Handbook of Green Chemistry–

Volume 6
Ionic Liquids

Edited by
Peter Wasserscheid and
Annegret Stark

Related Titles

Tanaka, K.

Solvent-free Organic Synthesis

Second, Completely Revised and Updated Edition
2009
ISBN: 978-3-527-32264-0

Lefler, J.

Principles and Applications of Supercritical Fluid Chromatography

2009
ISBN: 978-0-470-25884-2

Wasserscheid, P., Welton, T. (eds.)

Ionic Liquids in Synthesis

Second, Completely Revised and Enlarged Edition
2008
ISBN: 978-3-527-31239-9

Sheldon, R. A., Arends, I., Hanefeld, U.

Green Chemistry and Catalysis

2007
ISBN: 978-3-527-30715-9

Lindstrom, U. M. (ed.)

Organic Reactions in Water

Principles, Strategies and Applications
2008
ISBN: 978-1-4501-3890-1

Handbook of Green Chemistry

Volume 6
Ionic Liquids

Edited by
Peter Wasserscheid and Annegret Stark

WILEY-VCH Verlag GmbH & Co. KGaA

The Editor

Prof. Dr. Paul T. Anastas
Yale University
Center for Green Chemistry & Green Engineering
225 Prospect Street
New Haven, CT 06520
USA

Volume Editors

Prof. Dr. Peter Wasserscheid
LS f. Chem. Reaktionstechnik
University of Erlangen-Nürnberg
Egerlandstr. 3
91058 Erlangen
Germany

Dr. Annegret Stark
Institut für Technische Chemie und Umweltchemie
Friedrich-Schiller-Universität Jena
Lessingstr.12
07743 Jena
Germany

Handbook of Green Chemistry – Green Solvents
Vol. 4: Supercritical Solvents
ISBN: 978-3-527-32590-0
Vol. 5: Reactions in Water
ISBN: 978-3-527-32591-7
Vol. 6: Ionic Liquids
ISBN: 978-3-527-32592-4

Set II (3 volumes):
ISBN: 978-3-527-31574-1

Handbook of Green Chemistry
Set (12 volumes):
ISBN: 978-3-527-31404-1

All books published by Wiley-VCH are carefully produced. Nevertheless, authors, editors, and publisher do not warrant the information contained in these books, including this book, to be free of errors. Readers are advised to keep in mind that statements, data, illustrations, procedural details or other items may inadvertently be inaccurate.

Library of Congress Card No.:
applied for

British Library Cataloguing-in-Publication Data
A catalogue record for this book is available from the British Library.

Bibliographic information published by the Deutsche Nationalbibliothek
The Deutsche Nationalbibliothek lists this publication in the Deutsche Nationalbibliografie; detailed bibliographic data are available on the Internet at http://dnb.d-nb.de.

© 2010 WILEY-VCH Verlag GmbH & Co. KGaA, Weinheim

All rights reserved (including those of translation into other languages). No part of this book may be reproduced in any form – by photoprinting, microfilm, or any other means – nor transmitted or translated into a machine language without written permission from the publishers. Registered names, trademarks, etc. used in this book, even when not specifically marked as such, are not to be considered unprotected by law.

Typesetting Thomson Digital, Noida, India
Printing and Bookbinding betz-druck GmbH, Darmstadt
Cover Design Adam-Design, Weinheim

Printed in the Federal Republic of Germany
Printed on acid-free paper

ISBN: 978-3-527-32592-4

Contents

Ionic Liquids and Green Chemistry – an Extended Preface *XIII*
About the Editors *XXI*
List of Contributors *XXIII*

Part I Green Synthesis *1*

1 The Green Synthesis of Ionic Liquids *3*
Maggel Deetlefs and Kenneth R. Seddon
1.1   The *Status Quo* of Green Ionic Liquid Syntheses *3*
1.2   Ionic Liquid Preparations Evaluated for Greenness *4*
1.3   Which Principles of Green Chemistry are Relevant to Ionic Liquid Preparations? *6*
1.4   Atom Economy and the *E*-factor *7*
1.4.1 Atom Economy *7*
1.4.2 The *E*-factor *8*
1.5   Strengths, Weaknesses, Opportunities, Threats (SWOT) Analyses *8*
1.6   Conductive Heating Preparation of 1-Alkyl-3-methylimidazolium Halide Salts *8*
1.7   Purification of 1-Alkyl-3-methylimidazolium Halide Salts *12*
1.7.1 SWOT Analysis: Conductively Heated Preparation of 1-Alkyl-3-Methylimidazolium Halide Salts and Their Subsequent Purification *14*
1.8   Ionic Liquid Syntheses Promoted by Microwave Irradiation *15*
1.8.1 Microwave-assisted Versus Traditional Ionic Liquid Preparations *18*
1.8.2 SWOT Analysis: Microwave-promoted Syntheses of Ionic Liquids *18*
1.9   Syntheses of Ionic Liquids Promoted by Ultrasonic Irradiation *20*
1.9.1 SWOT Analysis: Ultrasound-promoted Syntheses of Ionic Liquids *22*
1.10 Simultaneous Use of Microwave and Ultrasonic Irradiation to Prepare Ionic Liquids *23*
1.10.1 SWOT Analysis: Simultaneous Use of Microwave and Ultrasonic Irradiation to Prepare Ionic Liquids *24*
1.11 Preparation of Ionic Liquids Using Microreactors *25*
1.11.1 SWOT Analysis: Preparation of Ionic Liquids Using Microreactors *27*

1.12	Purification of Ionic Liquids with Non-halide Anions 28
1.12.1	Purification of Hydrophobic *Versus* Hydrophilic Ionic Liquids 28
1.12.2	SWOT Analyses: Purification of Hydrophobic and Hydrophilic Ionic Liquids 29
1.13	Decolorization of Ionic Liquids 31
1.13.1	SWOT Analysis: Decolorization of Ionic Liquids 31
1.14	Conclusion 34
	References 36
Part II	**Green Synthesis Using Ionic Liquids** 39
2	**Green Organic Synthesis in Ionic Liquids** 41
	Peter Wasserscheid and Joni Joni
2.1	General Aspects 41
2.1.1	The Extremely Low Vapor Pressure of Ionic Liquids 43
2.1.2	Stability of Ionic Liquids in Organic Reactions 44
2.1.3	Liquid–Liquid Biphasic Organic Reactions 46
2.1.3.1	Tunable Solubility Properties 47
2.1.3.2	Product Isolation from Organic Reactions with Ionic Liquids 49
2.1.4	Reactive or Catalytic Ionic Liquids in Organic Synthesis 51
2.2	Friedel–Crafts Alkylation 54
2.2.1	Introduction and Technical Background 54
2.2.2	Ionic Liquids in Friedel–Crafts Reaction - the Unique Selling Point 55
2.2.3	Liquid–Liquid Biphasic Catalysis 56
2.2.4	Supported Ionic Liquid Phase (SILP) Friedel–Crafts Catalysis 57
	References 59
3	**Transition Metal Catalysis in Ionic Liquids** 65
	Peter Wasserscheid
3.1	Solubility and Immobilization of Transition Metal Complexes in Ionic Liquids 65
3.2	Ionic Liquid–Catalyst Interaction 67
3.2.1	Activation of Transition Metal Complexes by Lewis Acidic Ionic Liquids 68
3.2.2	*In Situ* Carbene Complex Formation 68
3.3	Distillative Product Isolation from Ionic Catalyst Solutions 70
3.4	New Opportunities for Biphasic Catalysis 72
3.5	Green Aspects of Nanoparticle and Nanocluster Catalysis in Ionic Liquids 75
3.6	Green Aspects of Heterogeneous Catalysis in Ionic Liquids 77
3.7	Green Chemistry Aspects of Hydroformylation Catalysis in Ionic Liquids 79
3.7.1	Feedstock Solubility 79
3.7.2	Catalyst Solubility and Immobilization 80

3.7.3	Use of Phosphite Ligands in Ionic Liquids *81*	
3.7.4	Halogen-containing Ionic Liquids Versus Halogen-free Ionic Liquids in Hydroformylation *81*	
3.7.5	Hydroformylation in scCO$_2$–Ionic Liquid Multiphasic Systems *82*	
3.7.6	Reducing the Amount of Ionic liquid Necessary – the Supported Ionic Liquid Phase (SILP) Catalyst Technology in Hydroformylation *83*	
3.8	Conclusion *85*	
	References *85*	

4	**Ionic Liquids in the Manufacture of 5-Hydroxymethylfurfural from Saccharides. An Example of the Conversion of Renewable Resources to Platform Chemicals** *93*
	Annegret Stark and Bernd Ondruschka
4.1	Introduction *93*
4.1.1	Areas of Application for HMF and its Derivatives *95*
4.1.1.1	Direct Uses of HMF *95*
4.1.1.2	Derivatives of HMF *96*
4.1.2	Summary: Application of HMF and Its Derivatives *98*
4.2	HMF Manufacture *99*
4.2.1	General Aspects of HMF Manufacture *99*
4.2.2	Methods of Manufacture of HMF from Fructose *100*
4.2.3	Methods of Manufacture of HMF from Sugars Other Than Fructose *104*
4.2.4	Deficits in HMF Manufacture *105*
4.3	Goals of Study *105*
4.4	HMF Manufacture in Ionic Liquids – Results of Detailed Studies in the Jena Laboratories *105*
4.4.1	Temperature *106*
4.4.2	Concentration and Time *106*
4.4.3	Effect of Water *108*
4.4.4	Effect of Purity *109*
4.4.5	Effect of the Choice of Ionic Liquid *111*
4.4.6	Other Saccharides *112*
4.4.7	Continuous Processing of HMF *114*
4.5	Conclusion *117*
	References *118*

5	**Cellulose Dissolution and Processing with Ionic Liquids** *123*
	Uwe Vagt
5.1	General Aspects *123*
5.2	Dissolution of Cellulose in Ionic Liquids *127*
5.3	Rheological Behavior of Cellulose Solutions in Ionic Liquids *129*
5.4	Regeneration of the Cellulose and Recycling of the Ionic Liquid *131*
5.5	Cellulosic Fibers *131*
5.6	Cellulose Derivatives *134*

5.7	Fractionation of Biomass with Ionic Liquids 134
5.8	Conclusion and Outlook 135
	References 135

Part III	Ionic Liquids in Green Engineering 137
6	**Green Separation Processes with Ionic Liquids 139**
	Wytze (G. W.) Meindersma, Ferdy (S. A. F.) Onink, and André B. de Haan
6.1	Introduction 139
6.2	Liquid Separations 141
6.2.1	Extraction 141
6.2.1.1	Metal Extraction 141
6.2.1.2	Extraction of Aromatic Hydrocarbons 145
6.2.1.3	Proteins 151
6.2.2	Extractive Distillation 153
6.2.2.1	Conventional Process 153
6.2.2.2	Ionic Liquids in Extractive Distillation 155
6.2.2.3	Conclusions 157
6.3	Environmental Separations 158
6.3.1	Desulfurization and Denitrogenation of Fuels 158
6.3.1.1	Conventional Desulfurization 158
6.3.1.2	Desulfurization with Ionic Liquids 158
6.3.1.3	Oxidative Desulfurization 162
6.3.1.4	Conclusions 163
6.4	Combination of Separations in the Liquid Phase with Membranes 163
6.4.1	Conclusions 164
6.5	Gas Separations 164
6.5.1	Conventional Processes 164
6.5.2	CO_2 Separation with Standard Ionic Liquids 165
6.5.3	CO_2 Separation with Functionalized Ionic Liquids 165
6.5.4	CO_2 Separation with Ionic Liquid (Supported) Membranes 166
6.5.5	Olefin–Paraffin Separations with Ionic Liquids 168
6.5.6	Conclusions 168
6.6	Engineering Aspects 168
6.6.1	Equipment 168
6.6.2	Hydrodynamics 169
6.6.3	Mass Transfer 171
6.6.4	Conclusions 172
6.7	Design of a Separation Process 172
6.7.1	Introduction 172
6.7.2	Application of COSMO-RS 173
6.7.3	Conclusions 174
6.8	Conclusions 175
	References 176

7	**Applications of Ionic Liquids in Electrolyte Systems** *191*
	William R. Pitner, Peer Kirsch, Kentaro Kawata, and Hiromi Shinohara
7.1	Introduction *191*
7.2	Electrolyte Properties of Ionic Liquids *193*
7.3	Electrochemical Stability *196*
7.4	Dye-sensitized Solar Cells *198*
	References *200*
8	**Ionic Liquids as Lubricants** *203*
	Marc Uerdingen
8.1	Introduction *203*
8.2	Why Are Ionic Liquids Good Lubricants? *204*
8.2.1	Wear and Friction Behavior *204*
8.2.2	Pressure Behavior *210*
8.2.3	Thermal Stability *210*
8.2.4	Viscosity Index and Pour Point *213*
8.2.5	Corrosion *215*
8.2.6	Electric Conductivity *215*
8.2.7	Ionic Greases *216*
8.3	Applications, Conclusion and Future Challenges *217*
	References *218*
9	**New Working Pairs for Absorption Chillers** *221*
	Matthias Seiler and Peter Schwab
9.1	Introduction *221*
9.2	Absorption Chillers *222*
9.3	Requirements and Challenges *223*
9.3.1	Thermodynamics, Heat and Mass Transfer *224*
9.3.2	Crystallization Behavior *224*
9.3.3	Corrosion Behavior *225*
9.3.4	Viscosity *225*
9.3.5	Thermal Stability *225*
9.4	State of the Art and Selected Results *226*
9.5	Abbreviations *228*
	References *228*
Part IV	**Ionic Liquids and the Environment** *233*
10	**Design of Inherently Safer Ionic Liquids: Toxicology and Biodegradation** *235*
	Marianne Matzke, Jürgen Arning, Johannes Ranke, Bernd Jastorff, and Stefan Stolte
10.1	Introduction *235*
10.1.1	The T-SAR Approach and the "Test Kit" Concept *236*
10.1.2	Strategy for the Design of Sustainable Ionic Liquids *238*

10.2	(Eco)toxicity of Ionic Liquids 239
10.2.1	Influence of the Side Chain 243
10.2.2	Influence of the Head Group 254
10.2.3	Influence of the Anion 255
10.2.4	Toxicity of Ionic Liquids as a Function of the Surrounding Medium 257
10.2.5	Combination Effects 259
10.2.6	(Quantitative) Structure–Activity Relationships and Modes of Toxic Action 261
10.2.7	Conclusion 263
10.3	Biodegradability of Ionic Liquids 265
10.3.1	Introduction 265
10.3.2	Testing of Biodegradability 266
10.3.3	Results from Biodegradation Experiments 268
10.3.3.1	Biodegradability of Ionic Liquid Anions 269
10.3.3.2	Biodegradability of Imidazolium Compounds 283
10.3.3.3	Pyridinium and 4-(Dimethylamino)pyridinium Compounds 284
10.3.3.4	Biodegradability of Other Head Groups 285
10.3.4	Misleading Interpretation of Biodegradation Data 286
10.3.5	Metabolic Pathways of Ionic Liquid Cations 288
10.3.6	Abiotic Degradation 290
10.3.7	Outlook 290
10.4	Conclusion 290
10.4.1	Toxicity and (Eco)toxicity of Ionic Liquids 291
10.4.2	Biodegradability of Ionic Liquids 293
10.4.3	The Goal Conflict in Designing Sustainable Ionic Liquids 293
10.4.4	Final Remarks 294
	References 295

11	**Eco-efficiency Analysis of an Industrially Implemented Ionic Liquid-based Process – the BASF BASIL Process** 299
	Peter Saling, Matthias Maase, and Uwe Vagt
11.1	The Eco-efficiency Analysis Tool 299
11.1.1	General Aspects 299
11.2	The Methodological Approach 299
11.2.1	Introduction 300
11.2.2	What is Eco-efficiency Analysis? 302
11.2.3	Preparation of a Specific Life-cycle Analysis for All Investigated Products and Processes 303
11.3	The Design of the Eco-efficiency Study of BASIL 303
11.4	Selected Single Results 304
11.4.1	Energy Consumption 304
11.4.2	Global Warming Potential (GWP) 306
11.4.3	Water Emissions 307
11.4.4	The Ecological Fingerprint 307
11.4.5	Cost Calculation 308

11.5	The Creation of the Eco-efficiency Portfolio *309*
11.6	Scenario Analysis *311*
11.7	Conclusion *312*
11.8	Outlook *313*
	References *314*

12 Perspectives of Ionic Liquids as Environmentally Benign Substitutes for Molecular Solvents *315*
Denise Ott, Dana Kralisch, and Annegret Stark

12.1	Introduction *315*
12.2	Evaluation and Optimization of R&D Processes: Developing a Methodology *317*
12.2.1	Solvent Selection Tools *317*
12.2.2	LCA Methodology *318*
12.2.3	The ECO Method *319*
12.2.3.1	The Key Objectives *320*
12.2.3.2	The Evaluation and Optimization Procedure *321*
12.3	Assessment of Ionic Liquid Synthesis – Case Studies *322*
12.3.1	Synthesis of Ionic Liquids: Extract from the Optimization Procedure *324*
12.3.2	Validation of *EF* as an Indicator for Several Impact Categories of the LCA Methodology *326*
12.3.3	Comparison of the Life Cycle Environmental Impacts of the Manufacture of Ionic Liquids with Molecular Solvents *327*
12.4	Assessment of the Application of Ionic Liquids in Contrast to Molecular Solvents *329*
12.4.1	Case Study: Diels–Alder Reaction *329*
12.4.1.1	Evaluation of the Solvent Performance *330*
12.4.1.2	Evaluation of the Energy Factor *EF* *330*
12.4.1.3	Evaluation of the Environmental and Human Health Factor *EHF* - Examples *332*
12.4.1.4	Evaluation of the Cost Factor *CF* *332*
12.4.1.5	Alternative Ionic Liquid Choices *334*
12.4.1.6	Decision Support *334*
12.5	Conclusions *335*
	References *336*

Index *341*

Ionic Liquids and Green Chemistry – an Extended Preface

Green Chemistry, or Sustainable Chemistry, deals with the development of chemicals (auxiliaries, intermediates, solvents, consumer products) and processes with the aim of designing for minimal environmental impact while providing maximum technical performance, as outlined in the Principles of Green Chemistry by Paul Anastas [1]. As such, Green Chemistry is part of Sustainability, meaning economically profitable production at lowest possible ecological intrusion, taking into account the social development of society (Three Pillars of Sustainability). Due to the globalization of markets, the social effects of industrial actions are difficult to assess, as many soft factors, such as politics, lobbying activities, funding strategies and artificial market regulations impact strongly, and are known to be unpredictable. Hence natural scientists have opted to concentrate on the consolidation of the equilibrium of two pillars only, i.e. ecology and economy.

Ionic liquids are salts which are characterized – due to their special distribution of charges and due to their special shape of ions – by melting points below 100 °C. Figure 1 displays typical cations and anions forming ionic liquids.

Ionic liquids represent a new class of non-molecular, liquid materials with unique property profiles. These unique properties originate from a complex interplay of Coulombic, hydrogen bonding and van der Waals interactions. To understand and to utilize these complex ion interactions is the heart of ionic liquid science and leads to the optimization of ionic liquid structures for specific applications. Figure 2 displays a number of properties that can be combined in ionic liquids in a unique manner.

In addition to the large number of modified and task-specific ions forming ionic liquids, the concept of using ionic liquids brings along some general features that have to do with the general properties of ionic liquids. Thus, some properties of ionic liquids are tunable in a wide range and others are more or less intrinsic to the approach. Table 1 shows an overview of typical ionic liquid property ranges. It gives typical values for selected properties and also known upper and lower limits. This overview is intended to give the less experienced reader quick access to some important ionic liquid facts. Much more detailed information about the specific properties of ionic liquids can be found, for example, in reference [2].

From the early 1990s, the development of ionic liquids as solvents for organic synthesis and homogeneous catalysis evolved and these studies went in parallel with

Handbook of Green Chemistry, Volume 6: Ionic Liquids
Edited by Peter Wasserscheid and Annegret Stark
Copyright © 2010 WILEY-VCH Verlag GmbH & Co. KGaA, Weinheim
ISBN: 978-3-527-32592-4

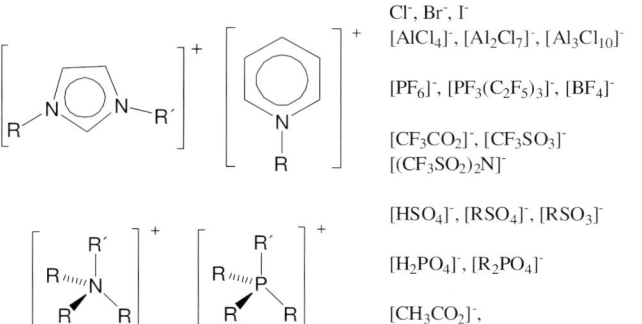

Figure 1 Typical cations and anions forming ionic liquids.

efforts to reduce the environmental impact of chemical synthesis and production (Green Chemistry). As the growing understanding of the unique properties of ionic liquids justified expectations that some improvements to existing technology could be made, the development of ionic liquids and Green Chemistry over the last 20 years went hand in hand.

The term Green Chemistry has been used extensively in the literature, and even a journal bearing this title has been established. However, upon examination of the literature, it becomes clear that in most cases synthetic strategies have been labeled with the term Green Chemistry rather uncritically. Ionic liquids are a prime example of this practice: due to their generally low vapor pressure (reducing the risk of exposure of workers and gaseous emissions), low flammability (reducing the risk of explosion) and often low toxicity when compared with conventional, volatile organic solvents, ionic liquids were often touted as Green Solvents.

From today's point of view, some of the early expectations proved to be unfounded, or at least not valid for the whole class of ionic liquids. Of course, it was found that just like any other physico-chemical property, toxicity, explosivity and volatility can also be

Figure 2 Properties of ionic liquids that can be exploited to design new and greener processes and products.

Table 1 Typical property ranges of ionic liquids [2].

Property	Lower limit example	Typical range of most ionic liquids	Upper limit example
Density	$[C_6C_1pyr][DCA]$ $= 0.92\,g\,l^{-1a}$	$1.1–1.6\,g\,l^{-1a}$	$[C_2mim]Br–AlBr_3{}^d$: $1/2 = 2.2\,g\,l^{-1a}$
Viscosity	$[C_2mim]Cl–AlCl_3{}^d$: $1/2 = 14\,mPa\,s^a$	$40–800\,mPa\,s^a$	$[C_4mim]Cl$ (supercooled) $= 40\,890\,mPa\,s^a$
Thermal stability	$[C_2mim][OAc]$ $\approx 200\,°C^b$	$230–300\,°C^b$	$[C_2mim][NTf_2]$ $= 400\,°C^b$
Hydrolytic stability	$[BF_4]^-$, $[PF_6]^-$	Also heterocyclic cations can hydrolyze under extreme conditions	$[NTf_2]^-$, $[OTf]^-$, $[CH_3SO_3]^-$
Base stability	$[Al_2Cl_7]^-$, $[HSO_4]^-$	All 1,3-dialkylimidazolium ionic liquids are subject to deprotonation	$[PR_4]^+$, $[OAc]^-$
Corrosion	$[NTf_2]^-$, $[OTf]^-$	Most ILs are corrosive towards Cu; additives available	Cl^-, HF formed from $[MF_x]^-$ hydrolysis
Price	$[HNR_3][HSO_4] \approx 3\,€\,kg^{-1c}$	$25–250\,€\,kg^{-1}$	$[C_4dmim][NTf_2]$ $\approx 1000\,€\,kg^{-1}$

[a]At room temperature.
[b]TGA experiments at $10\,K\,min^{-1}$.
[c]Estimation made for a production scale of 1000 kg and for a purity >98%.
[d]$[C_2mim]Br–AlBr_3$: 1/2 or $[C_2mim]Cl–AlCl_3$ denotes the ratio of 1-ethyl-3-methylimidazolium halide and aluminium halide.

designed into the structure of ionic liquids, but likewise, they can be carefully avoided.

Thus ionic liquids are not intrinsically green! They are a class of fascinating, new liquid materials providing unique combinations of properties. By making use of those unique properties, more efficient and greener processes and products can be realized. As a consequence of this statement, the greenness of an ionic liquid always has to be defined by its performance in a specific application –as also for any other performance chemical. The performance of an ionic liquid in a given application is highly dependent upon the structure of the ions forming the liquid. That means that some ionic liquids may be "green" in a given application whereas others, less favorably chosen ones, may be "not green". In addition to this performance argument, it must be noted that the greenness of any chemical is a function of many additional factors, including quantity, risk of release into the environment, location and type of environment, toxicological properties and biodegradation. These factors cannot be assessed *a priori*, but only in the context of a specific technical application at a given scale.

Designing green ionic liquids and also processes, applications and products based on ionic liquid technology requires a profound knowledge of a broad base of multidisciplinary aspects. In many cases that are discussed today, research is still at an early stage. This is because the use of ionic liquids in general is a very young scientific topic that has attracted massive scientific attention only in the last 10 years. Given the huge number of proposed Green Chemistry concepts involving ionic liquids and in the light of this short period, it is understandable why there are often too few data available to predict with certainty whether a certain application will really profit in its greenness from the use of an ionic liquid.

To illustrate this statement, one can look at ionic liquids in organic chemistry. Most organic reactions have been tested in ionic liquids to date, but this has often been done by simply transferring traditional reaction conditions to this new class of solvents. In many cases, interesting results were obtained; in others, the outcome was disappointing; in yet other cases, effects were first attributed to the ionic liquid solvent that later turned out to be due to impurities in the ionic liquid or due to ionic liquid decomposition products. From all these studies, it is evident, however, that the full utilization of the ionic liquid's potential in organic synthesis requires a detailed knowledge of the ionic liquid–substrate interactions which influence significantly rates and/or selectivities. This detailed knowledge is just about to be generated from ongoing and recent fundamental kinetic and spectroscopic work.

It should also be noted that it is not easy to derive the green advantage of a certain ionic liquid in a given application from an initial "proof of concept" research. Much information that is necessary to evaluate the greenness of an ionic liquid application is only generated in a more advanced state of development in which attention can be given to aspects such as recycling efficiency, ionic liquid recovery or the degree of ionic liquid degradation over time. However, this type of information is absolutely crucial for evaluating the greenness of every ionic liquid application.

As opposed to existing reviews and monographs, this book aims to summarize the current state of the art of Green Chemistry using ionic liquids. Authors from industry and academia have contributed their points of view on various aspects, starting with a thorough assessment of the synthesis of ionic liquids by Maggel Deetlefs and Ken Seddon of the Queen's University of Belfast Ionic Liquid Laboratories. Using SWOT analysis, the synthesis (by conductive heating, microwave or ultrasound-assisted) and purification of ionic liquids is analyzed with respect to the 12 Principles of Green Chemistry.

A large section of the book is dedicated to the application of ionic liquids to synthesis, with subsections dealing with organic synthesis, transition metal catalysis, the conversion of saccharides to platform chemicals and the processing of cellulose by direct dissolution in ionic liquids. In their contribution on organic synthesis, Peter Wasserscheid and Joni Joni of the University of Erlangen-Nuremberg discuss important aspects of green organic synthesis with ionic liquids and exemplify the latter for Friedel–Crafts alkylation reaction. In the section on transition metal catalysis, Peter Wasserscheid details the advantages of multiphase transition metal catalysis in ionic liquids. Both liquid–liquid biphasic catalysis and supported ionic liquid phase (SILP) catalysis are highlighted with special emphasis on catalytic hydroformylation as an illustrative example. A summary of the state of the art of saccharide transformation to 5-hydroxymethylfurfural, a potential platform chemical of future biorefineries, is presented by Annegret Stark and Bernd Ondruschka of the Friedrich-Schiller University in Jena, showing the advantages of biphasic liquid–liquid processing. Uwe Vagt of BASF SE details the achievements of direct dissolution and processing of cellulose from ionic liquids, and compares them with currently used production technology.

It is clear that Green Chemistry with ionic liquids is much more than just synthesis and catalysis. Improved separation processes, more efficient devices or optimized products can be realized by using ionic liquids and this can add to a greener chemistry as the ionic liquid applied may help to e.g. save, generate or store energy. Moreover, some non-synthetic aspects of Green Chemistry come naturally into play when evaluating the overall sustainability of a chemical process that includes not only reaction steps but also product purification and the operation of process machinery. Hence the second large section of this book is dedicated to Green Chemical Engineering. Here, Wytze Meindersma, Ferdy Onink and André de Haan summarize various separation techniques involving ionic liquids, including extraction of various liquid, solid and gaseous solutes, desulfurization methods, and separation over membranes. The use of ionic liquids in greener electrochemical applications, in particular dye-sensitized solar cells, are presented by Will Pitner and colleagues of Merck in Germany and Japan. Ionic liquids have also shown great promise for energy saving as advanced lubricants, as highlighted by Marc Uerdingen of Solvent Innovation. Matthias Seiler and Peter Schwab of Evonik Degussa demonstrate the potential of ionic liquids in smart heat pumps when compared with conventional fluids as working pairs for absorption chillers.

A very detailed knowledge of the toxicological properties of ionic liquids is required not only to adhere to REACH requirements for the introduction of novel ionic liquids into the market, but also for the conscious design of ionic liquid structures to reduce risk of environmental damage in the case of spillage. The team of authors headed by Stefan Stolte of the University of Bremen introduce the T-SAR concept and a test kit approach for the broad assessment of various toxicological properties and degradation pathways of ionic liquids to derive a structure–activity relationship, from which conclusions are drawn on how to design innocuous ionic liquids.

The assessment of the eco-efficiency of ionic liquids is discussed from both academic and industrial points of view. The group of Dana Kralisch of the Friedrich Schiller University at Jena has developed a tool which allows for a simplified life-cycle analysis (ECO method), which has been applied to the optimization of the synthesis of ionic liquids and their application in the Diels–Alder synthesis. Peter Saling and colleagues of BASF SE focus on the comparison of the performance of ionic liquids in the BASIL process, in which the formation of an ionic liquid as a side-product of the preparation of dialkoxyphenylphosphines increases the rate of reaction and facilitates phase separation.

In this book, we have tried to give a broad readership easy access to Green Chemistry studies involving ionic liquids. In the course of editing this book, we have realized that the number of different abbreviations and nomenclatures for ionic liquids is often a problem for the the non-expert to enter the field. Therefore, we have decided to use in this book a uniform system of abbreviations throughout all contributions to avoid any confusion in this respect. The list of abbreviations is given in Table 2.

Table 2 Abbreviations for ionic liquid ions used in this book.

Anion	Name	Cation	Name
[OAc]$^-$	Acetate	[C$_n$mim]$^+$	1-Alkyl-3-methylimidazolium
		[C$_n$C$_1$im]$^+$	
[CF$_3$CO$_2$]$^-$	Trifluoroacetate	[C$_n$dmim]$^+$	1-Alkyl-2,3-dimethylimidazolium
[OTf]$^-$	Trifluoromethanesulfonate	[Allylmim]$^+$	1-Allyl-3-methylimidazolium
[CF$_3$SO$_3$]$^-$			
[OTs]$^-$	p-Toluenesulfonate	[H-mim]$^+$	1-H-3-Methylimidazolium
[CH$_3$SO$_3$]$^-$	Methanosulfonate	[1-C$_4$py]$^+$	N-Butylpyridinium
[C$_4$F$_9$SO$_3$]$^-$	Nonafluorobutylsulfonate	[1-C$_4$,4-C$_1$py]$^+$	1-Butyl-4-methylpyridinium
[NTf$_2$]$^-$	Bis(fluoromethanesulfonyl)amide	[1-C$_4$,3-C$_1$py]$^+$	1-Butyl-3-methylpyridinium
[N(SO$_2$CF$_3$)$_2$]$^-$			
[DCA]$^-$	Dicyanamide	[H-Py]$^+$	N-H-Pyridinium
[N(CN)$_2$]$^-$			
[BF$_4$]$^-$	Tetrafluoroborate	[P$_{w,x,y,z}$]$^+$	Tetraalkylphosphonium
[TCB]$^-$	Tetracyanoborate	[N$_{w,x,y,z}$]$^+$	Tetraalkylammonium
[B(CN)$_4$]$^-$			
[BBB]$^-$	Bis[1,2-benzenediolato(2–)-O,O']borate	[Me$_3$NH]$^+$	Trimethylammonium
[SbF$_6$]$^-$	Hexafluoroantimonate	[Et$_3$NH]$^+$	Triethylammonium
[HSO$_4$]$^-$	Hydrogensulfate	[S$_{xyz}$]$^+$	Trialkylsulfonium
[PF$_6$]$^-$	Hexafluorophosphate	[C$_n$thia]$^+$	N-Alkylthiazolium
[FAP]$^-$	Tris(pentafluoroethyl)trifluorophosphate	[C$_n$C$_1$pyr]$^+$	N-Alkyl-N-methylpyrrolidinium
[PF$_3$(C$_2$F$_5$)$_3$]$^-$			
[(RO)$_2$PO$_2$]$^-$	Dialkylphosphate	[C$_4$C$_1$pip]$^+$	1-Butyl-1-methylpiperidinium
[DMP]$^-$	Dimethylphosphate	[C$_n$quin]$^+$	N-Alkylquinolinium
[DEP]$^-$	Diethylphosphate	[C$_n$picol]$^+$	N-Alkylpicolinium
[H$_2$PO$_4$]$^-$	Hydrogenphosphate	[C$_n$C$_1$morph]$^+$	N-Alkyl-N-methylmorpholinium
[TCM]$^-$	Tricyanomethide	[TGA)]$^+$	Tetramethylguanidinium
[C(CN)$_3$]$^-$			
[CTf$_3$]$^-$	Tris(trifluoromethanesulfony)methide		
[MeSO$_4$]$^-$	Methylsulfate		
[CH$_3$SO$_4$]$^-$			
[EtSO$_4$]$^-$	Ethylsulfate		
[C$_2$H$_5$SO$_4$]$^-$			
[C$_8$H$_{17}$SO$_4$]$^-$	Octylsulfate		
[n-C$_8$H$_{17}$OSO$_3$]$^-$			
[SCN]$^-$	Thiocyanate		

This book has been produced in a large and cooperative effort by many contributors. First, we would like to thank all authors for having brought in their expertise. Likewise, our thanks go to the Wiley-VCH staff for their support, their patience and their encouragement.

We are confident that this book will offer useful information for a broad and diverse readership interested in realizing Green Chemistry with the help of advanced solvent concepts. It is our hope that the book may contribute to the further promotion of ionic liquids in Green Chemistry both in industry and in academia, but also to a more critical assessment of this approach in the various potential fields of application. The book aims to stimulate further research and to deepen the collaboration between chemists, chemical engineers, mechanical engineers and toxicologists in the field of Green Chemistry with ionic liquids. We are convinced that ionic liquids as a scientific tool can help to realize a Greener Chemistry but much remains to be achieved and to be developed towards this goal. We hope that this book will help to attract young scientists to this important field of research for our future.

References

1 Anastas, P.T. and Kirchhoff, M.M. (2002) *Acc. Chem. Res.*, **35**, 686.
2 Wasserscheid, P. and Welton, T. (eds) (2007) *Ionic Liquids in Synthesis*, Wiley-VCH Verlag GmbH, Weinheim, (two volumes, 724 pages).

October 2009

Annegret Stark
Jena
Peter Wasserscheid
Erlangen

About the editors

Series Editor

Paul T. Anastas joined Yale University as Professor and serves as the Director of the Center for Green Chemistry and Green Engineering there. From 2004–2006, Paul was the Director of the Green Chemistry Institute in Washington, D.C. Until June 2004 he served as Assistant Director for Environment at the White House Office of Science and Technology Policy where his responsibilities included a wide range of environmental science issues including furthering international public-private cooperation in areas of Science for Sustainability such as Green Chemistry. In 1991, he established the industry-government-university partnership Green Chemistry Program, which was expanded to include basic research, and the Presidential Green Chemistry Challenge Awards. He has published and edited several books in the field of Green Chemistry and developed the 12 Principles of Green Chemistry.

Volume Editors

Annegret Stark studied pharmaceutical chemistry at the University of Applied Sciences in Isny, Germany. She conducted her diploma thesis in 1997 in the labs of R.D. Singer at St. Mary's University in Halifax, Nova Scotia, who inspired her to take up a researcher's career in the field of ionic liquids. After finishing her PhD in K.R. Seddon's research group at the Queen's University of Belfast, Northern Ireland, in 2001, she moved on to South Africa for a SASOL-sponsored postdoc in the group of H.G. Raubenheimer at Stellenbosch University (2001–2003).

Since 2003, she heads her own research group at the Institute for Technical Chemistry and Environmental Chemistry (B. Ondruschka) of the Friedrich-Schiller University in Jena, Germany. Her research focus lies, on the one hand, on the elucidation of structure-induced interactions between ionic liquids and solutes, and the resulting effects on the reactivity of these. On the other hand, she is interested in the application of microreaction technology, e.g. in the conversion of highly reactive intermediates. Both, ionic liquids and microreaction technology, are exploited as tools with the goal to provide sustainable chemical and engineering concepts.

Since October 2009, she has been an interim professor for Technical Chemistry at the TU Chemnitz, Germany.

Peter Wasserscheid studied chemistry at the RWTH Aachen. After receiving his diploma in 1995 he joined the group of Prof. W. Keim at the Institute of Technical and Macromolecular Chemistry at the RWTH Aachen for his PhD thesis. In 1998 he moved to BP Chemicals in Sunbury/GB for an industrial post-doc for six months. He returned to the Institute of Technical and Macro-molecular Chemistry at the RWTH Aachen where he completed his habilitation entitled "Ionic liquids – a new Solvent Concept for Catalysis". In the meantime, he became co-founder of Solvent Innovation GmbH, Cologne, one of the leading companies in ionic liquid production and application (since December 2007 a 100% affiliate of Merck KGaA, Darmstadt). In 2003 he moved to Erlangen as successor of Prof. Emig and since then is heading the Institute of Reaction Engineering. In 2005 he also became head of the department "Chemical and Bioengineering" of the University Erlangen-Nuremberg. P. Wasserscheid has received several awards including the Max-Buchner-award of DECHEMA (2001), the Innovation Award of the German Economy (2003, category "start-up") together with Solvent Innovation GmbH and the Leibniz Award of the German Science Foundation (2006). His key research interests are the reaction engineering aspects of multiphase catalytic processes with a particular focus on ionic liquid reaction media. The Wasserscheid group belongs to the top research teams in the development and application of ionic liquids in general, and in developing the ionic liquid technology for catalytic applications in special. For various reaction types the group has successfully demonstrated greatly enhanced performance of ionic liquid based catalyst systems *vs.* conventional systems.

Peter Wasserscheid has a scientific track record of more than 130 publications in peer-reviewed scientific journals plus many papers in the form of proceedings. Moreover, he is a co-inventor of more than 40 patents, most of them in the field of ionic liquids.

List of Contributors

Jürgen Arning
University of Bremen
UFT – Centre for Environmental
Research and Technology
Department 3: Bioorganic Chemistry
Leobener Strasse
28359 Bremen
Germany

Maggel Deetlefs
The Queen's University of Belfast
The QUILL Research Centre
David Keir Building
Stranmillis Road
Belfast BT9 5AG
Northern Ireland, UK

André B. de Haan
Eindhoven University of Technology
Department of Chemical Engineering
and Chemistry/Process Systems
Engineering
5600 MB Eindhoven
The Netherlands

Bernd Jastorff
University of Bremen
UFT – Centre for Environmental
Research and Technology
Department 3: Bioorganic Chemistry
Leobener Strasse
28359 Bremen
Germany

Joni Joni
Friedrich-Alexander-Universität
Erlangen-Nürnberg
Lehrstuhl für Chemische
Reaktionstechnik
Egerlandstrasse 3
91058 Erlangen
Germany

Kentaro Kawata
Merck Ltd Japan
New Technology Office
4084 Nakatsu
Aikawa-machi
Aiko-gun
Kanagawa 243-0303
Japan

Peer Kirsch
Merck Ltd Japan
New Technology Office
4084 Nakatsu
Aikawa-machi
Aiko-gun
Kanagawa 243-0303
Japan

Dana Kralisch
Friedrich-Schiller-Universität Jena
Institut für Technische Chemie und
Umweltchemie
Lessingstrasse 12
07743 Jena
Germany

Matthias Maase
BASF SE
67056 Ludwigshafen
Germany

Marianne Matzke
University of Bremen
UFT – Centre for Environmental
Research and Technology
Department 10: Ecology
Leobener Strasse
28359 Bremen
Germany

Wytze (G. W.) Meindersma
Eindhoven University of Technology
Department of Chemical Engineering
and Chemistry/Process Systems
Engineering
5600 MB Eindhoven
The Netherlands

Bernd Ondruschka
Friedrich-Schiller-Universität Jena
Institut für Technische Chemie und
Umweltchemie
Lessingstrasse 12
07743 Jena
Germany

Ferdy (S. A. F.) Onink
Eindhoven University of Technology
Department of Chemical Engineering
and Chemistry/Process Systems
Engineering
5600 MB Eindhoven
The Netherlands

Denise Ott
Friedrich-Schiller-Universität Jena
Institut für Technische Chemie und
Umweltchemie
Lessingstrasse 12
07743 Jena
Germany

William R. Pitner
Merck KGaA
PC R&D Ionic Liquids
Frankfurter Strasse 250
65293 Darmstadt
Germany

Johannes Ranke
University of Bremen
UFT – Centre for Environmental
Research and Technology
Department 3: Bioorganic Chemistry
Leobener Strasse
28359 Bremen
Germany

Peter Saling
BASF SE
67056 Ludwigshafen
Germany

Peter Schwab
Evonik Degussa GmbH
Care and Surface Specialties
Goldschmidtstrasse 100
45127 Essen
Germany

Kenneth R. Seddon
The Queen's University of Belfast
The QUILL Research Centre
David Keir Building
Stranmillis Road
Belfast BT9 5AG
Northern Ireland, UK

Matthias Seiler
Evonik Degussa GmbH
Process Technology and Engineering
Rodenbacher Chaussee 4
63457 Hanau
Germany

Hiromi Shinohara
Merck Ltd Japan
New Technology Office
4084 Nakatsu
Aikawa-machi
Aiko-gun
Kanagawa 243-0303
Japan

Annegret Stark
Friedrich-Schiller-Universität Jena
Institut für Technische Chemie und Umweltchemie
Lessingstrasse 12
07743 Jena
Germany

Stefan Stolte
University of Bremen
UFT – Centre for Environmental Research and Technology
Department 3: Bioorganic Chemistry
Leobener Strasse
28359 Bremen
Germany

Marc Uerdingen
Solvent Innovation GmbH
Nattermannallee 1
50829 Köln
Germany

Uwe Vagt
BASF SE
67056 Ludwigshafen
Germany

Peter Wasserscheid
Friedrich-Alexander-Universität Erlangen-Nürnberg
Lehrstuhl für Chemische Reaktionstechnik
Egerlandstrasse 3
91058 Erlangen
Germany

Part I
Green Synthesis

1
The Green Synthesis of Ionic Liquids
Maggel Deetlefs and Kenneth R. Seddon

1.1
The *Status Quo* of Green Ionic Liquid Syntheses

One of the greatest problems plaguing ionic liquid science is that the terms "green" and "ionic liquid(s)" are often used synonymously. This, however, is both incorrect and deceptive because implicit in the designer classification of ionic liquids [1] is the ability to prepare the salts to possess distinctly non-green characteristics (*viz.* toxic, explosive [2], etc.). What is more, the preparation and purification of the salts are also often extremely dirty, requiring the use of large volumes of harmful organic solvents. It therefore follows that a *bona fide* green ionic liquid synthesis has only occurred if both the ionic liquid product and its preparation comply with all 12 principles of green chemistry as listed below [3, 4].

The Twelve Principles of Green Chemistry
1. It is better to prevent waste than to treat or clean up waste after it has formed.
2. Synthetic methods should be designed to maximize the incorporation of all materials used in the process into the final product.
3. Wherever practicable, synthetic methodologies should be designed to use and generate substances that possess little or no toxicity to human health and the environment.
4. Chemical products should be designed to preserve efficacy of function while reducing toxicity.
5. The use of auxiliary substances (e.g. solvents, separation agents *etc.*) should be made unnecessary wherever possible and innocuous when used.
6. Energy requirements should be recognized for their environmental and economic impacts and should be minimized. Synthetic methods should be conducted at ambient temperature and pressure.

7. A raw material of feedstock should be renewable rather than depleting wherever technically and economically practicable.
8. Unnecessary derivatization (blocking group, protection/deprotection, temporary modification of physical/chemical processes) should be avoided whenever possible.
9. Catalytic reagents (as selective as possible) are superior to stoichiometric reagents.
10. Chemical products should be designed so that at the end of their function they do not persist in the environment and break down into innocuous degradation products.
11. Analytical methodologies need to be further developed to allow for real-time, in-process monitoring and control prior to the formation of hazardous substances.
12. Substances and the form of a substance used in a chemical process should be chosen so as to minimize the potential for chemical accidents, including releases, explosions and fires.

To date, many excellent reviews [5–8] have summarized the *status quo* of ionic liquid syntheses, but until the approach discussed here, no publication has exclusively evaluated the state-of-the-art for green ionic liquid syntheses. In addition (and as far as we know), there is also currently no system available to determine whether an existing or planned ionic liquid preparation is green. We therefore thought it was timely not merely to write another ionic liquid synthesis review, but instead to develop a simple, universal method to assess whether an ionic liquid preparation is green or not. In this chapter, we provide a detailed look at our recent approach [9] to gauge the greenness of ionic liquid preparations.

1.2
Ionic Liquid Preparations Evaluated for Greenness

The ionic liquid preparations and purifications that were assessed for greenness are shown in Figure 1.1 and Figure 1.2, respectively. It is important to note that we only considered laboratory scale preparations *i.e.* on scales <2 kg. In particular, we concentrated on the synthesis of ionic liquids containing 1-alkyl-3-methylimidazolium cations, $[C_n \text{mim}]^+$, since the most synthetic data are available in the literature for these types of salts [10]. We also included an evaluation of the pros and cons of using differing energy sources to promote ionic liquid syntheses, *viz.* conductive (conventional) heating, microwave irradiation, ultrasonic irradiation and simultaneous microwave and ultrasonic irradiation.

Although our greenness assessment approach was specifically applied to the methodologies shown in Figures 1.1 and Figure 1.2, in principle the method of analysis can be applied to any type of ionic liquid preparation, such as alkylsulfate [11, 12] and carbene routes [13]. We therefore also evaluated the greenness of the use microstructured reactors for ionic liquid syntheses (Section 1.11).

1.2 Ionic Liquid Preparations Evaluated for Greenness | **5**

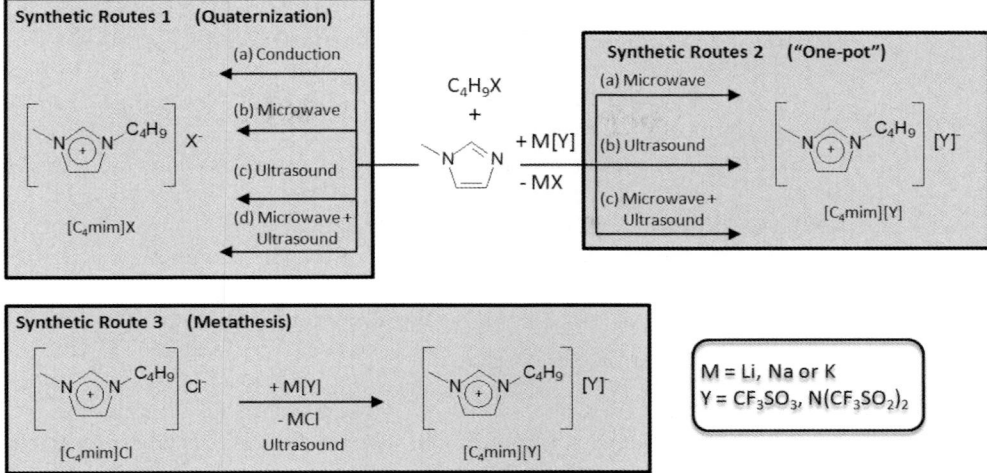

Figure 1.1 Typical ionic liquid synthetic routes [14].

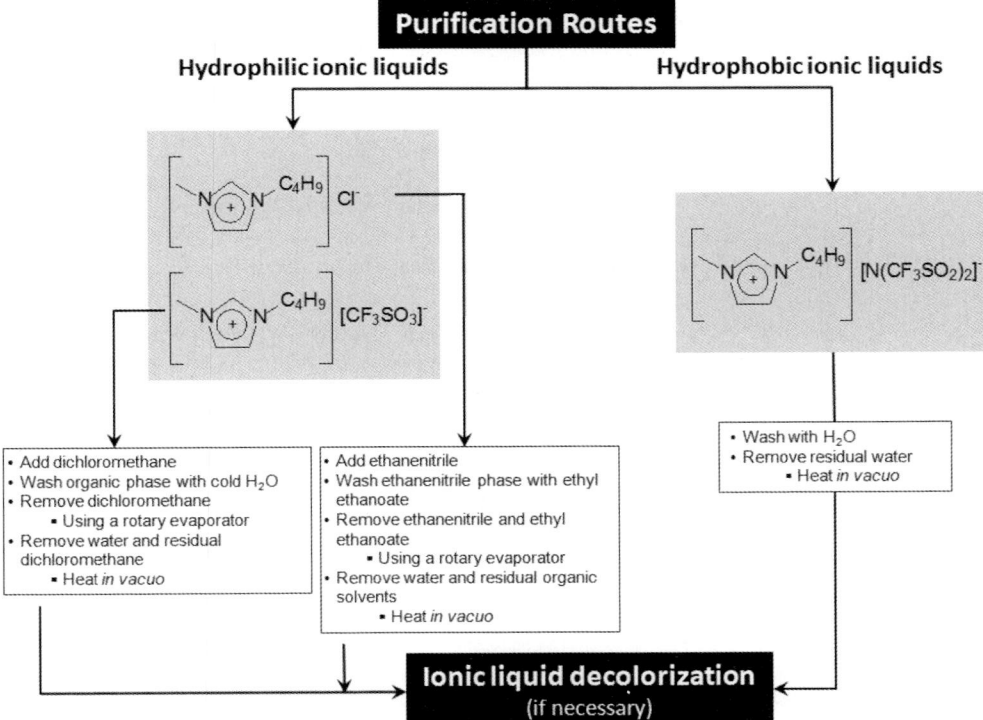

Figure 1.2 Typical ionic liquid purification routes [14].

1.3
Which Principles of Green Chemistry are Relevant to Ionic Liquid Preparations?

The 12 principles of green chemistry [3, 4] and also the 12 principles of green engineering [15] have been formalized over the past two decades by Professor Paul Anastas. Recently, both sets of lengthy principles were condensed by Professor Martyn Poliakoff and co-workers into mnemonics for easy communication, *viz.* PRODUCTIVELY and IMPROVEMENTS [16], as listed below.

The Twelve Principles of Green Chemistry		The Twelve Principles of Green Engineering	
P	Prevents Waste	I	Inherently non hazardous and safe
R	Renewable materials	M	Minimize material diversity
O	Omit derivatization steps	P	Prevention instead of treatment
D	Degradable chemical products	R	Renewable material and energy inputs
U	Use safe synthetic methods	O	Output-led design
C	Catalytic reagents	V	Very simple
T	Temperature, pressure ambient	E	Efficient use of mass, energy, space and time
I	In-process monitoring		
V	Very few auxiliary substances	M	Meet the need
E	*E*-factor, maximise feed in product	E	Easy to separate by design
L	Low toxicity of chemical products	N	Networks for exchange of local mass and energy
Y	Yes it's safe		
		T	Test the life cycle of the design
		S	Sustainability throughout product life cycle

The first step in conducting our greenness assessments involved a close examination of the 12 principles of green chemistry, which revealed that only eight of the 12 principles were relevant to the ionic liquid syntheses and purifications under scrutiny. The 4th, 7th, 9th and 10th principles are irrelevant to the ionic liquid preparations discussed here, since the 4th, 7th and 10th principles of green chemistry apply to the ionic liquid product itself rather than the procedure, and the 9th principle is associated with catalytic methodologies and hence is also not relevant to the typical ionic liquid preparations discussed here. On the other hand, the 8th, 11th and 12th principles of green chemistry apply to all the ionic liquid syntheses discussed here since:

1. Derivatization (8th principle) is not used in any of the ionic liquid syntheses.
2. In-process monitoring (11th principle), is not applicable to conventional laboratory syntheses, although it does apply to the special case of lab-on-a-chip [17, 18] (see Section 1.11).
3. Ionic liquids, by virtue of their negligible vapor pressure, minimize the potential for chemical accidents such as fires and explosions (12th principle).

To summarize, only the 1st, 2nd, 3rd, 5th, 6th, 8th, 11th and 12th principles of green chemistry are relevant to the synthetic procedures discussed here. Although

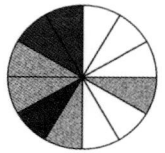
the principles of green engineering are of course important in determining the greenness of a given chemical reaction, they are more relevant to very large-scale chemical processes and therefore were not included in laboratory-scale (<2 kg) greenness assessments conducted here. As more data become available for industrial-scale ionic liquid preparations, we will expand our greenness assessment methodology to include the 12 principles of green engineering too.

Although, strictly, the 12 principles of green chemistry apply only to synthetic procedures, the 1st, 5th, 6th and 12th principles are relevant to ionic liquid purification procedures and, therefore, we assessed the purification of the salts according to these three principles also.

1.4
Atom Economy and the E-factor

Applying the 12 principles of green chemistry to evaluate the green credentials of ionic liquid syntheses and purifications is very important, but it would have been remiss of us not to include atom economy [19] and the E-factor [20, 21] in the assessments too.

1.4.1
Atom Economy

In simple terms, atom economy (also known as atom efficiency) measures, for a given reaction, how many of the atoms present in the starting materials form part of the final product and it also assumes 100% yield. Atom economy is reported as a percentage value with those reactions closest to 100% reflecting superior atom economies. This means that reactions with very low yields may possess 100% atom economies. For example, the preparation of 1-alkyl-3-methylimidazolium halide salts using quaternization [Routes 1(a)–(d), Figure 1.1] are 100% atom efficient regardless of the yield obtained, since no by-products are formed. On the other hand, ionic liquids prepared using one-pot [Routes 2(a)–(c), Figure 1.1] or metathesis reactions (Route 3, Figure 1.1) will be <100% atom efficient since a stoichiometric amount of MCl waste is generated.

$$\text{Atom economy (\%)} = \frac{\text{molecular weight of desired product (s)}}{\text{molecular weight of all reagents}} \times 100$$

The problem with employing atom economy to evaluate the green credentials of a reaction is that it does not take into account that some reactions with favorable stoichiometries require large excesses of reagents, give poor yields and often generate large amounts of unwanted products. This means that in terms of atom economy, a reaction with a very low yield can still be described as 100% atom efficient. As a result, the E-factor concept was introduced.

1.4.2
The E-factor

In the *E*-factor equation, *all* compounds used in a chemical process that are not contained in the final product are classified as waste. The nature of the *E*-factor equation dictates that the greenest chemical reactions have *E*-factor values close to zero. In brief, the *E*-factor gives a much truer reflection of the greenness of a chemical reaction than atom economy, since all generated waste is accounted for. The 'ideal' *E*-factor is reflected in the 2nd principle of green chemistry [3, 4], which states: "*Synthetic methods should be designed to maximize the incorporation of all materials used in the process into the final product*".

$$E\text{-factor} = \frac{\text{amount of waste produced in the process (kg)}}{\text{amount of the desired product(s) produced in the process (kg)}}$$

Although every chemical reaction will have its own unique atom economy and *E*-factor value, to simplify discussion, the atom economies and *E*-factors associated with ionic liquid preparations (Figure 1.1) and purifications (Figure 1.2) have been designated as low/medium/high and poor/good/excellent, respectively, since the literature has rarely given enough detail to allow definitive values to be assigned.

1.5
Strengths, Weaknesses, Opportunities, Threats (SWOT) Analyses

In order to simultaneously assess the green credentials of the selected synthesis and purification procedures of ionic liquids in terms of the 12 principles of green chemistry, atom economy and the *E*-factor, we have applied a common tool used in strategic planning, namely Strengths, Weaknesses, Opportunities, Threats (SWOT) analyses [22]. For ionic liquid preparation and purification methods, SWOT analyses involve specifying the objective(s) of a given procedure (e.g. obtaining an ionic liquid that is >99% pure) and identifying the internal and external factors that are favorable and unfavorable to achieving that objective. The SWOT analyses thus give an overview of the advantages or disadvantages of ionic liquid syntheses and purifications. In brief, the SWOT analyses provided here (represented skeletally in Figure 1.3) give a measure of the balance between good and bad for some common ionic liquid preparation and purification procedures and indicate potential directions for cleaning-up the methods.

1.6
Conductive Heating Preparation of 1-Alkyl-3-methylimidazolium Halide Salts

Even the most basic review of the ionic liquid synthesis literature reveals that the preparations of 1-alkyl-3-methylimidazolium halide salts have remained essentially

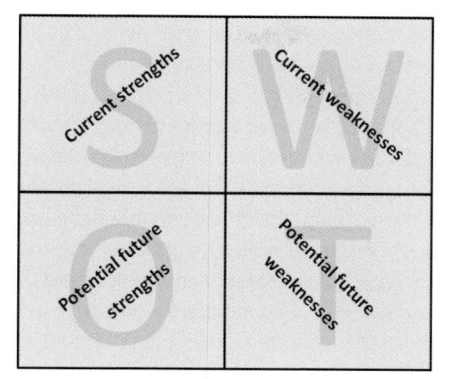

Figure 1.3 Skeletal Strengths Weaknesses Opportunities Threats (SWOT) analysis.

unchanged since they were first reported in 1982 [23, 24]; most reported syntheses still involve reaction of 1-methylimidazole with an excess of 1-haloalkane and are promoted using conductive heating [Route 1(a), Figure 1.1]. From a green chemistry perspective, much room for improvement exists for the syntheses of 1-alkyl-3-methylimidazolium halide salts, particularly regarding the use of more efficient energy sources and reducing/eliminating the need for organic solvents during both synthesis and purification; all these issues are discussed below.

The literature shows that most 1-alkyl-3-methylimidazolium halide salt preparations are executed using traditional heating under reflux, although nowadays an atmosphere of dry dinitrogen is sometimes used (Figure 1.4) since it has been found

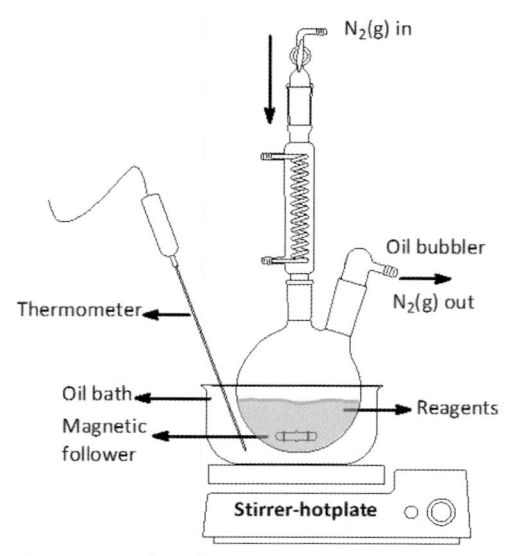

Figure 1.4 Traditional synthesis arrangement for preparing 1-butyl-3-methylimidazolium halide salts.

to promote the production of colorless [8] ionic liquids and also prevents the formation of hydrated halide salts from adventitious water [25] (see Section 1.7).

Almost without exception, reported preparations of 1-alkyl-3-methylimidazolium halide salts use an excess of 1-haloalkane [26], which means that the reactions are not in line with the 2nd principle of green chemistry, which states: *"Synthetic methods should be designed to maximize the incorporation of all materials used in the process into the final product"*. It also means that although the preparations of the halide salts are 100% atom efficient, with every atom of the starting materials incorporated into the final product, their *E*-factors are very poor, since excess 1-haloalkane is required to promote completion of the reactions at a reasonable rate. It also follows that the smaller the excess of 1-haloalkane, the lower the *E*-factor value will be, provided that no organic solvent is used during the preparation. For example, the *E*-factor for the preparation of [C$_4$mim]Cl (molar mass = 174.1 g mol^{-1}) is 0.106 and 0.005 when a 20% and 1% molar excess of 1-chlorobutane (molar mass = 92.56 g mol^{-1}), respectively, are employed.

$$E\text{-factor (20 mol \% excess C}_4\text{H}_9\text{Cl)} = \frac{0.2 \text{ mol} \times 92.56 \text{ g mol}^{-1}}{1 \text{ mol} \times 174.1 \text{ g mol}^{-1}} = 0.106$$

$$E\text{-factor (1 mol \% excess C}_4\text{H}_9\text{Cl)} = \frac{0.01 \text{ mol} \times 92.56 \text{ g mol}^{-1}}{1 \text{ mol} \times 174.1 \text{ g mol}^{-1}} = 0.005$$

Some reported preparations of 1-alkyl-3-methylimidazolium halide salts also make use of an organic solvent [27, 28] to reduce the viscosity of the reaction mixture and thus improve mass transfer, but also to control the reaction temperature and prevent product scrambling [29]. However, the molecular solvent employed, and also the excess of 1-haloalkane employed, require removal and subsequent disposal, which also does not comply with the 1st principle of green chemistry, which states: *"It is better to prevent waste than to treat or clean up waste after it has formed"*. Therefore, the practice of employing an organic solvent during preparation of 1-alkyl-3-methylimidazolium halide salts further increases the *E*-factor and is highly undesirable. For example, if 100 g of toluene (molar mass = 92.14 g mol^{-1}) is used during the synthesis of [C$_4$mim]Cl (and using a 20 mol% excess of 1-chlorobutane), the *E*-factor of increases from 0.106 to 0.681:

$$E\text{-factor (20 mol \% excess C}_4\text{H}_9\text{Cl} + 100 \text{ g toluene)}$$
$$= \frac{(0.2 \text{ mol} \times 92.56 \text{ g mol}^{-1} + 100 \text{ g})}{1 \text{ mol} \times 174.1 \text{ g mol}^{-1}} = 0.681$$

The 5th principle of green chemistry states that: *"The use of auxiliary substances (e.g. solvents, separation agents, etc.) should be made unnecessary wherever possible and innocuous when used"*, which means that ideally, no solvent should be used during preparation, but, if a solvent is to be used, it must be green and recycled. In order to align with the 1st, 2nd and 5th principles of green chemistry and have low *E*-factor

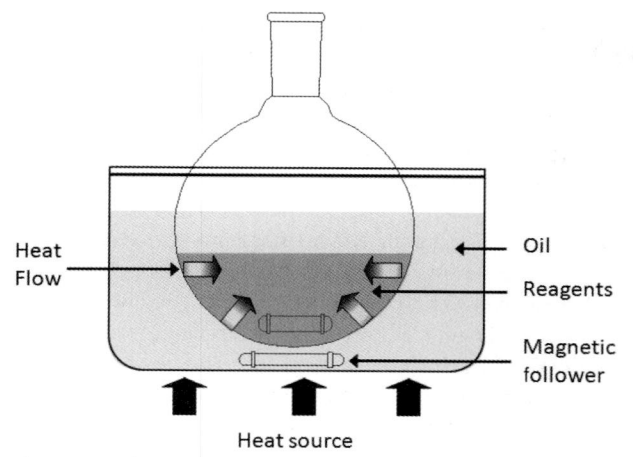

Figure 1.5 Schematic representation of conventional heating using an oil bath as external heat source.

values, neither excess 1-haloalkane nor organic solvent should be employed during the synthesis of 1-alkyl-3-methylimidazolium halide salts.

The use of conductive heating (usually an oil bath or heating mantle) to prepare 1-alkyl-3-methylimidazolium halide salts also does not align with the 6th principle of green chemistry, which states: "*Energy requirements should be recognized for their environmental and economic impacts and should be minimized. Synthetic methods should be conducted at ambient temperature and pressure*". Therefore, since conductive heating is slow, it is also energy inefficient because the transfer of heat from the heat source to the reaction mixture depends on the thermal conductivities of all the materials that must be penetrated such as the flask and solvent (Figure 1.5). In addition, the traditionally heated reactions are kinetically limited by the boiling point of the 1-haloalkane employed and can take several days, whereas microwave promoted reactions do not share these limitations (see below). The conductively heated preparations of 1-alkyl-3-methylimidazolium halide salts are also not performed at room temperature but usually at ~75 °C, which does not align with the second part of the 6th principle of green chemistry.

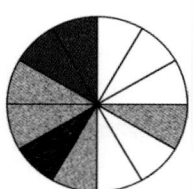

When all the above-mentioned negatives that are associated with the traditional syntheses of 1-alkyl-3-methylimidazolium halide ionic liquids are considered, it is safe to say that their preparations are not green. Indeed, of the eight relevant principles of green chemistry, they only comply with the three principles (8th, 11th and 12th) which apply to all the ionic liquid syntheses discussed here anyway. On the other hand, the *E*-factors for the preparations fall between good and excellent provided that excess 1-haloalkane is kept to a minimum and no harmful organic solvent is used and/or recycled during the preparation. Overall, however, the conductively heated preparations of 1-alkyl-3-methylimidazolium halide salts possess a lot of room for improvement, as indicated by their SWOT analysis (see Section 1.7.1).

1.7
Purification of 1-Alkyl-3-methylimidazolium Halide Salts

The purification of 1-alkyl-3-methylimidazolium halide salts (Figure 1.6) is generally dirty, with poor E-factors, since the excess 1-haloalkane, unconverted 1-methylimidazole and any solvent used during their preparation require removal and are therefore out of line with the 5th principle of green chemistry.

In our laboratories, removal of the excess 1-haloalkane and unconverted 1-methylimidazole is typically achieved by first adding a solvent such as ethanenitrile to the crude ionic liquid mixture and then repeatedly washing this phase with ethyl ethanoate. Once the washing process is complete, the two organic solvents are removed under reduced pressure (using a rotary evaporator). Since the amount of organic solvent used for the purification and also excess reagents are classified as waste in the E-factor equation, the more of each that is used, the higher the E-factor for a given purification procedure will be.

In less experienced groups, the purification of 1-alkyl-3-methylimidazolium halide salts is usually performed on the bench and therefore the salts absorb water from the atmosphere that also requires removal. It must be noted, however, that this adventitious water cannot be completely removed from ionic liquids containing halide anions (especially chloride) as they form, typically, very stable hydrates (Figure 1.7 [25]).

Therefore, if water-free ionic liquids containing halide ions are desired, their preparation and purification must be conducted under strictly anhydrous conditions from start to finish, for example, using Schlenk techniques. Nevertheless, some adventitious water can be removed by heating 1-alkyl-3-methylimidazolium halide salts under reduced pressure at ~70 °C for several hours. Almost needless to say,

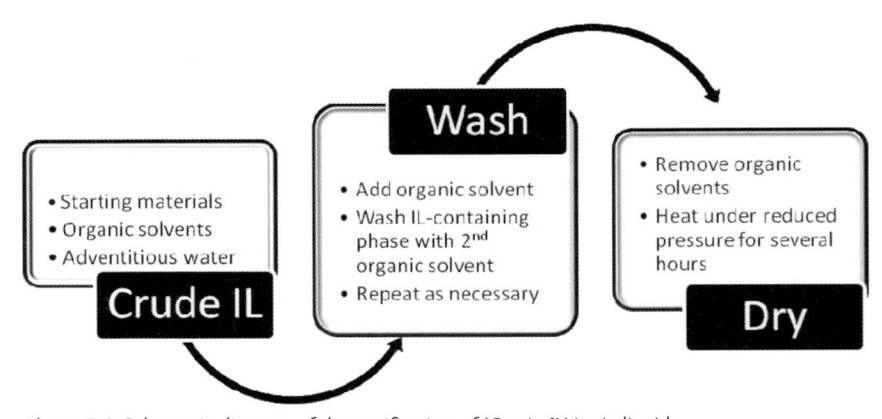

Figure 1.6 Schematic diagram of the purification of [C$_4$mim]X ionic liquids.

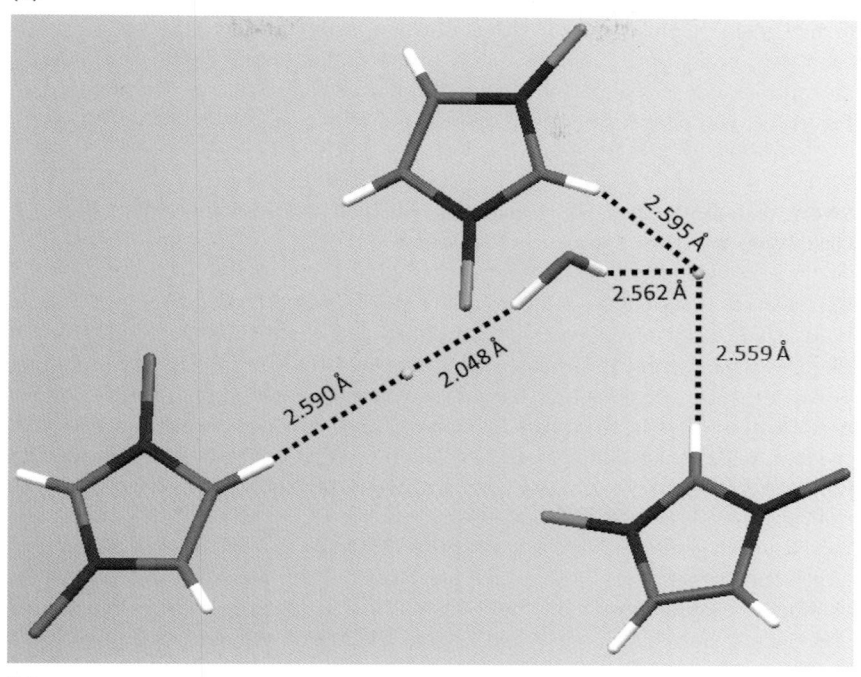

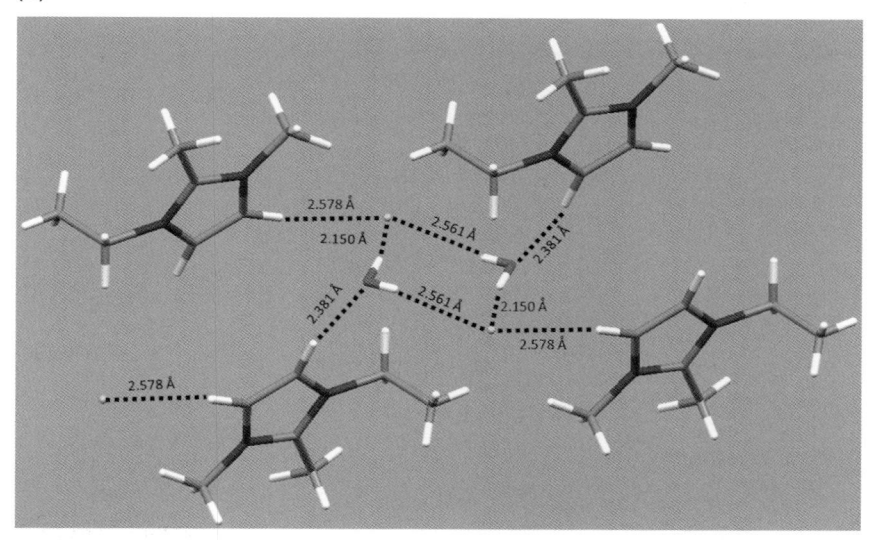

Figure 1.7 Structures of 1,3-dimethylimidazolium chloride hemihydrate (a) and 1,2-dimethyl-3-ethylimidazolium chloride hemihydrate (a) showing hydrogen bonding (dashed lines) [25] as determined by single crystal X-ray diffraction.

removing water in this way requires a large energy input and thus contributes further to the green inefficiency (*via* non-compliance with the 6th principle of green chemistry) of the purification of 1-alkyl-3-methylimidazolium halide salts. Indeed, the purification of 1-alkyl-3-methylimidazolium halide salts, as summarized in Figure 1.6, complies with none of the relevant principles of green chemistry.

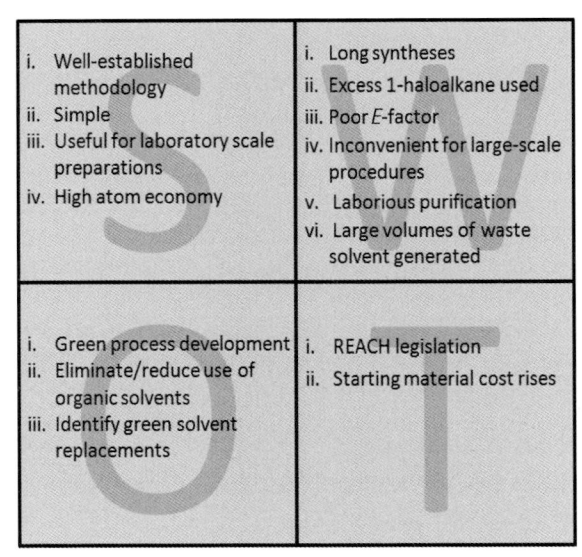

1.7.1
SWOT Analysis: Conductively Heated Preparation of 1-Alkyl-3-Methylimidazolium Halide Salts and Their Subsequent Purification

Examination of the SWOT analysis for the preparation of 1-alkyl-3-methylimidazolium halide salts using conductive heating and their subsequent purification (Figure 1.8) shows that the overall methodology is dirty. Nevertheless, the procedure possesses some strengths, *viz.* the processes are well established, simple, can be performed in even the most basic of laboratories to produce the salts on a small to medium scale (<1 kg) and possess 100% atom economies. On the other hand, both the preparations and purifications share the weakness of poor E-factors, since extensive purification is necessary to remove the excess of the 1-haloalkane employed, which generates large volumes of solvent waste. Further weaknesses are that large-scale preparations (>1 kg) and purifications are laborious and the use of conductive heating to promote the reactions, and also remove some residual water and organic solvents, is slow and energy inefficient. All the above-mentioned weaknesses show that opportunities exist to develop a greener methodology to

i. Well-established methodology ii. Simple iii. Useful for laboratory scale preparations iv. High atom economy	i. Long syntheses ii. Excess 1-haloalkane used iii. Poor E-factor iv. Inconvenient for large-scale procedures v. Laborious purification vi. Large volumes of waste solvent generated
i. Green process development ii. Eliminate/reduce use of organic solvents iii. Identify green solvent replacements	i. REACH legislation ii. Starting material cost rises

Figure 1.8 SWOT analysis: conductively heated preparation of 1-alkyl-3-methylimidazolium halide salts and their subsequent purification.

prepare pure 1-alkyl-3-methylimidazolium halide salts using conductive heating and thus reduce the costs of their manufacture, especially on large scales.

At present, the implementation of REACH (Registration, Evaluation, Authorization and restriction of CHemicals [30]) legislation is a threat to the continued low-cost manufacture of many chemicals, but especially for new products, since they require a large initial financial investment. Therefore, research scientists aiming to develop any new industrial-scale ionic liquid synthesis need to be extremely familiar with REACH legislation in order to achieve their aim. A further threat that exists for the synthesis of 1-alkyl-3-methylimidazolium halide salts using conductive heating is that starting material costs will undoubtedly rise and, therefore, the challenge facing chemists is to render preparations greener and concomitantly cheaper.

1.8
Ionic Liquid Syntheses Promoted by Microwave Irradiation

It has long been recognized that much more efficient energy sources exist to promote chemical reactions compared with conductive heating. In particular, microwave irradiation [31–35] has started to attract attention for the preparation of ionic liquids [Routes 1(b) and 2(a), Figure 1.1]. Some ionic liquids that have been prepared to date using microwave-assisted methodologies are shown in Table 1.1.

Microwave heating is governed by two mechanisms [34], dipole rotation and ionic conduction (Figure 1.9). Microwave heating by dipole rotation commonly occurs when molecules with high dielectric constants or polarities constantly and very rapidly realign to the fluctuating microwave field. Heating by an ionic conduction mechanism occurs because the ions present in a reaction medium (e.g. ionic liquid reaction mixture) begin to move under the influence of the electric field of the microwave irradiation, resulting in an increased collision rate and, therefore, the rapid conversion of kinetic energy into heat. Furthermore, microwaves penetrate a reaction medium directly as opposed to conductively heated reactions, provided the reaction vessel consists of a microwave-transparent material such as quartz and therefore, unlike conductive heating, preparation speeds are not limited by all the thermal conductivities of the materials to be penetrated (see Figure 1.5).

The promotion of 1-alkyl-3-methylimidazolium halide syntheses using microwave irradiation is favored by an ionic conduction heating mechanism since ionic liquids absorb microwave irradiation excellently. As an ionic liquid product is generated during a synthetic procedure, it absorbs microwave energy, which speeds up the reaction. Obviously, the more ionic liquid product has formed, the faster the reaction will proceed. In brief, microwave irradiation represents a far more efficient mode of heating to prepare ionic liquids than conductively heated syntheses. In addition, all microwave-assisted ionic liquid preparations align with the 6th principle of green chemistry by virtue of reduced preparation times and vastly reduced energy consumption (see Section 1.8.1).

At present, the implementation of microwave irradiation to prepare ionic liquids continues to evolve, but a question that remains unanswered is: How much greener

Table 1.1 Some ionic liquids prepared using microwave irradiation.

Ionic liquid			References
[imidazolium $-C_nH_{2n+1}$] X^-	[pyridinium-C_nH_{2n+1}] X^- $n = 3, 4, 6, 8$ or 10 $X = Cl, Br$ or I	[picolinium-C_nH_{2n+1}] X^- $n = 3$ or 4 $X = Cl$ or Br	[31, 33]
[thiazolium-C_nH_{2n+1}] X^-		[pyrrolidinium-C_nH_{2n+1}] X^-	[31]
	$n = 4, 6, 8$ or 10 $X = Br$ or I		
[ephedrinium-C_nH_{2n+1}] X^-		[ephedrinium-C_nH_{2n+1}] $[Y]^-$	[36]
	$n = 4, 8, 10$ or 16 $X = Br$ $Y = PF_6$ or CF_3SO_3		

are microwave-promoted preparations than traditional methods? From a very simplistic standpoint, tremendous energy savings have already been demonstrated for laboratory-scale ionic liquid syntheses [31, 33], since microwave-promoted reactions occur far more rapidly than the same preparations performed using conductive heating. For example, the synthesis of [C$_4$mim]Cl performed in a microwave oven at 300 W and taking 20 min to complete uses 99 W h (0.33 h × 300 W) of energy, whereas the same preparation uses 14 400 W h (24 h × 600 W) if performed employing an isomantle that operates at 600 W for 24 h [33]. Simplistically, this represents a 145-fold energy efficiency improvement when using microwave irradiation compared with conductive heating. This significant energy saving strongly aligns with the 6th principle of green chemistry by keeping the energy input of the syntheses to a minimum.

1.8 Ionic Liquid Syntheses Promoted by Microwave Irradiation | 17

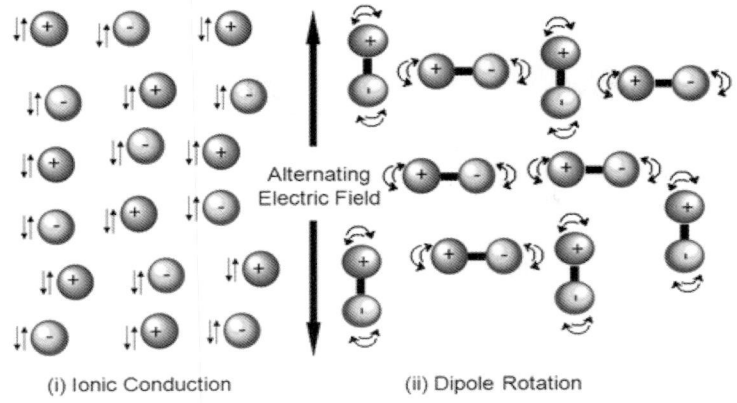

Figure 1.9 Schematic of microwave heating mechanisms: (i) ionic conduction and (ii) dipole rotation [34].

The first papers describing microwave syntheses of ionic liquids employed domestic microwave ovens that offered no temperature and pressure control and thus gave irreproducible results [35, 37, 38]. Since then, more reliable results have been obtained using commercial microwave reactors (Figure 1.10), which have allowed the temperature and pressure of reactions to be moderated, rendering the procedures far safer and more reproducible than their domestic predecessors [39].

It is worth noting that whereas the first reports describing microwave-assisted ionic liquid syntheses focused on their preparation *via* quaternization [Route 1(b), Figure 1.1], later reports [38, 40, 41] showed that a one-pot approach [Route 2(a), Figure 1.1] could also be used. Since the microwave-assisted procedures require much smaller 1-haloalkane excesses (~1 mol%) compared with conventional

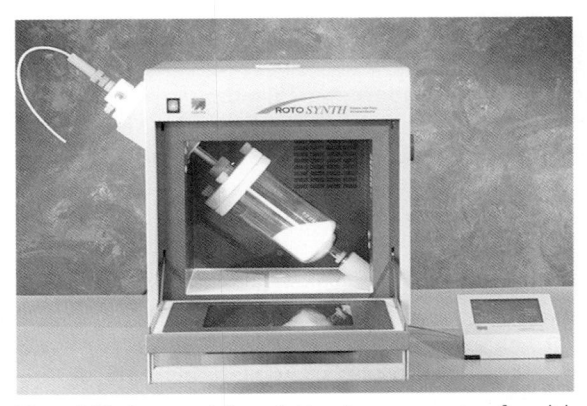

Figure 1.10 Commercially available microwave reactors for solid- and solution-phase batch synthesis. Photograph courtesy of Milestone srl, Sorisole, BG, Italy.

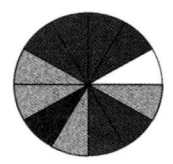

preparations (up to 400 mol% [26]), they align well with the 1st principle of green chemistry and their E-factors are far superior to those of traditional synthesis routes, especially if no organic solvent is used during the synthesis. In other words, if no organic solvent is used during synthesis, the preparations also align with the 2nd principle of green chemistry and the best E-factors for the preparations are obtained. In brief, the microwave-assisted preparations of 1-alkyl-3-methylimidazolium halide salts comply with seven of the eight relevant principles of green chemistry (1st, 2nd, 5th, 6th, 8th, 11th and 12th) whereas the analogous one-pot microwave-assisted syntheses align with six (1st, 5th, 6th, 8th, 11th and 12th). It must be noted that although the one-pot microwave-assisted ionic liquid syntheses generate a stoichiometric amount of MX waste (e.g. NaCl), the waste salt is far less harmful/toxic and straightforward to dispose of than 1-haloalkane waste and therefore the methodology aligns reasonably well with the 1st principle of green chemistry too.

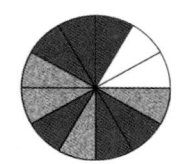

1.8.1
Microwave-assisted Versus Traditional Ionic Liquid Preparations

The superior E-factors and energy efficiencies of microwave-assisted syntheses of ionic liquids show that they are far greener than conductively heated preparations. However, the "degree of greenness" depends on (i) the type of quaternization being performed, since the kinetics of the reaction typically follow the order $C_nH_{2n+1}Cl < C_nH_{2n+1}Br < C_nH_{2n+1}I$, (ii) the 1-haloalkane excess required to achieve complete conversion to product, (iii) the scale of the reaction and (iv) if any solvent is necessary to reduce the viscosity and/or reduce product scrambling [29]. The proviso for the use of solvent in terms of green chemistry is that the solvent itself will have to be green (e.g. ethanol) and/or recycled to uphold the overall green credentials of a given preparation.

The preparation of ionic liquids on scales >2 kg also requires special consideration, as this is moving from bench scale to semi-pilot plant scale. However, at the semi-pilot plant scale, the 12 principles of green engineering [15] must also be considered, but as already mentioned, this falls outside the scope of this review. Nevertheless, current indications are that continuous flow reactors (Figure 1.11) represent the way forward to producing ionic liquids on large scales [37, 42].

1.8.2
SWOT Analysis: Microwave-promoted Syntheses of Ionic Liquids

The SWOT analysis for the preparation of ionic liquids using microwave irradiation is shown in Figure 1.12. The greatest green strengths of performing microwave-assisted ionic liquid preparations on a laboratory scale are the energy reduction by virtue of reduced reaction times and low 1-haloalkane excesses. Further strengths are that (i) different microwave reactors [35] can be employed to execute the preparations

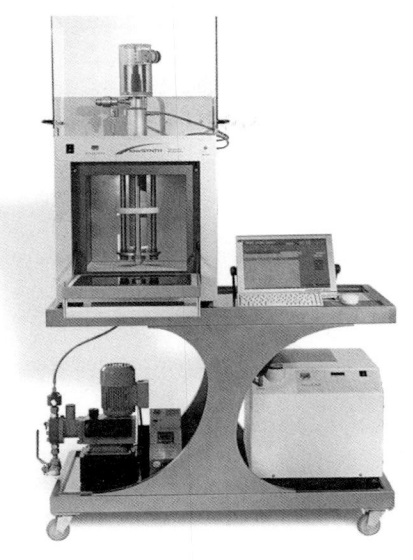

1.8 Ionic Liquid Syntheses Promoted by Microwave Irradiation

Figure 1.11 Commercially available microwave reactors for continuous flow synthesis. Photograph courtesy of Milestone srl, Sorisole, BG, Italy.

on different scales [37], (ii) syntheses can be performed solvent free, (iii) the reactions are 100% atom efficient and (iv) they have excellent (low) E-factor values compared with traditional preparations. All the before mentioned strengths render the microwave-promoted syntheses of ionic liquids as very green indeed.

i. Rapid ii. Energy efficient iii. Flexible reaction scales iv. Solvent-free synthesis v. High atom economy vi. Excellent E-factor	i. Lack of energy efficiency data ii. Expensive apparatus iii. Ionic liquid discoloration iv. Ionic liquid decomposition v. Mass transfer issues vi. Scale-up very expensive
i. Green process development ii. Quantitative energy input data iii. Reactor design iv. Intellectual property	i. Expensive operation ii. *In-situ* reaction monitoring iii. Safety controls

Figure 1.12 SWOT analysis for the preparation of ionic liquids using microwave irradiation.

The weaknesses associated with microwave-assisted preparation of ionic liquids are that:

1. Lack of energy efficiency data of syntheses makes an exact comparison with other methodologies difficult.
2. Microwave reactors are expensive compared with traditional synthesis apparatus.
3. Discolored ionic liquids are sometimes obtained at temperatures >75 °C [43, 44].
4. Ionic liquids decompose if overheated [29, 31, 45].
5. The high viscosity of ionic liquids can produce mass transport problems during synthesis.
6. The scale-up of the syntheses is very expensive.

Despite the above-mentioned weaknesses, opportunities exist for the development of green microwave reactor technology to produce ionic liquids. One major opportunity is green process development by establishing quantitative energy input data to assess definitively the greenness of microwave-assisted preparations versus other methodologies. Another opportunity is the design and manufacture of custom-made reactors to produce ionic liquids on large scales [42]. Both of these opportunities may generate valuable intellectual property and could also reduce the cost of producing ionic liquids on large scales.

The greatest threat to the development of microwave technology to produce ionic liquids on large scales by continuous flow is that it proves too expensive. Further threats include the inability, *via* design constraints, to incorporate *in situ* reaction monitoring (which aligns with the 11th principle of green chemistry) and the inability (albeit unlikely) to incorporate safety controls in industrial reactors to produce ionic liquids on large scales (which aligns with the 12th principle of green chemistry).

1.9
Syntheses of Ionic Liquids Promoted by Ultrasonic Irradiation

At about the same time that the first reports describing the microwave-assisted preparations of ionic liquids were published, reports also began to appear describing ultrasound-promoted preparations of ionic liquids [46, 47]. Some of the ionic liquids that have been prepared to date using ultrasonic irradiation are shown in Table 1.2.

The promotion of chemical reactions with ultrasound is due to a physical phenomenon known as cavitation, which is the formation, growth and implosive collapse of bubbles in a liquid [51]. The collapse or implosion of such bubbles results in some fascinating physical effects, which include the formation of localized "hotspots" in an elastic liquid and reduction of particle size [52, 53]. In terms of ionic liquid synthesis, the formation of hotspots favors quaternizations [Route 1(c), Figure 1.1], whereas the reduction of particle size favors both quaternizations and metathesis [Route 3, Figure 1.1] [47], since improved mass transport overcomes the viscosity issues generally associated with ionic liquid syntheses. Together, hotspot formation, particle size reduction and the reduced preparation times compared with traditional methods all speed up ionic liquid syntheses and thus represent a

1.9 Syntheses of Ionic Liquids Promoted by Ultrasonic Irradiation

Table 1.2 Some ionic liquids prepared using ultrasonic irradiation.

Ionic liquid	References
Imidazolium salt $[C_nH_{2n+1}]$, $n = 3, 4, 6$ or 8; $X = Cl, Br$ or I. Bis-imidazolium salt with C_nH_{2n+1}, $n = 4, 6$ or 8; $X = Br$ or I, counter-ion $2X^-$	[47]
Imidazolium $[Y]^-$ with C_nH_{2n+1}, $n = 4, 5$ or 6; $Y = BF_4, PF_6, BPh_4$ or CF_3SO_3. Pyridinium $[Y]^-$ with C_nH_{2n+1}, $n = 4, 5, 6$ or 8; $Y = PF_6$. Pyrrolidinium $[Y]^-$ with C_nH_{2n+1}, $n = 4$ or 8; $Y = PF_6$	[46, 48]
Pyridinium chloride with HO– and Cl– substituents; Pyridinium $[BF_4]^-$ with HO– and Cl– substituents	[49]

significant green advantage, especially if the preparations are performed solvent free. Indeed, it is worth noting that at present, the majority of papers describing ultrasound-promoted syntheses of ionic liquids have focused on 'one-pot' reactions [Route 2(b), Figure 1.1].

Despite the apparent green advantages of ultrasound-assisted ionic liquid preparations, a phenomenon which renders these preparations inadequate is that, almost without exception, ionic liquids discolor and decompose when exposed to ultrasonic irradiation for the time required to obtain acceptable conversions [50, 51]. Needless to say, this decomposition of ionic liquids is a severe disadvantage to the successful and widespread implementation of the technology and, from a green chemistry perspective, does not align with the 1st principle of green chemistry since very dirty purification and decolorization of the salts are required, which also give very poor *E*-factors. In addition, in a recent review, the highest reported tabulated yield to produce 1-alkyl-3-methylimidazolium halide and similar salts without halide anions are 95% and 90%, respectively [52]. This contrasts with yields of >99%

obtained by thermally induced and microwave-assisted preparations, rendering ultrasound-assisted ionic liquid syntheses far less green.

Although ultrasound-assisted preparations of 1-alkyl-3-methylimidazolium ionic liquids [Routes 1(c) and 2(b), Figure 1.1] comply with five of the eight relevant principles of green chemistry (namely the 5th, 6th, 8th, 11th and 12th), they do not follow the 1st and 2nd principles, since they give poor yields (compared with thermal and microwave routes) and the ionic liquid products require extensive purification. As a result, ultrasound-assisted ionic liquid syntheses also have poor E-factors and therefore are not as green as their microwave-assisted analogues.

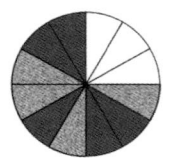

1.9.1
SWOT Analysis: Ultrasound-promoted Syntheses of Ionic Liquids

The SWOT analysis for the preparation of ionic liquids promoted by ultrasonic irradiation (Figure 1.13) reveals that its unique strength (compared with traditional and microwave-assisted syntheses) is improved mass transport. However, almost without exception, the preparations have very poor E-factors (high values) due to the discolored ionic liquid products, requiring extensive purification and decolorization efforts that produce large volumes of organic solvent and solid waste (see above). In addition to the major weakness of a poor E-factor, other weaknesses of ultrasound-assisted ionic liquid syntheses include the lack of quantitative energy efficiency data, the cost of the apparatus and lack of demonstrated large-scale production. It must be

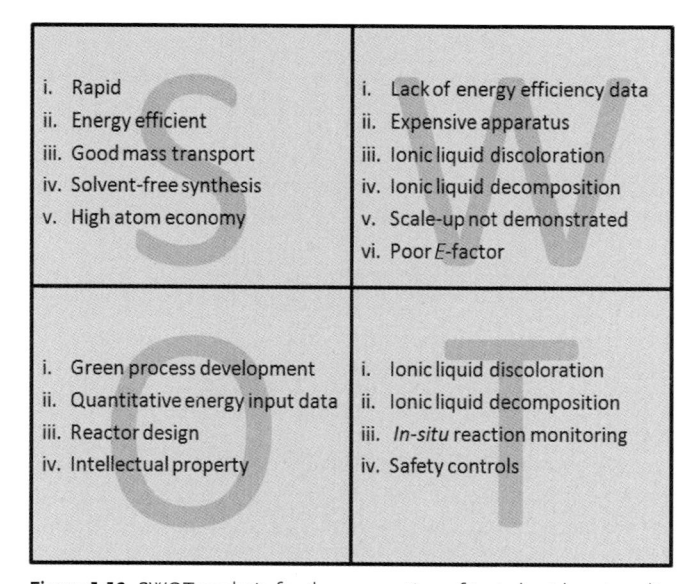

i. Rapid ii. Energy efficient iii. Good mass transport iv. Solvent-free synthesis v. High atom economy	i. Lack of energy efficiency data ii. Expensive apparatus iii. Ionic liquid discoloration iv. Ionic liquid decomposition v. Scale-up not demonstrated vi. Poor E-factor
i. Green process development ii. Quantitative energy input data iii. Reactor design iv. Intellectual property	i. Ionic liquid discoloration ii. Ionic liquid decomposition iii. *In-situ* reaction monitoring iv. Safety controls

Figure 1.13 SWOT analysis for the preparation of ionic liquids using ultrasonic irradiation.

pointed out that ionic liquid discoloration is far more severe under ultrasonic conditions than under microwave irradiation and, furthermore, the salts decompose under even mild ultrasonic conditions [51].

The greatest opportunities that exist for ultrasound-promoted ionic liquid syntheses are (i) to develop green procedures that do not lead to the discoloration and decomposition of ionic liquids and (ii) to determine the true energy efficiency of the preparations compared with other methodologies by collecting quantitative energy input data, thus allowing the opportunity for (iii) reactor design and (iv) generating valuable intellectual property.

It must be reiterated that ultrasound-assisted ionic liquid preparation will probably only find commercial application if its severe shortcomings (or threats in terms of the SWOT analysis) are overcome by future research and development efforts. Such research and development efforts will, similarly to microwave-assisted preparations, have to include safety controls and *in situ* reaction monitoring to align with the 11th and 12th principles of green chemistry.

1.10
Simultaneous Use of Microwave and Ultrasonic Irradiation to Prepare Ionic Liquids

Recent studies have shown that the simultaneous use of microwave and ultrasonic irradiation promote ionic liquid syntheses[Routes 1(d) and 2(c), Figure 1.1] [40, 52–54]. Some ionic liquids that have been prepared to date using this technology are shown in Table 1.3, which were obtained using one-step [Route 1(d), Figure 1.1], two-step [Route 1(a) then Route 3, Figure 1.1] as well as one-pot [Route 2(c), Figure 1.1] procedures.

The simultaneous use of microwave and ultrasonic irradiation to prepare ionic liquids should offer the cumulative benefits of the individual irradiations, *viz.* excellent coupling of microwaves with ionic liquids plus improved mass transport. Therefore, the time and energy saved using microwave and ultrasonic irradiation simultaneously would represent a significant green advantage, especially if the syntheses are performed solvent free. However, using this technology to prepare 1-alkyl-3-methylimidazolium halide ionic liquids [Route 1(d), Figure 1.1] gives very poor yields (<5%) and can also take significantly longer than using microwave or ultrasound irradiation alone [40]. Therefore, the combined use of microwave and ultrasonic irradiation to prepare 1-alkyl-3-methylimidazolium halide salts has very poor *E*-factors and requires significantly more energy than using the individual types of irradiation alone, rendering the syntheses dirty. In addition, a major problem that is anticipated for all ionic liquid syntheses using microwave and ultrasonic irradiation simultaneously is that discolored ionic liquid products will be obtained, requiring extensive decolorization, which will further negatively affect the *E*-factors.

Although the simultaneous microwave/ultrasound-assisted preparations of 1-alkyl-3-methylimidazolium halide [Route 1(d) Figure 1.1] and non-halide [Route 2 (c), Figure 1.1] salts comply with the five of the eight relevant principles of green

Table 1.3 Some ionic liquids prepared using simultaneous microwave and ultrasonic irradiation.

Ionic liquid	Reference
$\left[\text{N}\underset{+}{\bigcirc}\text{N}-C_nH_{2n+1}\right][Y]^-$ $\quad$ $n = 4$ or 8 $\quad$ $Y = BF_4, PF_6, CF_3SO_3$ or NTf_2 $\qquad$ $\left[\text{N}^+-C_8H_{17}\right][Y]^-$ (pyrrolidinium) $\quad$ $Y = BF_4, PF_6, CF_3SO_3$ or NTf_2	[40]
$\left[\text{Py}^+\text{-}C_8H_{17}\right][Y]^-$ $\quad$ $Y = BF_4, PF_6, CF_3SO_3$ or NTf_2	[53]

chemistry (namely the 5th, 6th, 8th, 11th and 12th), they do not follow the 1st and 2nd principles, since they give poor yields (compared with thermal and microwave routes) and the ionic liquid products require extensive purification. In addition, the E-factors of the preparations are poor due to the need to decolorize the ionic liquid products.

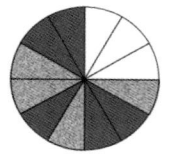

1.10.1
SWOT Analysis: Simultaneous Use of Microwave and Ultrasonic Irradiation to Prepare Ionic Liquids

The simultaneous use of microwave and ultrasonic irradiation to prepare ionic liquids offers both the best and worst of the individual technologies and the SWOT analysis shows that this indeed true (Figure 1.14).

The expected strengths of the technology include (i) rapid preparations with (ii) good energy efficiencies compared with traditional methods, (iii) improved mass transport (*versus* conductive heating and microwave-assisted preparations), (iv) the potential to perform the transformations solvent free and (v) high atom economies. At present, however, the limited number of studies describing syntheses have shown that the technology works best for one-pot preparations [Route 2(c), Figure 1.1] and is less successful for preparing ionic liquids containing halide anions [40]. The SWOT analysis clearly shows, without laboring the point, that at present, combined microwave/ultrasound irradiation technology has no obvious advantages or disadvantages.

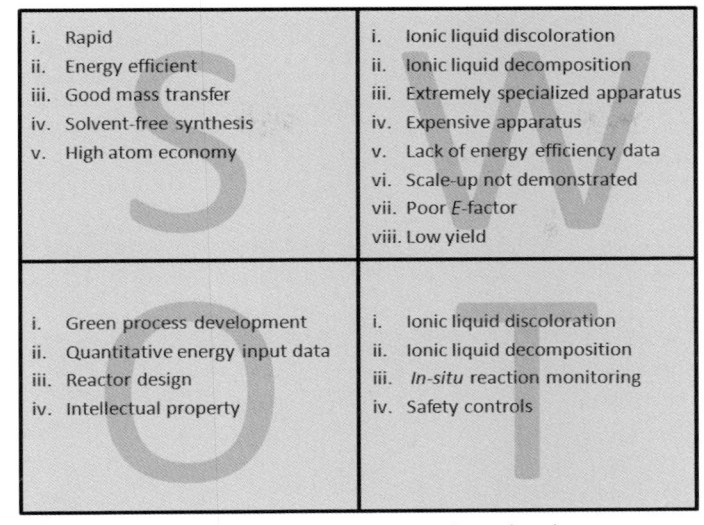

i. Rapid ii. Energy efficient iii. Good mass transfer iv. Solvent-free synthesis v. High atom economy	i. Ionic liquid discoloration ii. Ionic liquid decomposition iii. Extremely specialized apparatus iv. Expensive apparatus v. Lack of energy efficiency data vi. Scale-up not demonstrated vii. Poor *E*-factor viii. Low yield
i. Green process development ii. Quantitative energy input data iii. Reactor design iv. Intellectual property	i. Ionic liquid discoloration ii. Ionic liquid decomposition iii. *In-situ* reaction monitoring iv. Safety controls

Figure 1.14 SWOT analysis for the preparation of ionic liquids using simultaneous ultrasound and microwave irradiation.

1.11
Preparation of Ionic Liquids Using Microreactors

An exciting novel technology to prepare ionic liquids that has very recently emerged is the use of microstructured reactors (or microreactors), which function under continuous flow conditions [55–60]. In very simple terms, and as their name suggests, microstructured reactors are small reactors that can fit in the palm of the hand and their discerning feature is that they possess multiple parallel channels (microchannels) with diameters of $\sim$10–100 μm [61–63]. Two microreactors that have been used to prepare 1-ethyl-3-methylimidazolium ethylsulfate ([C$_2$mim][EtSO$_4$]) [56, 58] and 1-butyl-3-methylimidazolium bromide ([C$_4$mim]Br) [59] respectively, have their microchannels arranged in parallel either on a stack of plates (Figure 1.15 [58, 64]) or within a tube [56].

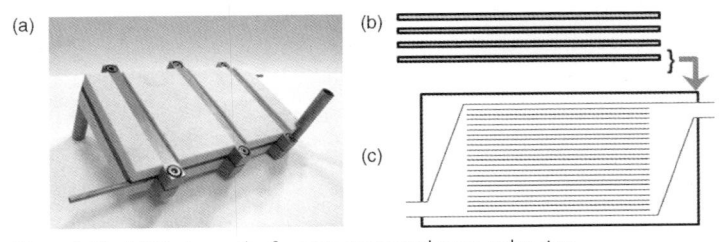

Figure 1.15 (a) Photograph of a microstructured reactor showing the arrangement of (b) four parallel plates and (c) a schematic of a single plate. Photograph courtesy of IMM, Mainz [64].

Table 1.4 Ionic liquids reported to date using microstructured reactors

Ionic liquid	References
[emim][EtSO$_4$] (1-ethyl-3-methylimidazolium ethylsulfate)	[56–58, 66]
[bmim]Br and [hmim]Br (1-butyl-3-methylimidazolium bromide; 1-hexyl-3-methylimidazolium bromide)	[59, 60]
1-(2-phenylethyl)-3-methylimidazolium bromide	[60]

Using the above-mentioned tubular- and stacked plate-type microstructured reactors, respectively, 1-ethyl-3-methylimidazolium ethylsulfate has been obtained with an excellent conversion rate of $X = 0.998$ [58], and 1-butyl-3-methylimidazolium bromide was obtained with a purity of >99% [56], demonstrating both the product yield and purity advantages of using microreactors [60]. In other words, ionic liquids can be prepared quickly and virtually without the need for purification by employing custom-made microstructured reactors. More examples of ionic liquids that have been prepared to date using microstructured reactors are shown in Table 1.4.

The key to the green advantages of using microstructured reactors is their microchannels, which allow for much better mass and heat transfer rates to prepare ionic liquids much faster than is possible with any other batch-type preparative method. In addition, excellent energy efficiency and reduced operational costs are achieved in the microstructured reactors, because heat exchange is far superior to that in traditional reactors, hotspot formation is prevented, higher reaction temperatures are possible and reaction volumes are reduced [58]. Since the salts are also obtained in high purities, purification using harmful organic solvents is not needed, which means that these types of preparations have excellent E-factors and are very green indeed. In fact, of all the preparative methodologies assessed here, the use of microstructured reactors to prepare the salts is the only technique that complies with all eight relevant principles of green chemistry, namely the 1st, 2nd, 3rd, 5th, 6th, 8th, 11th and 12th.

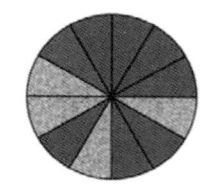

In practical terms, the advantages of using microstructured reactors become even clearer when it is considered that 9.3 kg per day of >99% pure 1-butyl-3-methylimidazolium bromide [59] is realistically achievable. To the best of our

knowledge, not even the most efficient batch-type industrial-scale ionic liquid preparations can compete with this greenness and efficiency. It is also worth highlighting that the high pressure stability of some microstructured reactors (up to 100 bar) means that they can permit continuous flow even at viscosities up to 10 000 MPa. Yet another advantage of this technology is that custom-made reactors are currently commercially available [64] and patented [65].

It must be pointed out that we have assumed that the E-factors associated with the preparation of ionic liquids using microreactors are excellent because of the high purity of the products obtained (e.g. >99% [59]). However, due to the nature of the apparatus (i.e. reagent delivery to a microreactor *via* pumps), reaction stoichiometries are not given in the literature. We have therefore presumed that these stoichiometries are 1 : 1 based on reported product purities of >99%. If, however, reaction stoichiometries are not 1 : 1, then we must remove one of the green credentials, namely the 2nd principle of green chemistry, which means that only seven of the eight relevant principles of green chemistry will apply (see Section 1.11).

1.11.1
SWOT Analysis: Preparation of Ionic Liquids Using Microreactors

The SWOT analysis for the preparation of ionic liquids using microstructured reactors (Figure 1.16) unambiguously shows that the strengths of the technology (as discussed in Section 1.11) heavily outweigh its weaknesses. Indeed, the only

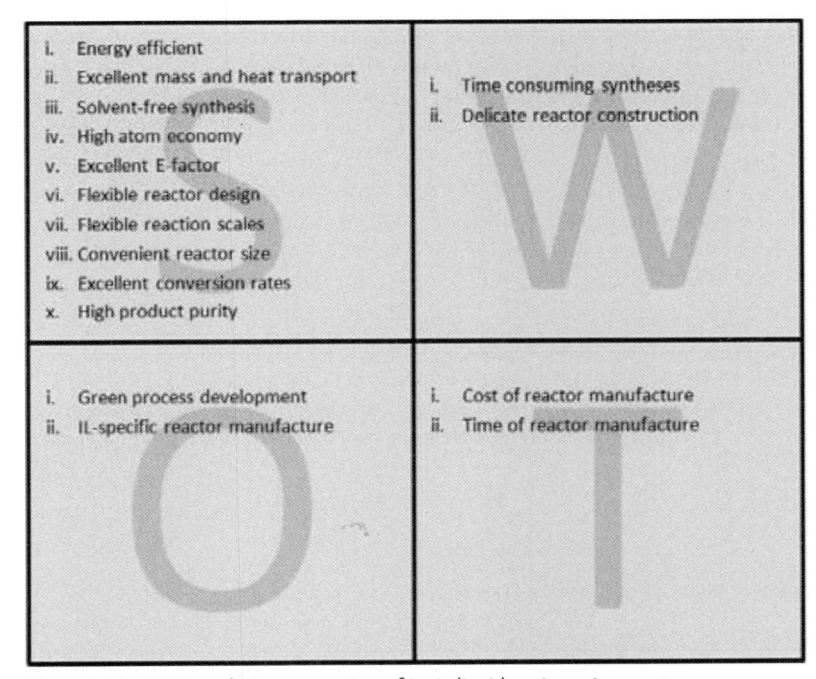

i. Energy efficient ii. Excellent mass and heat transport iii. Solvent-free synthesis iv. High atom economy v. Excellent E-factor vi. Flexible reactor design vii. Flexible reaction scales viii. Convenient reactor size ix. Excellent conversion rates x. High product purity	i. Time consuming syntheses ii. Delicate reactor construction
i. Green process development ii. IL-specific reactor manufacture	i. Cost of reactor manufacture ii. Time of reactor manufacture

Figure 1.16 SWOT analysis: preparation of ionic liquids using microreactors.

weakness of the technology is that some reported ionic liquid preparations take a long time to achieve good conversions when working at lower temperatures [56]. In addition, manufacture of the reactors is delicate and specialized, requiring laser welding, for example [58], to prevent potential problems such as leaking.

The opportunities of the technology are green process development and the market niche for the design and manufacture of suites of microstructured reactors for specific ionic liquid preparations or sets of preparations. The only threat that we see for the future widespread implementation of the technology is that microstructured reactors are currently still specialized pieces of apparatus that are expensive and time-consuming to manufacture.

1.12
Purification of Ionic Liquids with Non-halide Anions

Ionic liquid purification procedures can be divided into two main categories: purification of hydrophobic ionic liquids and purification of hydrophilic ionic liquids (Figure 1.2). To simplify discussion for the former and latter categories, we have selected two stereotypical salts as representatives, 1-butyl-3-methylimidazolium bisether amide, [C_4mim][NTf_2], and 1-butyl-3-methylimidazolium trifluorometha-nesulfonate, [C_4mim][OTf]. It must be noted that 1-alkyl-3-methylimidzaolium halide salts fall in the above-mentioned hydrophilic category, but their purification and associated green performance have already been discussed in Section 1.7). As already discussed in Section 1.3, only the 1st, 5th 6th and 12th principles of green chemistry are relevant to ionic liquid purification. Regardless of whether a one- or two-step methodology is employed to prepare [C_4mim][NTf_2] and [C_4mim][OTf] (Figure 1.1) a stoichiometric amount of MX waste (usually NaCl) is generated. If not removed from the ionic liquid product, the presence of the metal halide waste will severely affect the physical properties of the ionic liquid [7, 14]. The removal of MX from both [C_4mim][NTf_2] and [C_4mim][OTf] is achieved by washing the crude ionic liquid product with water, although the ease with which this is achieved and the respective green performances vary considerably.

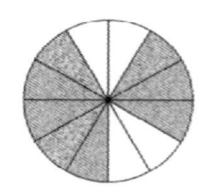

Although a stoichiometric amount of MX waste is generated during two-step ionic liquid syntheses, it is not toxic waste and therefore does not require incineration. Furthermore, the metal halide waste is also not contaminated with organics, which means that both one- and two-step syntheses to produce [C_4mim][NTf_2] and [C_4mim][OTf] closely align with the 1st principle of green chemistry.

1.12.1
Purification of Hydrophobic *Versus* Hydrophilic Ionic Liquids

In order to remove MX from hydrophilic [C_4mim][OTf], dichloromethane, a dense solvent, is usually added to the crude ionic liquid mixture. The dichloromethane is added since the ionic liquid preferentially dissolves therein, allowing repeated washing of the ionic liquid-containing phase with cold water to remove the MX waste more easily. The residual dichloromethane and also the water that remains in the ionic liquid as a result of the washing procedure require removal, which is achieved by first using a rotary evaporator and then drying the salt *in vacuo* at ∼70 °C for many hours, which is extremely energy inefficient. Almost needless to say, the use of dichloromethane to purify [C_4mim][OTf] is an extremely dirty practice since it is a toxic solvent that is detrimental to both humans and the environment [67]. Furthermore, the larger the volume of dichloromethane used, the poorer is the *E*-factor of the procedure.

The removal of MX from hydrophobic [C_4mim][NTf$_2$] is far easier than its removal from [C_4mim][OTf] because it does not readily mix with water and, therefore, washing the salt with water to extract MX is both faster and more efficient. This purification is also greener, since no organic solvent is required as is the case for hydrophilic salts to aid with the removal of MX. In principle, the water phase can also be recycled by distillation and reused, although some contaminated ionic liquid, now severely contaminated with MX, will remain, since even hydrophobic ionic liquids exhibit mutual solubility with water [68]. Moreover, the water must be removed in order to obtain the pure salt; this is achieved by drying the salt *in vacuo* at ∼70 °C for many hours, which is extremely energy inefficient.

In brief, the purification procedure of [C_4mim][OTf], is far less efficient than that of its cousin [C_4mim][NTf$_2$], since a significant amount of the ionic liquid is "lost" to the organic phase and higher levels of MX also remain in the purified ionic liquid. This is evidenced by the lower chloride content levels achievable for [C_4mim][NTf$_2$] [69] than for [C_4mim][OTf] [14, 70]. In addition, less energy is required to remove residual water from [C_4mim][NTf$_2$] than for [C_4mim][OTf] by heating the salts *in vacuo*, as the former should, by definition, hold less water than the latter. Although the purification of these ionic liquids does not comply with the relevant principles of green chemistry (namely the 1st, 5th, 6th and 12th), the hydrophilic ionic liquid processes are much greener than the hydrophilic processes, both in total isolated yield and in avoidance of organic solvents.

1.12.2
SWOT Analyses: Purification of Hydrophobic and Hydrophilic Ionic Liquids

The SWOT analyses for the purification of [C_4mim][NTf$_2$] and [C_4mim][OTf] show that the former procedure is the greener (Figure 1.17). This is mainly due to the purification of [C_4mim][OTf] requiring the use of dichloromethane (Section 1.12), which is an extremely harmful compound. Another major reason why the purification of [C_4mim][OTf] is dirty is because much greater losses of the ionic liquid to the water phase occur compared with the purification of [C_4mim][NTf$_2$] and it also requires longer heating *in vacuo* to remove residual water.

(a)

S
i. Simple
ii. Fast
iii. Efficient
iv. Good E-factor

W
i. Minimal ionic liquid losses to water phase
ii. Stoichiometric amounts of MX waste
iii. Heating *in vacuo*

O
i. Process optimization
ii. Scale-up

T
i. MX disposal

(b)

S
i. Simple

W
i. Poor E-factor
ii. Toxic dichloromethane used
iii. Large ionic liquid losses to water phase
iv. Stoichiometric amounts of MX waste
v. Extensive heating *in vacuo*
vi. Laboratory scale only

O
i. Process optimization
ii. Scale-up

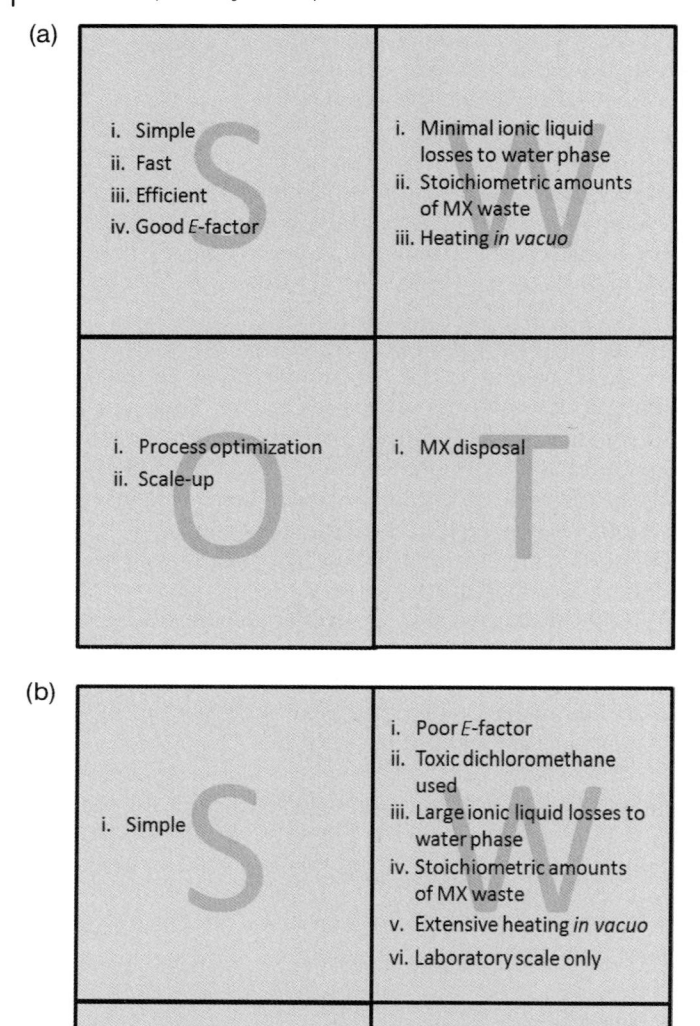

T
i. Waste disposal

Figure 1.17 SWOT analyses for the purification of (a) [C$_4$mim][NTf$_2$] and (b) [C$_4$mim][OTf].

1.13
Decolorization of Ionic Liquids

All practicing ionic liquid synthetic chemists know that the salts are sometimes obtained as pale yellow to black products. Ionic liquids containing halide anions are particularly susceptible to discoloration and the cause of the color can be transferred to subsequent ionic liquids prepared by metathesis (Route 3, Figure 1.1). Although chromophores are the suspected discoloration culprits [71], the true cause, and indeed the types of chromophores responsible, remain a mystery. To date, attempts to isolate the chromophores by column chromatography have failed to provide enough material for identification. This failure indicates the extremely low levels of the chromophores in ionic liquids (probably ppb levels of materials with molar extinction coefficients of $>10^6 \, l \, mol^{-1} \, cm^{-1}$). It is worth noting that even some of the most intensely colored ionic liquids are usually analytically pure to NMR and mass spectrometric techniques.

The color of an ionic liquid is usually not detrimental when using the salts as solvents (provided that they contain minimal levels of other impurities such as chloride and/or water), but colorless ionic liquids are essential for spectroscopic studies in order to eliminate interference from the suspected chromophoric impurity resonances that usually appear in the aromatic region of UV–Vis spectra.

Colorless ionic liquids may be obtained in two ways: careful reaction preparation and execution or post-synthesis, using decolorization. The former is the method of choice, achieved by purification of all starting materials and diligent monitoring of the reaction temperature to avoid overheating ($\sim<75\,°C$), which is known to produce colored ionic liquids. On the other hand, post-synthesis decolorization may be effected using a decolorizing column (Figure 1.18), a method that has recently been developed and which is scrutinized here for "greenness" in terms of the relevant principles of green chemistry (1st, 5th 6th and 12th) and also the *E*-factor.

1.13.1
SWOT Analysis: Decolorization of Ionic Liquids

It is almost unnecessary to do a SWOT analysis to establish that the decolorization of ionic liquids is not green (Figure 1.19). This is because from a green chemistry perspective, the toxicity of dichloromethane alone outweighs all strengths associated with the methodology and, since large volumes of the toxic solvent are also used during the process, it makes the decolorization procedure even dirtier, with an extremely poor *E*-factor.

It is important to note that the "column decolorization" of ionic liquids is not only environmentally unfriendly, but it will also probably only be applied in academia to obtain small quantities of ionic liquids destined for spectroscopic studies. Nevertheless, the method is simple and efficient on the laboratory scale but, in contrast, it uses a lot of dichloromethane and also produces solid waste (Celite, silica and activated charcoal) that need to be disposed of. On the upside, much research opportunity exists to render the methodology green and also to identify the nature

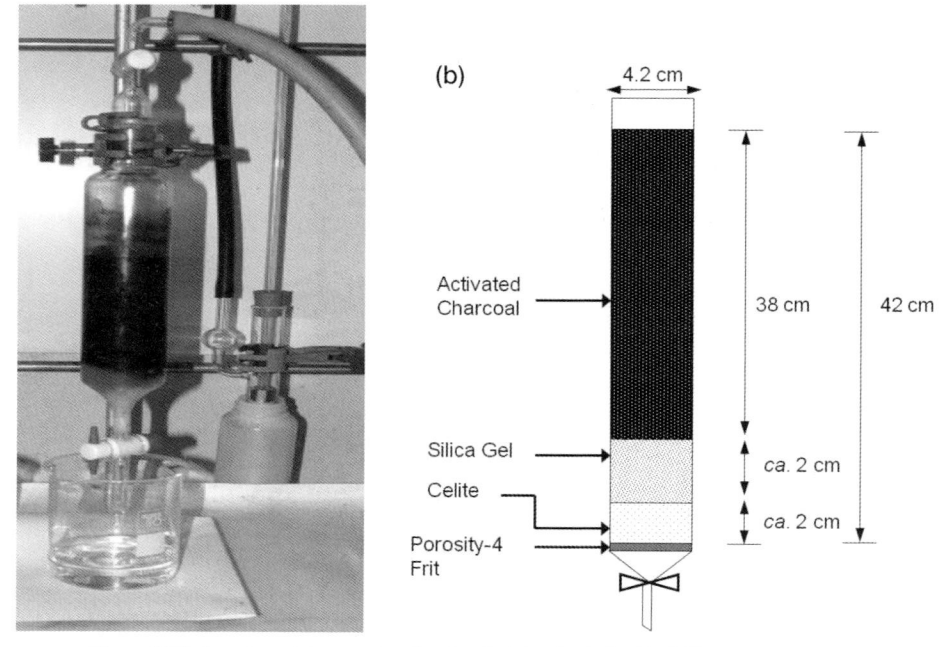

Figure 1.18 Experimental apparatus for decolorizing ionic liquids [71].

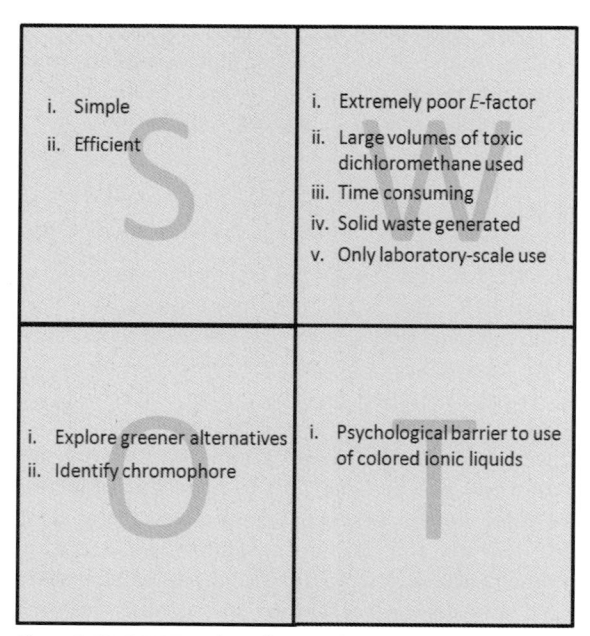

i. Simple ii. Efficient	i. Extremely poor E-factor ii. Large volumes of toxic dichloromethane used iii. Time consuming iv. Solid waste generated v. Only laboratory-scale use
i. Explore greener alternatives ii. Identify chromophore	i. Psychological barrier to use of colored ionic liquids

Figure 1.19 SWOT analysis for ionic liquid decolorization.

Table 1.5 Greenness summary of ionic liquid preparative methods.

Methodology	E-factor	Atom economy	Principles of green chemistry complied with (total number and type)[a]
Route 1(a)	Good–excellent	High	3 (8th, 11th, 12th)
Route 1(b)	Excellent	High	7 (1st, 2nd, 5th, 6th, 8th, 11th, 12th)
Route 1(c)	Poor	Low–high	5 (5th, 6th, 8th, 11th, 12th)
Route 1(d)	Poor	High	5 (5th, 6th, 8th, 11th, 12th)
Route 2(a)	Poor–good	Low–medium	7 (1st, 2nd, 5th, 6th, 8th, 11th, 12th)
Route 2(b)	Poor	Low–Medium	5 (5th, 6th, 8th, 11th, 12th)
Route 2(c)	Very poor	Low–medium	5 (5th, 6th, 8th, 11th, 12th)

(*Continued*)

Table 1.5 (*Continued*)

Methodology	E-factor	Atom economy	Principles of green chemistry complied with (total number and type)[a]
Route 3	Poor	Low–medium	3 (8th, 11th, 12th)
Microstructured reactors	Excellent[b]	Excellent	8 (1st, 2nd, 3rd, 5th, 6th, 8th, 11th, 12th)

[a] 4th, 7th, 9th and 10th principles of green chemistry do not apply.
[b] But see Section 1.11.

and identity of the chromophore. Until these opportunities exist, the threat to the methodology is the psychological barrier that exists for many ionic liquid researchers, especially those new to the field, to use colored ionic liquids.

1.14
Conclusion

The greenness of the synthetic procedures and purification methodologies assessed in this critical review are summarized in Table 1.5.

To maintain green credibility, ionic liquids must be green both in application and in their synthesis. The above discussions clearly illustrate that ionic liquid synthesis and purification can certainly be considered as green if microwave-assisted synthesis and microstructured reactors are employed. It is also clear that the purification of hydrophobic ionic liquids is intrinsically greener than that of hydrophilic ionic liquids.

With current trends towards the design of non-toxic, biodegradable ionic liquids, the new challenges become to develop improved purification procedures for hydrophilic ionic liquids alongside a universal requirement for the development of *in situ* on-line analytical monitoring for industrial-scale syntheses. Overall, the judgment provided here for the synthesis and purification of ionic liquids is "green, but not green enough".

Acknowledgments

We gratefully acknowledge the EPSRC for funding under their Portfolio Partnership Scheme (Grant No. EP/D029538/1) and also thank Milestone, particularly Dr Mauro

Iannelli, for the supply of their microwave units to QUILL. We are also grateful to Professor Paul J. Dyson for his insightful comments on the manuscript. M.D. thanks Dr J.D. Holbrey and Dr M. Nieuwenhuyzen for help in generating the images for Figure 1.7.

References

1 Freemantle, M. (1998) Designer solvents – ionic liquids may boost clean technology development. *Chemical & Engineering News*, **76** (30 March), 32–37.

2 Katritzky, A.R., Singh, S., Kirichenko, K., Holbrey, J.D., Smiglak, M., Reichert, W.M. and Rogers, R.D. (2005) 1-Butyl-3-methylimidazolium 3,5-dinitro-1,2,4-triazolate: a novel ionic liquid containing a rigid, planar energetic anion. *Chemical Communications*, 868–870.

3 Anastas, P.T. and Kirchhoff, M.M. (2002) Origins, current status and future challenges of green chemistry. *Accounts of Chemical Research*, **35**, 686–694.

4 Anastas, P.T. and Warner, J.C. (1998) *Green Chemistry: Theory and Practice*, Oxford University Press, New York.

5 Davis, J.H. Jr., Gordon, C.M., Hilgers, C. and Wasserscheid, P. (2003) Synthesis and purification of ionic liquids, in *Ionic Liquids in Synthesis*, 1st edn (eds P. Wasserscheid and T. Welton), Wiley-VCH Verlag GmbH, Weinheim, pp. 7–21.

6 Carmichael, A.J., Deetlefs, M., Earle, M.J., Fröhlich, U. and Seddon, K.R. (2003) Ionic liquids: Improved syntheses and new products, in *Ionic Liquids as Green Solvents: Progress and Prospects*, vol. 856 (eds R.D. Rogers and K.R. Seddon), ACS Symposium Series American Chemical Society, Washington, DC, pp. 14–31.

7 Stark, A. and Seddon, K.R. (2007) Ionic liquids, in *Kirk–Othmer Encyclopedia of Chemical Technology*, 5th edn, vol. 26 (ed. A. Seidel), John Wiley & Sons, Inc, Hoboken, NJ, pp. 836–920.

8 Gordon, C.M. and Muldoon, M.J. (2008) Synthesis of ionic liquids, in *Ionic Liquids in Synthesis*, vol. 1 (eds P. Wasserscheid and T. Welton), Wiley-VCH Verlag GmbH, Weinheim, pp. 7–25.

9 Deetlefs, M. and Seddon, K.R. (2010) Assessing the greenness of some typical ionic liquid preparations. *Green Chemistry*, submitted.

10 Deetlefs, M. and Seddon, K.R. (2010) *The Handbook of Ionic Liquid Synthesis*, Wiley-VCH Verlag GmbH, Weinheim, in preparation.

11 Wasserscheid, P., van Hal, R. and Boesmann, A. (2002) 1-Butyl-3-methylimidazolium ([bmim]) octylsulfate – an even "greener" ionic liquid. *Green Chemistry*, **4**, 400–404.

12 Wasserscheid, P., van Hal, R., Bosmann, A., Esser, J. and Jess, A. (2003) New ionic liquids based on alkylsulfate and alkyl oligoether sulfate anions: synthesis and applications, in *Ionic Liquids as Green Solvents: Progress and Prospects*, vol. 856 (eds R.D. Rogers and K.R. Seddon), ACS Symposium Series, American Chemical Society, Washington, DC, pp. 57–69.

13 Earle, M.J. and Seddon, K.R. (2001) Preparation of imidazole carbenes and the use thereof for the synthesis of ionic liquids. World Patent, WO 2001077081.

14 Seddon, K.R., Stark, A. and Torres, M.-J. (2000) Influence of chloride, water and organic solvents on the physical properties of ionic liquids. *Pure and Applied Chemistry*, **72**, 2275–2287.

15 Anastas, P.T. and Zimmerman, J.B. (2003) Design through the 12 principles of green engineering. *Environmental Science & Technology*, **37**, 94A.

16 Tang, S., Bourne, R., Smith, R. and Poliakoff, M. (2008) The 24 principles of

green engineering and green chemistry: IMPROVEMENTS PRODUCTIVELY. *Green Chemistry*, **10**, 268–269.

17 McHale, G., Hardacre, C., Ge, R.N.D., Allen, R.W.K., MacInnes, J.M., Bown, M.R. and Newton, M.I., (2008) Density–viscosity product of small-volume ionic liquid samples using quartz crystal impedance analysis. *Analytical Chemistry*, **80**, 5806–5811.

18 Ge, R., Allen, R.W.K., Aldous, L., Bown, M.R., Doy, N., Hardacre, C., MacInnes, J.M., McHale, G., Newton, M.I. and Rooney, D.W. (2009) Evaluation of a Microfluidic device for the electrochemical determination of halide content in ionic liquids. *Analytical Chemistry*, **81**, 1628–1637.

19 Trost, B.M. (1991) The Atom economy – a search for synthetic efficiency. *Science*, **254**, 1471–1477.

20 Sheldon, R.A. (1992) Organic-synthesis – past, present and future. *Chemistry & Industry*, **23**, 903–906.

21 Sheldon, R.A. (2007) The E-factor: fifteen years on. *Green Chemistry*, **9**, 1273–1283.

22 Appleby, R.C., (1994) *Modern Business Administration*, 6th edn, Pearson Education Limited, Harlow, England.

23 Fannin, A.A., Floreani, D.A., King, L.A., Landers, J.S., Piersma, B.J., Stech, D.J., Vaughn, R.L., Wilkes, J.S. and Williams, J.L. (1984) Properties of 1,3-dialkylimidazolium chloride-aluminum chloride ionic liquids. 2. Phase transitions, densities, electrical conductivities and viscosities. *The Journal of Physical Chemistry*, **88**, 2614–2621.

24 Wilkes, J.S., Levisky, J.A., Wilson, R.A. and Hussey, C.L. (1982) Dialkylimidazolium chloroaluminate melts: a new class of room-temperature ionic liquids for electro-chemistry, spectroscopy and synthesis. *Inorganic Chemistry*, **21**, 1263–1264.

25 Elaiwi, A.E. (1994) Mass spectrometry of organic and chlorometallated salts, DPhil thesis, University of Sussex, Brighton.

26 Smith, G.P., Dworkin, A.S., Pagni, R.M. and Zingg, S.P. (1989) Brønsted superacidity of HCl in a liquid chloroaluminate. $AlCl_3$.1-ethyl-3-methyl-1H-imidazolium chloride. *Journal of the American Chemical Society*, **111**, 525–530.

27 Bonhôte, P., Dias, A.-P., Papageorgiou, N., Kalyanasundaram, K. and Grätzel, M. (1996) Hydrophobic, highly conductive ambient-temperature molten salts. *Inorganic Chemistry*, **35**, 1168–1178.

28 Suarez, P.A.Z., Dullius, J.E.L., Einloft, S., De Souza, R.F. and Dupont, J. (1996) The use of new ionic liquids in two-phase catalytic hydrogenation reaction by rhodium complexes. *Polyhedron*, **15**, 1217–1219.

29 Jeapes, A.J., Thied, R.C., Seddon, K.R., Pitner, W.R., Rooney, D.W., Hatter, J.E. and Welton, T. (2001) Process for recycling ionic liquids. World Patent, WO 2001015175.

30 Health and Safety Executive (2008) *REACH (Registration, Evaluation, Authorisation & restriction of CHemicals)*, http://www.hse.gov.uk/reach/index.htm (Last accessed February 01, 2010).

31 Deetlefs, M. and Seddon, K.R. (2003) Improved preparations of ionic liquids using microwave irradiation. *Green Chemistry*, **5**, 181–186.

32 Varma, R.S. and Namboodiri, V.V. (2001) Solvent-free preparation of ionic liquids using a household microwave oven. *Pure and Applied Chemistry*, **73**, 1309–1331.

33 Khadilkar, B.M. and Rebeiro, G.L. (2002) Microwave-assisted synthesis of room-temperature ionic liquid precursor in closed vessel. *Organic Process Research & Development*, **6**, 826–828.

34 Leadbeater, N.E. and Torenius, H.M. (2006) Microwaves and ionic liquids, in *Microwaves in Organic Synthesis* (ed. A. Loupy), Wiley-VCH Verlag GmbH, Weinheim, pp. 327–361.

35 Varma, R.S. and Namboodiri, V.V. (2001) An expeditious solvent-free route to ionic liquids using microwaves. *Chemical Communications*, 643–644.

36 Vo Thanh, G., Pegot, B. and Loupy, A. (2004) Solvent-free microwave-assisted preparation of chiral ionic liquids from

(−)-N-methylephedrine. *European Journal of Organic Chemistry*, 1112–1116.
37. Boros, È., Seddon, K.R. and Strauss, C.R. (2008) Chemical processing with microwaves and ionic liquids. *Chimica Oggi-Chemistry Today*, **26** (November–December), 28–30.
38. Namboodiri, V.V. and Varma, R.S. (2002) An improved preparation of 1,3-dialkylimidazolium tetrafluoroborate ionic liquids using microwaves. *Tetrahedron Letters*, **43**, 5381–5383.
39. Roberts, B.A. and Strauss, C.R. (2005) Toward rapid, green, predictable microwave-assisted synthesis. *Accounts of Chemical Research*, **38**, 653–661.
40. Cravotto, G., Gaudino Emanuela, C., Boffa, L., Lévêque, J.-M., Estager, J. and Bonrath, W. (2008) Preparation of second generation ionic liquids by efficient solvent-free alkylation of N-heterocycles with chloroalkanes. *Molecules*, **13**, 149–156.
41. Xu, D.Q., Liu, B.Y., Luo, S.P., Xu, Z.Y. and Shen, Y.C. (2003) A novel and eco-friendly method for the preparation of ionic liquids. *Synthesis*, **7**, 2626–2629.
42. Erdmenger, T., Paulus, R.M., Hoogenboom, R. and Schubert, U.S. (2008) Scaling-up the synthesis of 1-butyl-3-methylimidazolium chloride under microwave irradiation. *Australian Journal of Chemistry*, **61**, 197–203.
43. Begg, C.G., Grimmett, M.R. and Wethey, P.D. (1973) The thermally induced rearrangement of 1-substituted imidazoles. *Australian Journal of Chemistry*, **26**, 2435–2438.
44. Farmer, V. and Welton, T. (2002) The oxidation of alcohols in substituted imidazolium ionic liquids using ruthenium catalysts. *Green Chemistry*, **4**, 97–102.
45. Abdul-Sada, A.K., Ambler, P.W., Hodgson, P.K.G., Seddon, K.R. and Stewart, N.J. (1995) Ionic liquids of imidazolium halide for oligomerization or polymerization of olefins. World Patent, WO 9521871.
46. Lévêque, J.-M., Luche, J.-L., Pétrier, C., Roux, R. and Bonrath, W. (2002) An improved preparation of ionic liquids by ultrasound. *Green Chemistry*, **4**, 357–360.
47. Namboodiri, V.V. and Varma, R.S. (2002) Solvent-free sonochemical preparation of ionic liquids. *Organic Letters*, **4**, 3161–3163.
48. Estager, J., Lévêque, J.-M., Cravotto, G., Boffa, L., Bonrath, W. and Draye, M. (2007) One-pot and solventless synthesis of ionic liquids under ultrasonic irradiation. *Synlett*, 2065–2068.
49. Zhao, S., Zhao, E., Shen, P., Zhao, M. and Sun, J. (2008) An atom-efficient and practical synthesis of new pyridinium ionic liquids and application in Morita–Baylis–Hillman reaction. *Ultrasonics Sonochemistry*, **15**, 955–959.
50. Li, X., Zhao, J., Li, Q., Wanga, L. and Tsang, S.C. (2007) Ultrasonic chemical oxidative degradations of 1,3-dialkylimidazolium ionic liquids and their mechanistic elucidations. *Dalton Transactions*, 1875–1880.
51. Oxley, J.D., Prozorov, T. and Suslick, K.S. (2003) Sonochemistry and sonoluminescence of room-temperature ionic liquids. *Journal of the American Chemical Society*, **125**, 11138–11139.
52. Lévêque, J.-M., Estager, J., Draye, M., Cravotto, G., Boffa, L. and Bonrath, W. (2007) Synthesis of ionic liquids using non-conventional activation methods: an overview. *Monatshefte fur Chemie*, **138**, 1103–1113.
53. Cravotto, G., Boffa, L., Lévêque, J.-M., Estager, J., Draye, M. and Bonrath, W. (2007) A speedy one-pot synthesis of second-generation ionic liquids under ultrasound and/or microwave irradiation. *Australian Journal of Chemistry*, **60**, 946–950.
54. Cravotto, G. and Cintas, P. (2007) The combined use of microwaves and ultrasound: improved tools in process chemistry and organic synthesis. *Chemistry - A European Journal*, **13**, 1902–1909.
55. Grosse Böwing, A. and Jess, A. (2005) Kinetics of single- and two-phase synthesis of the ionic liquid 1-butyl-3-

methylimidazolium chloride. *Green Chemistry*, **7**, 230–235.
56 Grosse Böwing, A. and Jess, A. (2007) Kinetics and reactor design aspects of the synthesis of ionic liquids – experimental and theoretical studies for the ethylmethylimidazole ethylsulfate. *Chemical Engineering Science*, **62**, 1760–1769.
57 Grosse Böwing, A., Jess, A. and Wasserscheid, P. (2005) Kinetik und Reaktionstechnik der Synthese ionischer Flüssigkeite. *Chemical Engineering & Technology* , **77**, 1430–1439.
58 Renken, A., Hessel, V., Löba, P., Miszczuk, R., Uerdingen, M. and Kiwi-Minsker, L. (2007) Ionic liquid synthesis in a microstructured reactor for process intensification. *Chemical Engineering and Processing* , **46**, 840–845.
59 Waterkamp, D.A., Heiland, M., Schlüter, M., Sauvageau, J.C., Beyersdorff, T. and Thöming, J. (2007) Synthesis of ionic liquids in micro-reactors – a process intensification study. *Green Chemistry*, **9**, 1084–1090.
60 Wilms, D., Klos, J., Kilbinger, A.F.M., Löwe, H. and Frey, H. (2009) Ionic liquids on demand in continuous flow. *Organic Process Research & Development*, **13**, 961–964.
61 Hessel, V., Löwe, H., Müller, A. and Kolb, G. (2005) *Chemical Micro Process Engineering*, Wiley-VCH Verlag GmbH, Weinheim.
62 Ehrfeld, W., Hessel, V. and Löwe, H. (2000) *Microreactors: New Technology for Modern Chemistry*, Wiley-VCH Verlag GmbH, Weinheim.
63 Kockmann, N., Brand, O. and Fedder, G.K. (2006) *Micro Process Engineering: Fundamentals, Devices, Fabrication and Applications*, Wiley-VCH Verlag GmbH, Weinheim.
64 Institut für Mikrotechnik Mainz (IMM) (2009) http://www.imm-mainz.de/ (accessed 29 July 2009).
65 Stark, A., Bierbaum, R. and Ondruschka, B. (2004) Verfahren zur Herstellung ionischer Flüssigkeiten (process for the manufacture of ionic liquids), DE 102004040016A1.
66 Löb, P., Hessel, V., Krtschil, U. and Löwe, H., (2006) Continuous micro-reactor rigs with capillary sections in organic synthesis: generic process flow sheets, practical experience and novel chemistry. *Chimica Oggi-Chemistry Today*, **24**, (November–December), 46–50.
67 Scorecard – The Pollution Information Site (2005) Dichloromethane http://www.scorecard.org/chemical-profiles/summary.tcl?edf_substance_id=75-09-2 (Last accessed August 20, 2009).
68 Deetlefs, M. and Seddon, K.R. (2006) Ionic liquids: fact and fiction. *Chimica Oggi-Chemistry Today*, **24** (March–April), 16–17, 20–23.
69 Driver, G.W. (2006) Physical chemistry and thermodynamics of ionic liquids and ionic liquid solutions, PhD thesis, Queen's University of Belfast.
70 Villagrán, C., Deetlefs, M., Pitner, W.R. and Hardacre, C. (2004) Quantification of halide in ionic liquids using ion chromatography. *Analytical Chemistry*, **76**, 2118–2123.
71 Earle, M.J., Gordon, C.M., Plechkova, N.V., Seddon, K.R. and Welton, T., (2007) Decolorization of ionic liquids for spectroscopy. *Analytical Chemistry*, **79**, 758–764.

Part II
Green Synthesis Using Ionic Liquids

2
Green Organic Synthesis in Ionic Liquids
Peter Wasserscheid and Joni Joni

2.1
General Aspects

Ionic liquids are fascinating liquid materials which can be applied as solvents and as catalytic liquids in organic synthesis. The properties of ionic liquids are primarily governed by a complex interplay of Coulombic, hydrogen bonding and van der Waals interactions [1]. This special interplay is highly sensitive to the chemical structure of both the ionic liquid's anion and cation. For many cation–anion combinations it leads to unique property profiles and to a liquid character that is distinctively different from that of neutral molecular solvents.

However, this distinct difference in solvent and catalyst properties does not necessarily mean that in every organic reaction ionic liquids behave very differently from neutral molecular solvents. Consequently, the performance of an organic synthetic protocol can be very different when carried out in an ionic liquid but there are also many reported cases in which the reaction in ionic liquids proceeds more or less like in a molecular, polar and aprotic solvent.

When it comes to an evaluation of organic synthesis in ionic liquids in the light of green chemistry this performance aspect is the crucial point to consider. Why?

In comparison with many of the commonly used molecular solvents, most ionic liquids are more difficult and more costly to prepare (see Chapter 1 and Section 5.2 in Chapter 5 for details). Moreover, it has been clearly shown that some ionic liquid structures (by far not all!) exhibit problematic toxicity and ecotoxicity properties (see Chapter 4 for details). Compared with methanol and toluene, ionic liquids will always be more expensive and more difficult to produce and to dispose of. Hence ionic liquids are *not intrinsically greener* than molecular solvents and a synthetic protocol to produce an organic compound will not become greener just by replacing a molecular solvent by an ionic liquid!

How can we realize green synthetic chemistry with the help of ionic liquids? It is the clever use of the unique property profile of a specific ionic liquid in a specific reaction that can create Green chemistry in ionic liquids if the resulting new protocol is better, more efficient and more sustainable. Moreover, the amount of ionic liquid that is necessary to achieve this superior performance should be as low as possible to

Handbook of Green Chemistry, Volume 6: Ionic Liquids
Edited by Peter Wasserscheid and Annegret Stark
Copyright © 2010 WILEY-VCH Verlag GmbH & Co. KGaA, Weinheim
ISBN: 978-3-527-32592-4

keep the negative influence of ionic liquid synthesis and disposal on the life cycle assessment to a minimum (see Section 5.2 in Chapter 5 for details).

Consequently, it is very unlikely that a transformation in ionic liquids is particularly "green" if the same reaction works with the same performance (same selectivity, same rate, same recycling procedure) in water, ethanol or even without a solvent (see Section 5.2 in Chapter 5 for a particularly convincing example).

To identify promising examples of green synthetic chemistry with ionic liquids, the most promising (but time-consuming) approach is to start from a detailed understanding of both the properties of the ionic liquid under investigation and the mechanism of the intended reaction. The need for a specific ionic liquid structure in green synthetic chemistry with ionic liquids depends completely on the desired ionic liquid effect. If the envisaged advantage is simply linked to a very general effect of the nature of the ionic liquid, many different ionic liquid structures can be used (e.g. vapor pressure reduction of highly volatile toxic reactants). In this case, the greenest (which means the most ecologically benign and the cheapest) ionic liquid should be applied. However, in most instances the ionic liquid needs to fulfill special functions (e.g. it needs to act as an acid or it should provide a very high solubility for one of the reactants) and in this case a special chemical structure of the ionic liquid is needed. In some special cases, the use of task-specific ionic liquid [2] or even toxic ionic liquids may be required to provide the desired improvement in performance. In this event, the overall result of using this problematic ionic liquid can still result in a significant green chemistry advantage if the ionic liquid-based process saves a lot of waste or reduces significantly raw material consumption or energy. The advantage of the application of an ionic liquid is even more obvious if the amount of the ionic liquid used is very small and – in the case of a toxic ionic liquid – its handling occurs in the protected environment of a closed chemical plant.

Obviously, all applications of a toxic ionic liquid leave room for further improvement with respect to green chemistry, namely to identify a less toxic or less complex ionic liquid which can fulfill the same task. In any case, the benefits offered by ionic liquid-based processes must greatly overcompensate the additional investment needed for the safe handling of toxic ionic liquids or efforts to synthesize a highly complex task-specific ionic liquid.

It is remarkable in this context that most organic reactions using ionic liquids that fulfill these requirements are characterized by the fact that the ionic liquid is more than just the solvent but acts as a kind of functional part of the reaction mixture. Often the ionic liquid is catalytically active or provides an environment that stabilizes reaction intermediates in such a way that the reaction is accelerated or the desired product is produced in higher selectivity.

By careful revision of the many hundreds of publications on organic syntheses in ionic liquids [3], four main arguments can be extracted to support the green nature of ionic liquid-based organic synthesis.

1. the negligible vapor pressure of ionic liquids;
2. the higher chemical and thermal stability of ionic liquids;

3. the possibility of creating liquid–liquid biphasic reaction systems including ionic liquid recycling;
4. the additional functionality that an ionic liquid can provide compared with organic solvents.

In the following sections we take a closer look to these arguments and their relevance for specific organic reactions.

2.1.1
The Extremely Low Vapor Pressure of Ionic Liquids

A special property of ionic liquids that is directly derived from their ionic nature their extremely low vapor pressure. Indeed, ionic liquids have in the past been described as non-volatile substances. Today we know from the pioneering work of Rebelo's group (in collaboration with Seddon's and Magee's groups) [4] and from many other subsequent physico-chemical investigations [5] that ionic liquids have a measurable vapor pressure and that some families of aprotic ionic liquids – in particular bis (trifluoromethylsulfonyl)amide salts – can even be distilled without decomposition under drastic conditions. As a consequence of these studies, the vapor pressure of an ionic liquid can no longer be considered to be zero. However, on a practical scale, the vapor pressures of many relevant ionic liquids have been determined to be in the range of those of liquid metals [6], which means that their vapor pressure is irrelevant for applications in organic synthesis.

Of course, the negligible vapor pressure of ionic liquids under ambient conditions is a strong green asset in contrast to volatile molecular solvents. Atmospheric pollution by an evaporating ionic liquid is extremely unlikely during organic synthesis. This is of special relevance for non-continuous organic reactions where complete recovery of a volatile organic solvent is usually difficult to integrate into the process.

Additionally, the vapor pressure of reactants and products is reduced when dissolved in an ionic liquid, which also decreases their potential for atmospheric pollution. This is an important aspect with regard to the operational safety of organic reactions as mixtures of volatile organics and air are always linked with explosion hazards. Thus an organic reaction in ionic liquids can be characterized in most cases by a much lower fire and explosion hazard compared with the same reaction in a volatile and flammable solvent. This by itself is an important advantage with respect to principle 12 of the 12 Principles of Green Chemistry [7]:

> *Substances and the form of a substance used in a chemical process should be chosen so as to minimize the potential for chemical accidents, inclusion releases, explosions and fires.*

This aspect is especially true for oxidation reactions where air or pure oxygen are used as oxidants. Here, the use of common organic solvents is often restricted by the potential formation of explosive mixtures between oxygen and the volatile organic solvent in the gas phase.

Additionally, the negligible vapor pressure of ionic liquids simplifies the distillative work-up of volatile products, especially in comparison with the use of low-boiling solvents. In chemical process schemes this advantage will save the energy requirements linked to the distillation of the solvent during product isolation. Moreover, common problems related to the formation of azeotropic mixtures in a volatile solvent and organic reaction mixtures can be avoided when using an ionic liquid.

Although the lack of an ionic liquid's vapor pressure benefits the concept of green chemistry, it could also leads to negative effects. This can be briefly illustrated with two examples:

1. The reduced vapor pressure of the reaction products in ionic liquids may increase the energy demand to isolate the product by distillation considerably (ionic liquid entrainer effect). In some cases, isolation by means of distillation may even become inappropriate and more complex separation methods (e.g. extraction) may be necessary.

2. The negligible vapor pressure prevents the purification and regeneration of ionic liquids by distillation. Again, alternative methods are available, such as extraction, chromatography or membrane processes. However, these alternatives can be significantly more tedious than a simple solvent distillation that is routinely carried out with classical volatile solvents.

2.1.2
Stability of Ionic Liquids in Organic Reactions

The stability of ionic liquids under the reaction conditions is a very important aspect as almost all green synthetic protocols will have to involve extensive ionic liquid recycling.

Ionic liquids have been found in many cases to show higher thermal stability than classical organic solvents. A particularly important argument in favor of ionic liquids is the fact that the high thermal stability in combination with their negligible vapor pressure allows for an extremely broad liquid range for pressureless synthetic application of up to 300 °C for some ionic liquid candidates (e.g. $[C_2mim][NTf_2]$ melts at 9 °C [8] and shows even over longer times thermal stability up to 300 °C [9]). Again, the absence of solvent pressure at high temperatures is an important safety issue and corresponds to Principle 6 in which it is claimed that synthetic methods should be conducted at ambient pressures.

From the reaction engineering point of view, this wide temperature window of pressureless organic chemistry allows for great kinetic control. Assuming that the reaction rate of an organic reaction doubles with a temperature increase of 10 °C (which is true for a reaction with an activation energy in the region of 60 kJ mol^{-1}), the rate of an organic reaction can be increased by a factor of 250×10^6 by going from 20 to 300 °C. Unfortunately, in most cases stability and selectivity issues strongly limit the practical use of this potential.

Another very important strength of ionic liquids is their generally very high stability towards acids and even super-acids. This particular feature explains the large number of very successful applications in the area of acid-catalyzed reaction in ionic liquids ([3]; see also Section 2.1). As protonation of a cation would ultimately result in a doubly charged cation, this reaction is improbable if there is not a distinctive basic functionality (e.g. in the case of amine-functionalized task-specific cations). On the anion side, all anions derived from strong or very strong acids are usually very stable against acidic media. Exceptions are, of course, anions which are themselves basic or unstable towards acids such as acetate and formate.

Apart from these positive general aspects, it is useful to mention a couple of limiting facts. In order to obtain a realistic view of ionic liquid stabilities based on today's knowledge, one has to admit that the thermal stabilities of ionic liquids have been seriously over-estimated in the past. Thermal stabilities of over 400 °C in early papers [8] do not reflect long-term stabilities but were obtained in thermogravimetric analysis (TGA) onset measurements with steep heating ramps (10 K min^{-1}). Thorough investigations on the long-term stability of ionic liquids indicate stability limits of up to 100 °C less for the same ionic liquid depending on its purity [10].

Moreover, the estimation of the thermal stability of a given ionic liquid in a certain organic reaction needs to take into account that the presence of nucleophilic or basic reactants can greatly reduce the thermal stability of ionic liquids. This is due not only to deprotonation of 1,3-dialkylimidazolium ions (forming carbene species that will further undergo consecutive reactions [11]) but also to thermal transalkylation from the ionic liquid's cation to the nucleophilic substrate or product. If basic reaction conditions are required for the intended organic reaction, only tetraalkylphosphonium ions can be recommended as the ionic liquid's cation at present. Tetraalkylphosphonium cations have recently been shown to display reasonably stabilities under even strongly basic conditions [12]. In contrast, all nitrogen-based cations suffer to some extent from either carbene formation, Hofmann elimination or rapid dealkylation [3].

Another very important stability aspect for organic synthesis in ionic liquids is their stability towards hydrolysis. Water is likely to be omnipresent in organic syntheses unless extreme precautions are taken to keep the system inert. Many of the reported examples of organic chemistry have applied [C_4mim][BF_4] and [C_4mim][PF_6], two ionic liquids which are known to undergo anion hydrolysis in the presence of water even under mild conditions [13]. This hydrolysis releases the highly toxic and very corrosive degradation product HF. Consequently, applications of these classes of ionic liquids can hardly be considered as green in any respect; even so, the presence of additional acidity has certainly led to increased reactivity and unusual selectivities in certain cases. To avoid these problems, hydrolysis-stable anions such as [NTf_2]$^-$, [TfO]$^-$, [CH_3SO_3]$^-$ and [$(RO)_2PO_2$]$^-$ should be used for organic reactions where water is not excluded. Note that alkylsulfate anions are also labile towards hydrolysis, forming an alcohol and hydrogensulfate as the hydrolysis products [14].

Finally, electrochemical stability of ionic liquids should be mentioned as a precondition for electrochemical organic reactions in ionic liquids. Only a few examples have been reported (e.g. [15]) so far. The reader interested in this specific aspect of organic chemistry is therefore referred to the more specialized literature on electrochemistry in ionic liquids [16] and to Section 4.2 in Chapter 4.

2.1.3
Liquid–Liquid Biphasic Organic Reactions

It is correct to state that almost all green chemistry approaches that build on ionic liquids in organic synthesis include ionic liquid recycling after reaction. This requires an appropriate method to separate the ionic liquid from the reaction product. Although distillation of the reaction product is the most obvious separation method (given the negligible vapor pressure of the ionic liquid), it is not always practicable due to the relatively high boiling points of many organic products.

A very elegant alternative method that can separate reaction products from the ionic liquid solvent already during the reaction is liquid–liquid biphasic reaction. In an ideal case, the ionic liquid is able to act as a catalyst (e.g. in case of a Friedel–Crafts alkylation or transesterification) or suitable compartment for the intended organic reaction, and the reaction benefits from the ionic reaction media in the form of high rate and excellent selectivity. Importantly, the ionic liquid should further display partial solubility for the substrates but poor solubility for the reaction products. Under these conditions, the product phase (also containing the unconverted reactants) can be isolated by simple decantation and the ionic liquid phase can be recycled (see Figure 2.1).

A very interesting extension of classical liquid–liquid biphasic systems with ionic liquids is supported ionic liquid phase (SILP) catalysis. The development of SILP catalysts was first aimed at minimizing the amount of ionic liquid applied in a liquid–liquid biphasic reaction. However, as stirring of this small amount of liquids becomes problematic, immobilization of this liquid on the support's surface in SILP was introduced. In SILP systems, the problem is overcome through immersing a support with very high internal surface area in the ionic liquid to provide a maximum

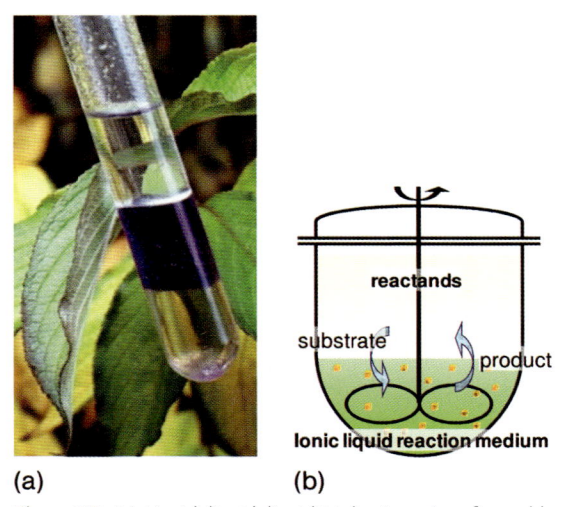

Figure 2.1 (a) Liquid–liquid–liquid triphasic system formed by nonane (top layer), water (middle layer with a blue dye) and [C_2mim][NTf_2]; (b) schematic diagram of a liquid–liquid biphasic reaction with an ionic liquid reaction medium.

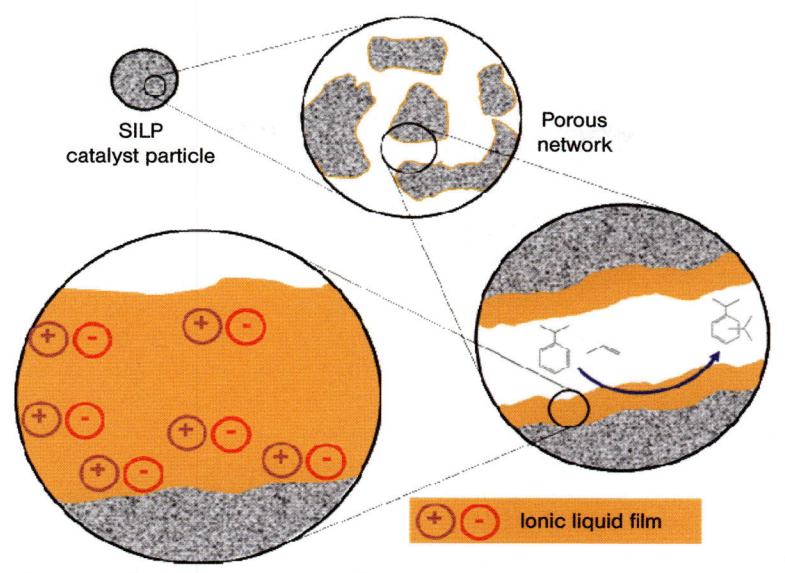

Figure 2.2 Schematic representation of a supported ionic liquid phase (SILP) catalyst material.

liquid–liquid exchange surface with a minimum amount of ionic liquid. In the end, a virtually heterogeneous catalyst system is obtained that is characterized by highly efficient use of the ionic liquid. Figure 2.2 shows a schematic representation of a SILP catalyst material.

In recent years, the SILP catalysis concept has been very successfully established for many transition metal-catalyzed transformations [17] (see Chapter 3 for details). However, also in classical organic synthesis a growing number of examples have appeared including Friedel–Crafts reaction with acidic SILP materials (see Section 2.2), SILP material-based enzymatic reactions [18] and SILP material-based organocatalysis [19].

2.1.3.1 Tunable Solubility Properties

One of the key factors controlling the reaction rate in an organic–ionic liquid biphasic reaction is the solubility of the substrate in the ionic liquid phase. Due to their very high structural variability, ionic liquids can be designed to provide a specific solubility difference between the substrate and product of a given reactant [20]. In recent times, selection of the ionic liquid has been based not only on the steeply growing set of experimental solubility data [21] but also on theoretical predictions. Based on initial work dealing with the selection of ionic liquids for separation technologies [22], several groups have used successfully the conductor-like screening model for real solvent (COSMO-RS) approach and the corresponding commercial software [23] to select or optimize their ionic liquid for optimum solubility properties in the context of organic synthesis. Relevant recent publications have applied the COSMO method to predict the solubility of alcohols [24] and flavonoids [25] in ionic liquids. Even the dependence of the ionic liquid's structure on the conversion and selectivity of enzymatic glycerolysis has been predicted using the COSMO-RS method [26].

The COSMO-RS method is able to predict the solubility of a compound in an ionic liquid in thermodynamic equilibrium. It starts from separate DFT calculations of the ionic liquid ions and the solute molecules. These calculations have to take into consideration different conformers [27]. Interestingly, data files can be combined so that, for example, a file generated for the $[C_2mim]^+$ ion can be used in combination with different anion files and substrate files to calculate solubilities in different systems. This makes the method fairly flexible and efficient for ionic liquids. The DFT results for each moiety are converted into a statistics of surface elements of the same charge density. This distribution – the so-called σ-profile – is used to calculate the activity coefficient of each component by pair-wise interacting the surface elements of the different components and minimizing the system energy in this process. As one result of this exercise, the activity coefficient of the solute in the cation–anion mixture forming the ionic liquid is obtained which allows one to calculate a large number of important system properties, such as solubility data, distribution coefficients and vapor pressures of volatile substances in the respective ionic liquid [23a]. Figure 2.3 shows a schematic representation of the process to predict partition coefficients between an ionic liquid phase and an organic phase using COSMO-RS.

Note that COSMO-RS only helps to determine the thermodynamic equilibrium data. However, in order to realize fast organic reactions in an organic–ionic liquid biphasic system, high substrate availability in the ionic liquid reaction phase is necessary. The latter requires both high thermodynamic solubility of the substrate in the ionic liquid and fast mass transfer of the substrates into the ionic reaction layer.

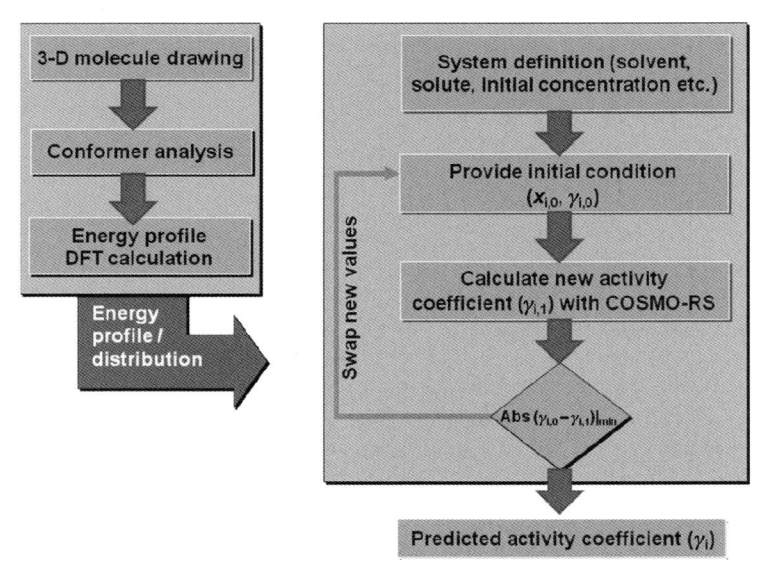

Figure 2.3 Schematic representation of predicting the partition coefficient between an ionic liquid phase and an organic phase using COSMO-RS.

Sufficiently fast mass transport of reactants into the ionic liquids phase is not a trivial prerequisite. This is true since the mass transport coefficient is dependent on the substrate size and the viscosity of the ionic liquid [28]. The typically high viscosity of ionic liquids will substantially slow down all mass transfer rates. Therefore, it is important to select a low-viscosity ionic liquid for fast organic reactions in order to avoid limitation of the overall rate by the rate of mass transfer. Moreover, effective stirring is a very important issue to make full use of the potential rate acceleration in ionic liquids in these multiphasic reactions.

Other important points of practical relevance for liquid–liquid biphasic organic reactions concern the ease of phase separation and the formation of side-products or feedstock impurities that tend to accumulate in the ionic liquid layer with increasing recycling. The latter may affect the ionic liquid's physico-chemical properties over longer run times.

2.1.3.2 Product Isolation from Organic Reactions with Ionic Liquids

Liquid–liquid biphasic operation without an additional volatile solvent is the ideal and greenest case of ionic liquid multiphasic operation. Unfortunately, this ideal scenario can only be realized in a few special cases. The BASF BASIL process is a particularly convincing example of this kind [29]. Here, the butylimidazolium chloride ionic liquid formed during the process forms a pronounced miscibility gap with the alkoxyphenylphosphine products of the process thus enabling very efficient product separation from the ionic liquid (see Section 6.1 in Chapter 6 for details).

Unfortunately, many products of organic reactions are polar themselves (e.g. alcohols, ketones, amines) and do not form spontaneously a second product phase on top of the ionic liquid reaction phase. In these cases, the use of an additional, volatile organic solvent to extract the product from the ionic liquid layer has been proposed by many authors (see [3] for various examples). This approach can seriously harm the green character of the overall process depending on the environmental impact (toxicity, ecotoxicity) of the applied additional solvent. As a green extraction medium for product isolation from ionic liquids, supercritical CO_2 (scCO_2) has been suggested by, for example, Reetz and co-workers for the isolation of products of an enzymatic transesterification in ionic liquids [30]. Note that for the evaluation of this very elegant method, the additional investment in high-pressure equipment and the energy requirement of CO_2 compression have to be taken into account.

It is important to note that not in all cases can a suitable extraction solvent be identified to isolate polar organic products from their ionic liquid reaction medium. In these cases, product crystallization or membrane technologies [31] remain as alternative isolation techniques. However, both methods are not ideal from a green chemistry perspective. Product precipitation usually requires extensive washing procedures to make sure that the product is obtained free of traces of the ionic liquid. For membrane separation processes, a strong dilution of the reaction mixture is usually required to lower the viscosity. Thus, both alternative routes again result in the extensive use of volatile solvents and add additional energy requirements for solvent recovery and product isolation from the solvent.

From all these considerations, it becomes clear that green chemistry in ionic liquids is very much about efficient product separation from the ionic liquid reaction medium. In cases where this separation step is straightforward, the chance of realizing greener organic synthesis with ionic liquids is significantly higher. Figure 2.4 shows

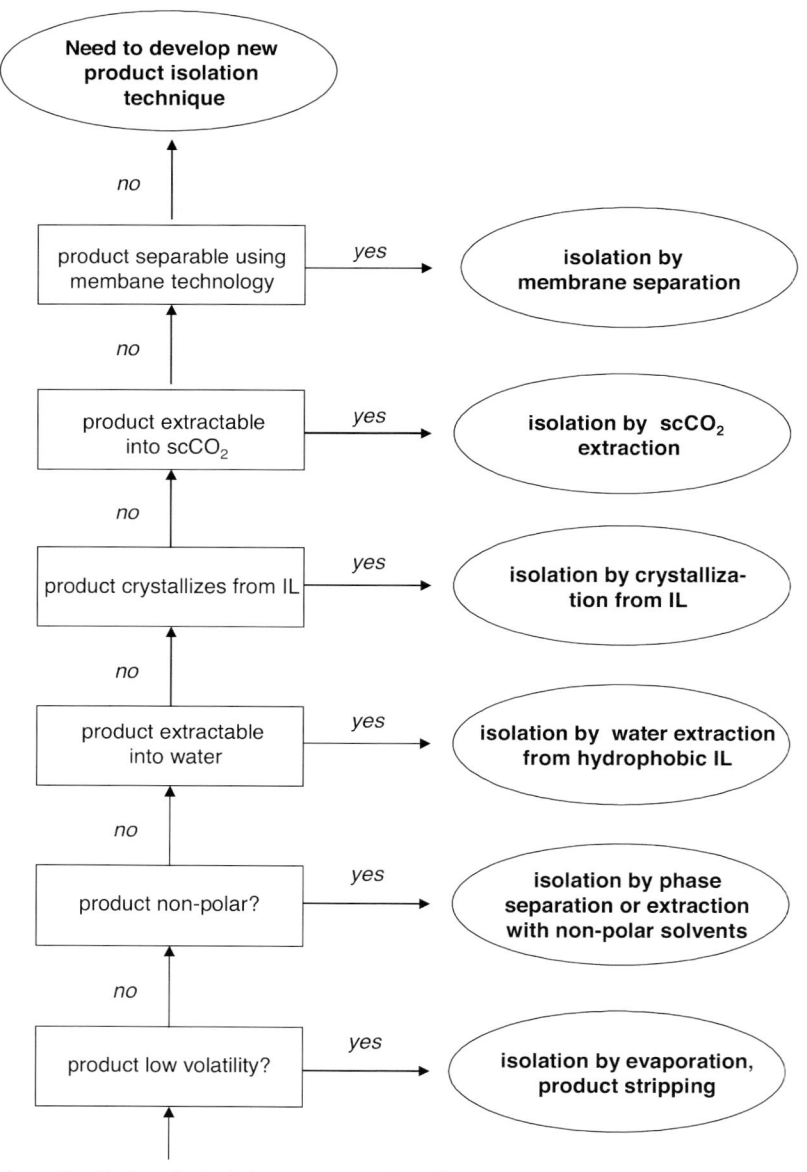

Figure 2.4 Options for isolating organic products from an ionic liquid reaction medium ranked from bottom to top in the order of increasing separation effort, which to a good approximation is the order of decreasing greenness. Adapted from [32].

an overview of the most important options for isolating organic products from an ionic liquid reaction medium. The different methods are ranked from bottom to top in order of increasing separation effort, which to a good approximation is the order of decreasing greenness.

2.1.4
Reactive or Catalytic Ionic Liquids in Organic Synthesis

It is clear from everything that has been discussed previously that the ionic liquid needs to offer a substantial and relevant advantage and contribution to justify its use in green organic chemistry. This point is often realized in a most convincing way if the ionic liquid significantly influences the reactivity of the reactants (catalytic effect) or if the ionic liquid itself is a reactive component in the intended application.

The most traditional way to implement catalytic activity in an ionic liquid is the use of acidic chloroaluminate ionic liquids [33]. Acidity and thus catalytic activity in these systems are a function of the [cation]Cl to $AlCl_3$ ratio, with ionic liquids of a molar [cation]Cl to $AlCl_3$ ratio <1 being called "basic" (free chloride is present in these systems) and systems with a ratio of >1 being called "acidic" (the acidic anions $[Al_2Cl_7]^-$ and $[Al_3Cl_{10}]^-$ are present in these systems). Chloroaluminate ionic liquids have proven to be excellent media for many organic reactions, including Friedel–Crafts alkylation reactions (for more details, see Section 2.2), Friedel–Crafts acylation reactions [34], cracking/isomerization reactions [35], chlorination [36], ether cleavage [37], Diels–Alder reactions [38] and reductions [39].

However, it is very important to note that the use of chloroaluminate ionic liquids in organic reactions suffers from two major disadvantages:

1. All chloroaluminate ionic liquids are very moisture sensitive. Any contact with water destroys the ionic liquid in an immediate hydrolysis reaction while liberating HCl as toxic and corrosive gas. Therefore, extreme care is necessary to exclude any source of water, which includes tedious drying of all reactants and the handling of the reaction in a completely inert atmosphere (e.g. in a glove-box). In the context of green chemistry, these restrictions are problematic as not all feedstock is available in a very dry form and all drying processes are cost and energy consuming. Moreover, it is problematic to ensure complete inertness of operation in the case of an incident, thus affecting the claim of designing inherently safer synthetic methodologies (see Principle 3 of the 12 Principles of Green Chemistry [7]).

2. In many cases, it has proven practically impossible to separate products containing electron-rich heteroatoms (e.g. alcohols, ketones, aldehydes, amines) from chloroaluminate ionic liquids without prior hydrolysis of the ionic liquid. This step destroys the acidic ionic liquid and, if such a product isolation step is necessary, the major advantage of the ionic liquid system compared with the simple use of $AlCl_3$ in homogeneous solution is lost. In contrast, after chloroaluminate hydrolysis, the aqueous waste contains additionally the ionic liquid cation, which may further complicate the downstream purification and disposal concept. Note that for products that do not have polar electron-donating functional

groups and are sufficiently non-polar to show a pronounced miscibility gap with the catalytic chloroaluminate ionic liquid (e.g. alkanes, alkenes, non-functionalized alkylated arenes), product isolation by decantation is straightforward and the ionic liquid can be reused without further treatment.

From these two important restrictions, it is predictable that chloroaluminate ionic liquids can contribute to green synthetic methods only in those cases where feedstock is used that is available in a dry form or is easy to dry and for feedstock and the products do not contain electron-rich heteroatoms. Despite these severe restrictions, there are still a significant number of industrially interesting reactions left, as will be demonstrated in more detail in Section 2.2.

The disadvantages associated with the use of chloroaluminate ionic liquids have led various authors to investigate alternative Lewis acidic ionic liquid media such as those based on zinc chloride [40], tin chloride [41], indium chloride [42], iron chloride [43] and scandium triflate [44]. These ionic liquids are considerably more water stable but also much less reactive than the above-described chloroaluminate systems.

Brønsted acidity can be incorporated in the ionic liquid's cation and anion (and even both). The range of acid strength that can be realized in ionic liquids ranges from very modest (e.g. the acidity of the proton in the 2-position of the 1,3-dialkylimidazolium cation [11]) to super-acidic (in the case of acidic chloroaluminate ionic liquids contacted with strong Brønsted acids, such as HCl [45]). Some selected examples of Brønsted acidic ionic liquids with different acidities are displayed in Figure 2.5. Note that the acidity of an ionic liquid cannot be easily extrapolated from the acidity of similar acids in an aqueous environment. In a water-free ionic liquid, the degree of Brønsted acidity is a function of the proton concentration and the proton coordination to the ionic liquid's ions. This is the reason why dissolving a strong

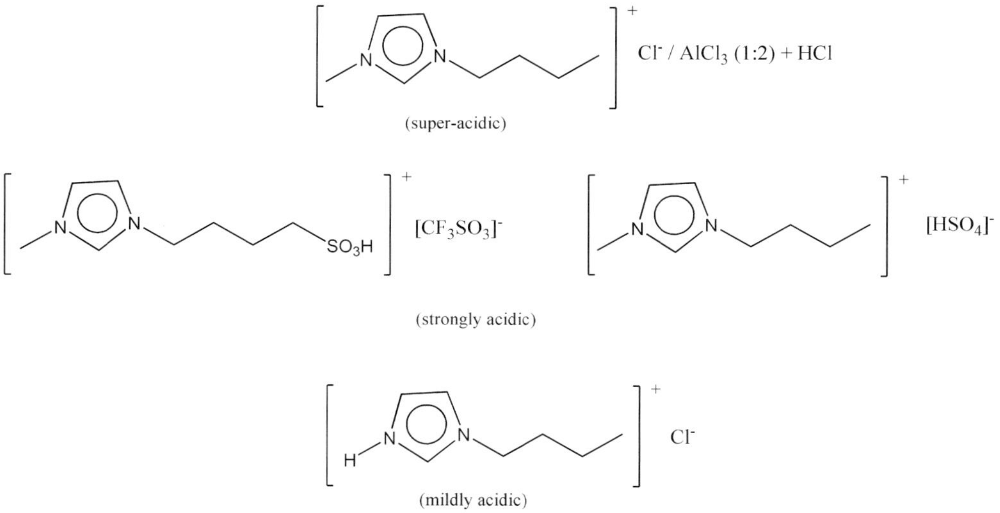

Figure 2.5 Examples of Brønsted acidic ionic liquids of different acid strengths.

Brønsted acid HY in an ionic liquid with the same anion Y⁻ always leads to a lower acidity compared with pure HY [46]. Examples of organic reactions catalyzed by Brønsted acidic ionic liquids include Friedel–Crafts alkylations [47] (see also Section 2.2), esterification reactions [48] and nitrations [49].

Another option that attracted attention only recently is the use of ionic liquids as a solvent or precursor for organocatalysts. In a recent example of this approach, Orsini *et al.* [50] generated electrochemically *N*-heterocyclic carbenes under galvanostatic control and used the latter *in situ* as organocatalysts for the benzoin condensation reaction. Yields were up to 84% and the authors claimed that this method is particularly green as it avoids the use of any volatile solvent or additional base. A similar strategy was applied by the same group for the *N*-acylation of chiral oxazolidin-2-ones [51] and for the synthesis of 2-azetidinones [52].

Apart from the possibility of incorporating or generating catalytic functions in an ionic liquid, the ionic liquid medium itself may act as a catalyst simply by accelerating an organic reaction by its solvent nature and its interaction with the dissolved reactants. Remarkable examples of this effect originated, for example, from systematic kinetic studies of nucleophilic reactions at cationic centers performed by Welton's group [53]. They reacted sulfonium electrophiles with three different amine nucleophiles in several molecular solvents and ionic liquids and found that the degree of hydrogen bonding – in particular between the electrophile and the solvent – provided the dominant effect in determining the rate of reaction. Earlier, the same group reported that the nucleophilicity of amines can be manipulated in ionic liquids by demonstrating that the amines under investigation were more nucleophilic in the tested ionic liquids (e.g. [C_4mim][OTf]) than in the studied organic solvent dichloromethane and acetonitrile [54]. A similar study investigated the ionic liquid's influence on the nucleophilicity of dissolved halides [55]. In this case, however, it was found that the nucleophilicities of all halides were lower in all ionic liquids (e.g. [C_4mim][BF_4], [C_4mim][PF_6], [C_4mim][OTf], [C_4mim][NTf_2]) than dichloromethane. Recently, detailed kinetic studies aimed at revealimg the influence of the ionic liquid solvent on the reaction rate of organic reactions were extended to Diels–Alder [56] and esterification [57] reactions. For both examples the solvent effects on the reaction rates were examined by using a linear solvation energy relationship based on the Kamlet–Taft solvent scales.

Finally, another interesting area in organic chemistry in which the chemical reactivity of an ionic liquid has been applied successfully should be briefly mentioned. It has been shown in recent years by many authors that functionalized and task specific ionic liquids (TSILs) can be designed and applied as new soluble supports for supported organic synthesis and combinatorial chemistry. The usefulness of these supports has been demonstrated, for example, for the synthesis of heterocyclic compounds, coupling reactions, peptide synthesis and sulfonamide preparations [58].

After this general introduction highlighting the potential role of ionic liquids in green organic reactions, one specific example will be selected to illustrate the above-mentioned aspects in more detail. For this purpose, the Friedel–Crafts alkylation reaction catalyzed by acidic ionic liquids has been selected. This reaction is important, in our view, owing to the number of high-quality publications available and the very clear green benefit of the use of an ionic liquid. Furthermore, examples of Friedel–Crafts alkylations with ionic liquids have been successfully demonstrated in continuous miniplants, and such work can be regarded as reasonable proof of concept to allow an estimation of the technical potential of this example.

2.2 Friedel–Crafts Alkylation

2.2.1 Introduction and Technical Background

The Friedel–Crafts alkylation reaction is an electrophilic substitution reaction that was developed by Charles Friedel and James Crafts in 1877 and has been studied in detail by Olah and co-workers [59]. The Friedel–Crafts alkylation involves the reaction of an alkylating agent such as an alkene, alcohol or alkyl halide with an aromatic compound to form an alkylated aromatic compound. With respect to green chemistry, alkenes are clearly preferred as the feedstock due to the better atom efficiency of the reaction.

The reaction is catalyzed by strong acids. Scheme 2.1 illustrates the Friedel–Crafts alkylation for the example of toluene isopropylation with propene. The scheme highlights two important selectivity issues commonly found in Friedel–Crafts alkylation, namely the control of the regioselectivity of the alkylation reaction and the prevention of undesired consecutive alkene addition to form higher polyalkylated aromatics.

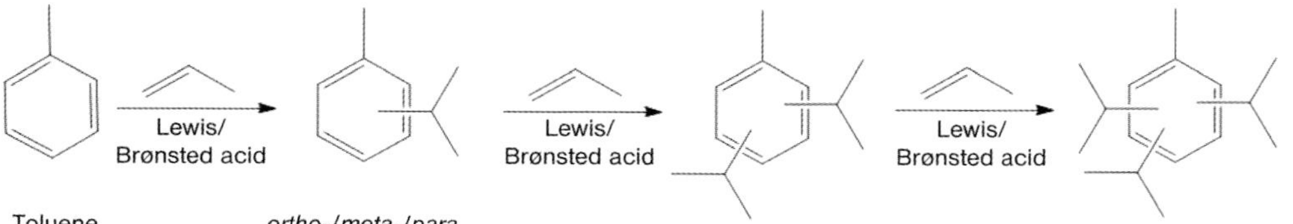

Toluene — *ortho-/meta-/para* Isopropyltoluene (cymene) — Diisopropyltoluenes — Triisopropyltoluenes

Scheme 2.1 Isopropylation of toluene as an example for the Friedel–Crafts alkylation reaction.

Alkylated aromatics are important intermediates in industrial chemistry, with applications ranging from pharmaceutical and agricultural chemicals to bulk chemicals such as styrene [60]. Industrial Friedel–Crafts alkylations are catalyzed by either homogeneous and heterogeneous catalyst systems. Over decades, the process has been carried out using homogeneous aluminum(III) chloride (e.g. the so-called Monsanto/Kellog technology that is still operated by many companies) or other homogeneously dissolved Lewis acid catalysts. In these processes, water or basic aqueous solutions have to be added to the reaction mixture for product isolation. This causes irreversible hydrolysis of the catalyst, releases toxic, corrosive gases and

Table 2.1 Common acid catalysts for Friedel–Crafts alkylation reactions [60]

Type of catalyst	Example	Acid type
Metal halides	$AlCl_3$, $AlBr_3$, BF_3, $ZrCl_4$, $FeCl_3$	Lewis acid
Metal alkyls and metal alkoxides	AlR_3, BR_3, ZnR_2, $Al(OPh)_3$	Lewis acid
Protonic acids	H_2SO_4, HF, H_3PO_4, HCl	Brønsted acid
Acidic oxides	Zeolites, BeO, Cr_2O_3, P_2O_5	Lewis acid
Supported acids	H_3PO_4–SiO_2, BF_3–Al_2O_3	Brønsted acid/Lewis acid
Exchange resins	Dowex 50, Amberlite IR 112	Brønsted acid

produces a large amount of wastewater contaminated with organic substances. With regard to green chemistry, it is this product isolation step that is most critical in the homogeneous process.

Since 1980, zeolites have developed into an important class of solid acids for heterogeneously catalyzed Friedel–Crafts alkylation. Mesoporous zeolites, such as H-ZSM-5, are particularly interesting and are technically applied, for example, in the production of ethylbenzene (Mobil Badger process) [61]. The technology has gradually replaced the conventional homogeneously catalyzed alkylation process [62]. Another group of interesting solid acids for Friedel–Crafts alkylation are supported Brønsted acids on a support, such as phosphoric acid on silica [63] and ion-exchange resins, such as Dowex. Table 2.1 gives an overview of technically important catalysts for the Friedel–Crafts alkylation reaction.

However, most of these heterogeneous acids require harsh reaction conditions due to their significantly lower reactivity compared with the best homogeneous counterparts. Operation under such harsh reaction conditions is often accompanied by unwanted coke formation [64] or oligomerization [65] followed by subsequent deactivation of the catalyst.

2.2.2
Ionic Liquids in Friedel–Crafts Reaction - the Unique Selling Point

Through dissolving the active acidic catalyst in an ionic liquid or by incorporating acidity as a function in one of its ions, acidic ionic liquids are obtained that themselves act as catalysts in the Friedel–Crafts alkylation. By far the most prominent and best studied examples concern the use of an acidic chloroaluminate ionic liquid [3a,e], but chloroferrates [66], scandium-based systems [67] and hydrogensulfate [46a] ionic liquids have also been studied as catalytic phases. More recent variations that are of particular interest for green Friedel–Crafts alkylations utilize the remarkably high solubility of $AlCl_3$ in bis(trifluoromethylsulfonyl)amide ionic liquids (up to 4.5 equiv. $AlCl_3$ per equivalent of [cation][NTf_2]). These systems were introduced by Brausch and co-workers [68] and have the enormous advantage that the expensive ionic liquid cation can be recycled even from an aqueous work-up of the ionic liquid. The latter may become necessary after extensive recycling of the catalyst phase, e.g. due to the build-up of polar heavy components in the ionic liquid. The ionic liquid recycling builds on the known miscibility gap between the bis(trifluoromethylsulfonyl)amide ionic liquid and water. After isolation and drying of the liquid bis(trifluoromethylsulfonyl)amide support, the system can be reloaded with $AlCl_3$ and used again as fresh catalyst.

Note that all reported Friedel–Crafts alkylation reactions in tetrafluoroborate or hexafluorophosphate ionic liquids have to be treated with extreme caution. In the reports, Lewis or Brønsted acids (which are never 100% water-free) are usually added to these hydrolysis-sensitive ionic liquids [69], quickly forming the corresponding HF which itself also acts as a strong catalyst. As this reactivity is combined with the irreversible destruction of the ionic liquid, such approach can hardly be called green or efficient. In some cases [70], the authors even claimed that $[C_4MIM][PF_6]$ alone can act as a catalyst for a Friedel–Crafts alkylation reaction. However, this claim is uncertain and it is very likely linked to the presence of HF impurities in the applied hexafluorophosphate melt.

What is the unique selling point of acidic ionic liquids with respect to greener Friedel–Crafts alkylations? A careful analysis of the literature reveals that a combination of four aspects is mainly responsible for the significant scientific and technical interest that has focused on this application:

1. The high concentration of highly acidic species that can be immobilized in ionic liquids and that leads to a highly reactive catalytic phase allowing for fast reactions under mild conditions. In fact, acidic chloroaluminate ionic liquids catalyze all reactions that are conventionally catalyzed by $AlCl_3$, but without suffering the disadvantage of the low solubility of $AlCl_3$ in many solvents.

2. The liquid nature of the catalytic phase eliminates the heat and mass transfer problems frequently encountered with heterogeneous Friedel–Crafts-catalysts.

3. The miscibility gap of most ionic liquids with alkylated aromatics allows for product isolation by simple decantation.

4. The decreasing solubility from monoalkylated aromatics to dialkylated aromatics (and further to trialkylated aromatics) in most ionic liquids allows in principal for the *in situ* extraction of the desired monoalkylated products from the catalytic ionic liquid phase.

2.2.3
Liquid–Liquid Biphasic Catalysis

The first liquid–liquid biphasic Friedel–Crafts alkylation reactions with room-temperature liquid ionic liquids were carried out by Wilkes and co-workers in 1986 [71]. This group investigated the reactions of benzene and toluene with various alkyl chlorides in acidic $[C_2mim]Cl$–$AlCl_3$ mixtures. The greener version of Friedel–Crafts alkylation using alkenes as alkylating agent (no HCl formation in contrast to the use of alkyl chlorides!) was first reported in the patent literature by Seddon's group, describing acidic chloroaluminate and chlorogallate ionic liquids as the liquid catalytic phase for the reaction of ethylene with benzene to give ethylbenzene [72]. While this patent claimed the reaction to be catalyzed by acidic

1,3-dialkylimidazolium, tetraalkylammonium, tetraalkylphosphonium and pyridinium melts, two later patents extended the range of suitable ionic liquids to trialkylammonium salts [73], which are very easily and cheaply obtained by the reaction of a trialkylamine with first gaseous HCl and then an excess of $AlCl_3$.

A study aimed at maximizing the amount of monoalkylbenzene by using a high alkene to benzene ratio was later reported by Xiao and Johnson [74]. They used two different chloroaluminate ionic liquids, [PyH]Cl–$AlCl_3$ and [S_{222}]Br–$AlCl_3$, in the alkylation of benzene with 1-pentene, 2-pentene and 1-octene.

Recently, a first detailed kinetic study of a liquid–liquid biphasic Friedel–Crafts alkylation reaction was reported by Joni et al. [75]. They studied the isopropylation of cumene and m-xylene in the ionic liquid [C_2mim]Cl–$AlCl_3$ in a semi-batch reactor. Kinetic models representing the alkylation reaction network have been established based on variations in reaction temperature and propylene partial pressure. Corrections of the measured product concentrations in the organic phase from COSMO-RS-based solubility calculations allowed the mass balance to be closed. It was also demonstrated that the reaction needs efficient stirring to obtain the full chemical reactivity, which was found to be limited by the availability of propene in the ionic liquid reaction phase.

Continuous liquid–liquid biphasic Friedel–Crafts alkylations have been demonstrated using acidic chloroaluminate ionic liquids and the [cation][NTf_2]–$AlCl_3$ reaction system [76]. The latter catalyst was applied in a loop reactor with an internal gravity separator for time on-stream of more than 10 h without catalyst deactivation or change in selectivity [77].

2.2.4
Supported Ionic Liquid Phase (SILP) Friedel–Crafts Catalysis

In order to optimize green Friedel–Crafts alkylations with ionic liquids even further, it was an obvious concept to reduce the amount of ionic liquids applied to a minimum. Hölderich and co-workers were the first to realize this step in 2000 [78] by adding small amounts of acidic chloroaluminate ionic liquids to various types of silica, alumina, TiO_2 and ZrO_2 supports. During the preparation, the formation of HCl was observed, probably due to the presence of water on the supports. Moreover, it is clear from the reported synthetic protocols that the acidic ionic liquid reacted chemically with the OH groups on the support surface, thus creating an ionic liquid-terminated solid. The materials obtained were tested for the alkylation of benzene, toluene, naphthalene and phenol with 1-dodecene in batch, continuous liquid-phase and continuous gas-phase systems. The catalytic activities of the immobilized ionic liquids were found to be higher than those of the conventional H-beta zeolite under the same conditions. Despite the fact that no leaching of $AlCl_3$ in the organic product phase was reported, some catalyst deactivation was concluded from the slight loss in conversion with time in the continuous liquid-phase reaction. Later, the same group reported the grafting of ionic liquids on the surface of a silica support by means of a chemical reaction of alkoxysilyl-functionalized cations with the support material [79].

Recently, this concept has been developed further by applying a defined pretreatment of the support. This procedure led to much better reproducibility of the catalytic

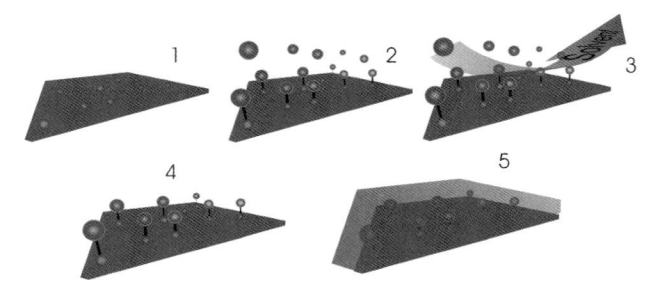

Figure 2.6 Schematic representation of the support pretreatment process post-calcination of the SiO_2 support. (1) Fresh calcined SiO_2 containing basic surface hydroxyl group; (2) loading of the support's surface with acidic chloroaluminate ionic liquid; (3) washing off excess and unreacted acidic species with fresh non-reacting solvent; (4) evacuation of washing solvent residue giving an interacting-free support surface ready for acidic ionic liquid immobilization; (5) utilization of the pretreated support for acidic ionic liquid loading.

results and much better reactivity in the isopropylation of cumene [80]. The pretreatment of the support comprises a washing step of the freshly calcined support with an acidic ionic liquid followed by removal of the excess ionic liquid in a solvent wash. The support pretreated and dried in this way is coated with a defined load of ionic liquid to obtain the active catalyst. The pretreatment procedure is shown schematically in Figure 2.6.

The same acidic SILP materials (including pretreatment) were also successfully applied in a continuous gas-phase alkylation of cumene and toluene using a four-stage reactor with gas chromatographic analysis after each stage [81] (see Figure 2.7).

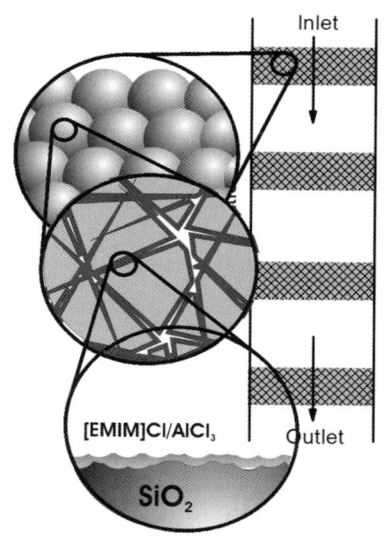

Figure 2.7 Schematic view of the acidic SILP Friedel–Crafts catalyst in a four-stage reactor applied for the continuous gas-phase isopropylation of toluene.

Variation of ionic liquid loading and the aromatic to propylene ratio showed that a moderate ionic liquid loading (20 vol.% with respect to support pore volume) with a considerably high aromatic to propylene ratio (7 : 1 molar ratio) were optimal for suppressing consecutive alkylation reactions. Moreover, it was found that a higher acidity of the ionic liquid film on the support has a beneficial effect on the selectivity to monoalkylated products. With very dry toluene and propene feed streams (guard beds), a fully stable 210 h time-on-stream operation could be realized showing constantly high selectivity to the monoalkylated product (>95%) and excellent selectivity to *m*-cymene within the cymenes (up to 80%). Based on these very recent and very promising results, a great technical potential for more sustainable and greener Friedel–Crafts alkylation processes can be anticipated for these acidic SILP catalysts [81].

References

1 Wasserscheid, P. and Welton, T. (eds.) (2007) *Ionic Liquids in Synthesis*, Wiley-VCH Verlag GmbH, Weinheim.

2 (a) Fei, Z., Geldbach, T.J., Zhao, D. and Dyson, P.J. (2006) *Chemistry - A European Journal*, **12**, 2122;(b) Davis, J.H. Jr. (2007) Synthesis of task-specific ionic liquids, in *Ionic Liquids in Synthesis* (eds P. Wasserscheid and T. Welton), Wiley-VCH Verlag GmbH, Weinheim, pp. 45–56.

3 (a) Welton, T. (1999) *Chemical Reviews*, **99**, 2071; (b) Welton, T. (2004) *Coordination Chemistry Reviews*, **248**, 2459; (c) Chiappe, C. (2007) *Ionic Liquids in Synthesis* (eds P. Wasserscheid and T. Welton), Wiley-VCH Verlag GmbH, Weinheim, pp. 265–291; (d) Earle, M.J. and Seddon, K.R. (2000) *Pure and Applied Chemistry*, **72**, 1391–1398; (e) Earle, M.J. (2007) *Ionic Liquids in Synthesis* (eds P. Wasserscheid and T. Welton), Wiley-VCH Verlag GmbH, Weinheim, pp. 292–368; (f) Afonso, C.A.M., Branco, L.C., Candelas, N.R., Gois, P.M.P., Lourenco, N.M.T., Mateus, N.M.M. and Roas, J.N. (2007) *Chemical Communications*, **26**, 2669; (g) Malhotra, S.V., Kumar, V. and Parmar, V.S. (2007) *Current Organic Synthesis*, **4** (4), 370; (h) Winkel, A., Reddy, P.V.G. and Wilhelm, R. (2008) *Synthesis*, **7**, 999; (i) Pavlinac, J., Zupan, M., Laali, K.K. and Stojan, S. (2009) *Tetrahedron*, **65** (29–30), 5625; (j) Toma, S., Meciarova, M. and Sebesta, R. (2009) *European Journal of Organic Chemistry*, **3**, 321.

4 Earle, M.J., Esperanca, J., Gilea, M.A., Lopes, J.N.C., Rebelo, L.P.N., Magee, J.W., Seddon, K.R. and Widegren, J.A. (2006) *Nature*, **439**, 831.

5 (a) Zaitsau, D.H., Kabo, G.J., Strechan, A.A., Paulechka, Y.U., Tschersich, A., Verevkin, S.P. and Heintz, A. (2006) *Journal of Physical Chemistry A*, **110**, 7303; (b) Leal, J.P., Esperanca, J., da Piedade, M.E.M., Lopes, J.N.C., Rebelo, L.P.N. and Seddon, K.R. (2007) *Journal of Physical Chemistry A*, **111**, 6176.

6 Armstrong, J.P., Hurst, C., Jones, R.G., Licence, P., Lovelock, K.R.J., Satterley, C.J. and Villar-Garcia, I.J. (2007) *Physical Chemistry Chemical Physics*, **9**, 982.

7 Anastas, P.T. and Kirchhoff, M.M. (2002) *Accounts of Chemical Research*, **35**, 686.

8 Bonhôte, P., Dias, A.-P., Papageourgiou, N., Kalyanasundaram, K. and Grätzel, M. (1996) *Inorganic Chemistry*, **35**, 1168.

9 Bösmann, A. and Wasserscheid, P.unpublished results.

10 Baranyai, K.J., Deacon, G.B., MacFarlane, D.R., Pringle, J.M. and Scott, J.L. (2004) *Australian Journal of Chemistry*, **57**, 145.

11 (a) Arduengo, A.J., Harlow, R.L. and Kline, M. (1991) *Journal of the American Chemical Society*, **113**, 361; (b) Arduengo, A.J., Dias, H.V.R. and Harlow, R.L. (1992) *Journal of the American Chemical Society*, **114**, 5530.

12 (a) Earle, M.J., Fröhlich, U., Huq, S., Katdare, S., Lukasik, R.M., Bogel, E., Plechkova, N.V. and Seddon K.R. (to Queen's University Belfast), World Patent WO 2006072785; (2006) *Chemical Abstracts*, **145**, 145001; (b) Earle, M.J., Seddon, K.R., Forsyth, S., Fröhlich, U., Gunaratne, N. and Katdare S. (to Queen's University Belfast), World Patent WO 2006072775; (2006) *Chemical Abstracts*, **145**, 145000; (c) Ramnial, T., Ino, D.D. and Clyburne, J.A.C. (2005) *Chemical Communications*, 325; (d) Ramnial, T., Taylor, S.A., Bender, M.L., Gorodetsky, B., Lee, P.T.K., Dickie, D.A., McCollum, B.M., Pye, C.C., Walsby, C.J. and Clyburne, J.A.C. (2008) *The Journal of Organic Chemistry*, **73** (3), 801.

13 Swatloski, R.P., Holbrey, J.D. and Rogers, R.D. (2003) *Green Chemistry*, **5**, 361.

14 (a) Wasserscheid, P., van Hal, R. and Bösmann, A. (2002) *Green Chemistry*, **4**, 400; (b) Himmler, S., Hörmann, S., van Hal, R., Schulz, P.S. and Wasserscheid, P. (2006) *Green Chemistry*, **8** (10), 887.

15 Lagrost, C., Carrie, D., Vaultier, M. and Hapiot, P. (2003) *Journal of Physical Chemistry A*, **107** (5), 745.

16 Endres, F. and El Abedin, S.Z. (2007) *Ionic Liquids in Synthesis* (eds P. Wasserscheid and T. Welton), Wiley-VCH Verlag GmbH, Weinheim, pp. 575–618.

17 (a) Riisager, A., Fehrmann, R., Haumann, M. and Wasserscheid, P. (2006) *European Journal of Inorganic Chemistry*, 695; (b) Riisager, A., Fehrmann, R., Haumann, M. and Wasserscheid, P. (2006) *Topics in Catalysis*, **40**, 91; (c) Riisager, A., Fehrmann, R., Haumann, M. and Wasserscheid, P. (2008) *Topics in Organometallic Chemistry*, **23**, 149.

18 (a) Jiang, Y., Guo, C., Xia, H., Mahmood, I. and Chunzhao, L.H. (2009) *Journal of Molecular Catalysis B-Enzymatic*, **58** (1–4), 103; (b) Lozano, P., De Diego, T., Sauer, T., Vaultier, M., Gmouh, S. and Iborra, J.L. (2007) *Journal of Supercritical Fluids*, **40** (1), 93.

19 (a) Li, P., Wang, L., Zhang, Y. and Yicheng, W.G. (2008) *Tetrahedron*, **64** (32), 7633; (b) Kaper, H., Antonietti, M. and Göttmann, F. (2008) *Tetrahedron Letters*, **49** (29–30), 4546; (c) Aprile, C., Giacalone, F., Gruttadauria, M., Marculescu, A.M., Noto, R., Revell, J.D. and Wennemers, H. (2007) *Green Chemistry*, **9** (12), 1328; (d) Gruttadauria, M., Riela, S., Aprile, C., Lo Meo, P., D'Anna, F. and Noto, R. (2006) *Advanced Synthesis and Catalysis*, **348** (1 + 2), 82–92; (e) Gruttadauria, M., Riela, S., Lo Meo, P., D'Anna, F. and Noto, R. (2004) *Tetrahedron Letters*, **45** (32), 6113.

20 Fremantle, M. (1998) *Chemical 37*.

21 (a) Cocalia, V.A., Visser, A.E., Rogers, R.D. and Holbrey, J.D. (2007) *Ionic Liquids in Synthesis* (eds P. Wasserscheid and T. Welton), Wiley-VCH Verlag GmbH, Weinheim, pp. 89–102; (b) Anderson, J.L., Amthony, J.L., Brennecke, J.F. and Maginn, E.J. (2007) *Ionic Liquids in Synthesis* (eds P. Wasserscheid and T. Welton), Wiley-VCH Verlag GmbH, Weinheim, pp. 103–129; (c) Kakiuchi, T. and Takashi (2008) *Analytical Sciences*, **24** (10) 1221; (d) Chirico, R.D., Diky, V., Magee, J.W., Frenkel, M., Marsh, K.N., Rossi, M.J., McQuillan, A.J., Lynden-Bell, R.M., Brett, C.M.A., Dymond, J.H., Goldbeter, A., Hou, J.-G., Marquardt, R., Sykes, B.D. and Yamanouchi, K. (2009) *Pure and Applied Chemistry*, **81** (5), 791.

22 Lei, Z., Arlt, W. and Wasserscheid, P. (2006) *Fluid Phase Equilibria*, **241**, 290.

23 (a) Klamt, A. and Eckert, F. (2000) *Fluid Phase Equilibria*, **172**, 43; (b) COSMOlogic GmbH, *Computational Chemistry and Fluid Thermodynamics*, www.COSMOlogic.de.

24 Navas, A., Ortega, J., Vreekamp, R., Marrero, E. and Palomar, J. (2009) *Industrial & Engineering Chemistry Research*, **48** (5), 2678.

References

25 Guo, Z., Lue, B.-M., Thomasen, K., Meyer, A.S. and Xu, X. (2007) *Green Chemistry*, **9** (12), 1362.

26 Chen, B., Guo, Z., Tan, T. and Xu, X. (2007) *Biotechnology and Bioengineering*, **99** (1), 18.

27 Jork, C., Kristen, C., Pieraccini, D., Stark, A., Chiappe, C., Beste, Y.A. and Arlt, W. (2005) *Journal of Chemical Thermodynamics*, **37** (6), 537.

28 Camper, D., Becker, C., Koval, C. and Noble, R. (2006) *Industrial & Engineering Chemistry Research*, **45** (1), 445.

29 (a) BASF AG, World Patent WO 03/062171; WO 03/062251; WO 05/061416; (b) Maase, M. (2004) *Chemie in Unserer Zeit*, 434; (c) Freemantle, M. (2003) *Chemical* (d) Rogers, R.D. and Seddon, K.R. *(2003) Inorganic Materials*, **2**, 363; (e) Seddon, K.R. (2003) *Science*, **302**, 792.

30 (a) Reetz, M., Wiesenhöfer, W., Francio, G. and Leitner, W. (2003) *Advanced Synthesis and Catalysis*, **345**, 1221; (b) Reetz, M.T., Wiesenhöfer, W., Franciò, G. and Leitner, W. (2002) *Chemical Communications*, 992.

31 (a) Wasserscheid, P., Kragl, U. and Kröckel, J. (to Solvent Innovation GmbH), World Patent WO 2003039719;(2003) *Chemical Abstracts*, **138**, 387435; (b) Kröckel, J. and Kragl, U. (2003) *Chemical Engineering & Technology*, **26** (11), 1166.

32 Wasserscheid, P. and Schulz, P.S. (2007) *Ionic Liquids in Synthesis* (eds P. Wasserscheid and T. Welton), Wiley-VCH Verlag GmbH, Weinheim, pp. 369–463.

33 Boon, J.A., Levisky, J.A., Pflug, J.L. and Wilkes, J.S. (1986) *The Journal of Organic Chemistry*, **51**, 480.

34 (a) Surette, J.K.D., Green, L. and Singer, R.D. (1996) *Chemical Communications*, 2753; (b) Stark, A., MacLean, B.L. and Singer, R.D. (1999) *Journal of The Chemical Society-Dalton Transactions*, 63.

35 Adams, C.J., Earle, M.J. and Seddon, K.R. (2000) *Green Chemistry*, **2**, 21.

36 Patell, Y., Winterton, N. and Seddon, K.R. (2000) World Patent WO 0037400.

37 Green, L., Hemeon, I. and Singer, R.D. (2000) *Tetrahedron Letters*, **41**, 1343–1345.

38 Lee, C.W. (1999) *Tetrahedron Letters*, **40**, 2461–2462.

39 Adams, C. J., Earle, M. J. and Seddon, K. R. (1999) *Chemical Communications*, 1043–1044.

40 Huang, J.F. and Sun, I.W. (2003) *Journal of the Electrochemical Society*, **150**, E299–E306.

41 Abbott, A.P., Capper, G., Davies, D.L., Rasheed, R.H. and Tambyrajah, V. (2002) *Green Chemistry*, **4**, 24–26.

42 Hardacre, C., Mcauley, B.J. and Seddon, K.R. (2003) World Patent WO03028883.

43 (a) Davey, P.N., Earle, M.J., Newman, C.P. and Seddon, K.R. (1999) World Patent WO 9919288; (b) Valkenberg, M.H., deCastro, C. and Hölderich, W.F. (2001) *Applied Catalysis A-General*, **215**, 185–190.

44 Zulfiqar, F. and Kitazume, T. (2000) *Green Chemistry*, **2**, 296.

45 (a) Smith, G.P., Dworkin, A.S., Pagni, R.M. and Zingg, S.P. (1989) *Journal of the American Chemical Society*, **111**, 525; (b) Ma, M. and Johnson, K.E. (1995) *Journal of the American Chemical Society*, **117**, 1508.

46 (a) Robert, T., Magna, L., Olivier-Bourbigou, H. and Gilbert, B. (2009) *Journal of the Electrochemical Society*, **156** (9), F115–F121; (b) Thomazeau, C., Olivier-Bourbigou, H., Magna, L., Luts, S. and Gilbert, B. (2003) *Journal of the American Chemical Society*, **125** (18), 5264; (c) Sesing, M. and Korth, W. (2002) *Green Chemistry*, **4**, 134.

47 Keim, W., Korth, W. and Wasserscheid, P. (2000) World Patent WO 0016902.

48 (a) Forbes, D.C. and Weaver, K.J. (2004) *Journal of Molecular Catalysis A-Chemical*, **214**, 129; (b) Gu, Y., Shi, F. and Deng, Y. (2004) *Journal of Molecular Catalysis A-Chemical*, **212**, 71.

49 (a) Earle, M.J., Katdare, S. P. and Seddon, K. R. (2002) World Patent WO0230865; (b) Earle, M.J., Katdare, S.P. and Seddon, K.R. (2004) *Organic Letters*, **6**, 707; (c) Qiao, K. and Yokoyama, C. (2004) *Chemistry Letters*, **33**, 808–809.

50 Orsini, M., Chiarotto, I., Elinson, M.N., Sotgiu, G. and Inesi, A. (2009) *Electrochemistry Communications*, **11** (5), 1013.

51 Chiarotto, I., Feeney, M.M.M., Feroci, M. and Inesi, A. (2009) *Electrochimica Acta*, **54** (5), 1638.

52 Feroci, M., Chiarotto, I., Orsini, M., Sotgiu, G. and Inesi, A. (2008) *Advanced Synthesis and Catalysis*, **350** (9), 1355.

53 Ranieri, G., Hallet, J.P. and Welton, T. (2008) *Industrial & Engineering Chemistry Research*, **47** (3), 638.

54 Crowhurst, L., Lancaster, N. L., Perez Arlandis, J. M. and Welton, T. (2004) *Journal of the American Chemical Society*, **126** (37), 11549.

55 Lancaster, N.L. and Welton, T. (2004) *The Journal of Organic Chemistry*, **69** (18), 5986.

56 Bini, R., Chiappe, C., Mestre, V.L., Pomelli, C.S. and Welton, T. (2008) *Organic and Biomolecular Chemistry*, **6** (14), 2522.

57 Wells, T.P., Hallet, J.P., Williams, C.K. and Welton, T. (2008) *The Journal of Organic Chemistry*, **73** (14), 5585.

58 Vaultier, M., Kirschning, A. and Singh, V. (2007) *Ionic Liquids in Synthesis* (eds P. Wasserscheid and T. Welton), Wiley-VCH Verlag GmbH, Weinheim, pp. 488–526, and references cited therein.

59 (a) Olah, G.A. (1963) *Friedel–Crafts and Related Reactions*, Interscience, New York; (b) Olah, G.A. (1973) *Friedel–Crafts Chemistry*, Wiley-Interscience, New York.

60 Keim, W. and Röper, M. (1995) Acylation and alkylation, in *Ulmann's Encyclopedia of Industrial Chemistry* (eds W. Gerhartz, Y.S. Yamamoto, F.T. Campbell, R. Pfefferkorn and J.F. Rounsaville), Wiley-VCH Verlag GmbH, pp. 185–220.

61 Sheldon, R.A., Arends, I. and Hanefeld, U. (2007) *Green Chemistry and Catalysis*, Wiley-VCH Verlag GmbH, Weinheim.

62 (a) Sheldon, R.A. and Downing, R.S. (1999) *Applied Catalysis A-General*, **189**, 163; (b) Tanabe, K. and Hölderich, W.F. (1999) *Applied Catalysis A-General*, **181**, 399.

63 Riisager, A., Fehrmann, R. and Wasserscheid, P. (2008) Supported liquid catalysts, in *Handbook of Heterogeneous Catalysis* (eds G. Ertl, H. Knözinger, F. Schüth and J. Weitkamp), 2nd edn, vol. 1, Wiley-VCH Verlag GmbH, Weinheim, pp. 631–644.

64 (a) Arroyo, P.A., Henriques, C.A., Sousa-Aguiar, E.F., Martinez, A. and Monteiro, J.L.F. (2000) *Studies in Surface Science and Catalysis*, **130C**, 2555; (b) Querini, C.A. and Roa, E. (1997) *Applied Catalysis A-General*, **163**, 199; (c) Sotelo, J.L., Uguina, M.A., Valverde, J.L. and Serrano, D.P. (1994) *Applied Catalysis A-General*, **114**, 273.

65 Kaeding, W.W. (1989) *Journal of Catalysis*, **120**, 409.

66 Sun, X. and Zhao, S. (2006) *Chinese Journal of Chemical Engineering*, **14** (3), 289–293.

67 Song, C.E., Shim, W.H., Roh, E.J. and Choi, J.H. (2000) *Chemical Communications*, 1695.

68 (a) Wasserscheid, P., Metlen, A. and Brausch, N. (2005) World Patent WO 2005014547; (b) Brausch, N., Metlen, A. and Wasserscheid, P. (2004) *Chemical Communications*, 1552.

69 Song, C.E., Shim, W.H., Roh, E.J. and Choi, J.H. (2000) *Chemical Communications*, 1695–1696.

70 Shen, H.-Y., Judeh, Z.M.A. and Ching, C.B. (2003) *Tetrahedron Letters*, **44**, 981–983.

71 Boon, J.A., Levisky, J.A., Pflug, J.L. and Wilkes, J.S. (1986) *The Journal of Organic Chemistry*, **51**, 480.

72 Hodgson, P.K.G., Morgan, M.L.M., Ellis, B., Abdul-Sada, A.A.K., Atkins, M.P. and Seddon, K.R. (1999) US Patent 5 994 602.

73 (a) Greco, C.C., Fawzy, S. and Lieh-Jiun, S. (1998) US Patent 5 824 832; (b) Wasserscheid, P., Ellis, B. and Fabienne, H. (2000) World Patent WO 0041809.

74 Xiao, L. and Johnson, K.E. (2004) *Canadian Journal of Chemistry*, **82**, 491–498.

75 Joni, J., Schmitt, D., Schulz, P.S., Lotz, T.J. and Wasserscheid, P. (2008) *Journal of Catalysis*, **258** (2), 401.

76 Ladnak, V., Joni, J. and Wasserscheid, P. (2007) DGMK Tagungsbericht (2007),

2007-2 (Preprints of the DGMK/SCI-Conference, Opportunities and Challenges at the Interface between Petrochemistry and Refinery), pp. 253–257.
77 Landnak, V., Hofmann, N., Brausch, N. and Wasserscheid, P. (2007) *Advanced Synthesis and Catalysis*, **349** (4–5), 719.
78 (a) DeCastro, C., Sauvage, E., Valkenberg, M.H. and Hölderich, W.F. (2000) *Journal of Catalysis*, **196**, 86; (b) Valkenberg, M.H., de Castro, C. and Hölderich, W.F. (2002) *Green Chemistry*, **4**, 88.
79 Valkenberg, M.H., de Castro, C. and Hölderich, W.F. (2001) *Studies in Surface Science and Catalysis*, **135**, 4629.
80 Joni, J., Haumann, M. and Wasserscheid, P. (2009) *Advanced Synthesis and Catalysis*, **351** (3), 423.
81 Joni, J., Haumann, M. and Wasserscheid, P. (2010) *Applied Catalysis A-General*, **372** (1), 8.

3
Transition Metal Catalysis in Ionic Liquids
Peter Wasserscheid

All points that were discussed in the previous chapter, Green Organic Synthesis in Ionic Liquids, in favor or disfavor of the use of ionic liquids also hold true in principle for the application of ionic liquid solvents in transition metal catalysis. However, new green chemistry aspects arise from the molecular interaction of the catalytic metal sites with the ionic liquid medium. As the catalytic metal sites are the centers of reaction, their chemical nature in the ionic liquid will greatly define the catalytic activity and selectivity and thus the greenness of the whole approach. Moreover, in many cases the catalytic metal compound is a particularly expensive part of the reaction mixture, which puts extra focus on the recycling of the ionic catalyst solution.

This chapter focuses on the additional green chemistry aspects that arise from the use of a catalytic transition metal in an ionic liquid reaction mixture. It takes into account that catalytically active metals can be present in an ionic liquid as homogeneously dissolved metal–ligand complexes, as nanoclusters/nanoparticles or as part of a classical heterogeneous catalyst that forms a suspension with the ionic liquid.

3.1
Solubility and Immobilization of Transition Metal Complexes in Ionic Liquids

Many transition metal complexes dissolve readily in ionic liquids in quantities that are suitable for catalysis (typically from 100 ppm to 2 mass%). This is due to the moderate polarity of most ionic liquids [1] and also due to their soft electronic character. Both properties combine in a way that allows the dissolution of many ionic or at least polar transition metal complexes. Sufficient solubility for a wide range of catalyst complexes is an obvious, but not a trivial, prerequisite for applying ionic liquids in homogenous catalysis. Some of the other green solvents, such as supercritical CO_2 (scCO_2) or perfluorinated solvents, suffer strongly from their limited ability to dissolve transition metal complexes. In these latter cases, this problem is solved by applying special ligand systems (e.g. ponytail ligands), which themselves are synthesized in expensive, multi-step procedures.

In this context, it is noteworthy that the full immobilization of transition metal complexes in ionic liquids, e.g. under the conditions of intensive extraction with a non-miscible product phase in a continuous process, also requires synthetic ligand modification in many cases. Ligands with sulfonato [2], guanidinium [3], imidazolium [4], pyridinium groups [5], 2-imidazolyl groups [6] and cobaltocenium [7] anchoring groups have been applied in this context (see Figure 3.1). In some cases, the ionic nature of the transition metal complex is sufficient to realize complete immobilization of the active catalyst in the ionic liquid. Examples from the literature include the $[H_4Ru_4(C_6H_6)_4][BF_4]_2$ cluster for arene hydrogenation [8], the *in situ*-generated [HNi(alkene)][AlCl$_4$] complex for alkene dimerization [9] and the complex [Pd(PPh$_3$)$_2$(Ar)][X] that is active in the Suzuki coupling reaction [10]. The design of metal complexes with ionic tags tailored for high catalytic activity and excellent immobilization in ionic liquids has been recently reviewed by Toma and co-workers [11].

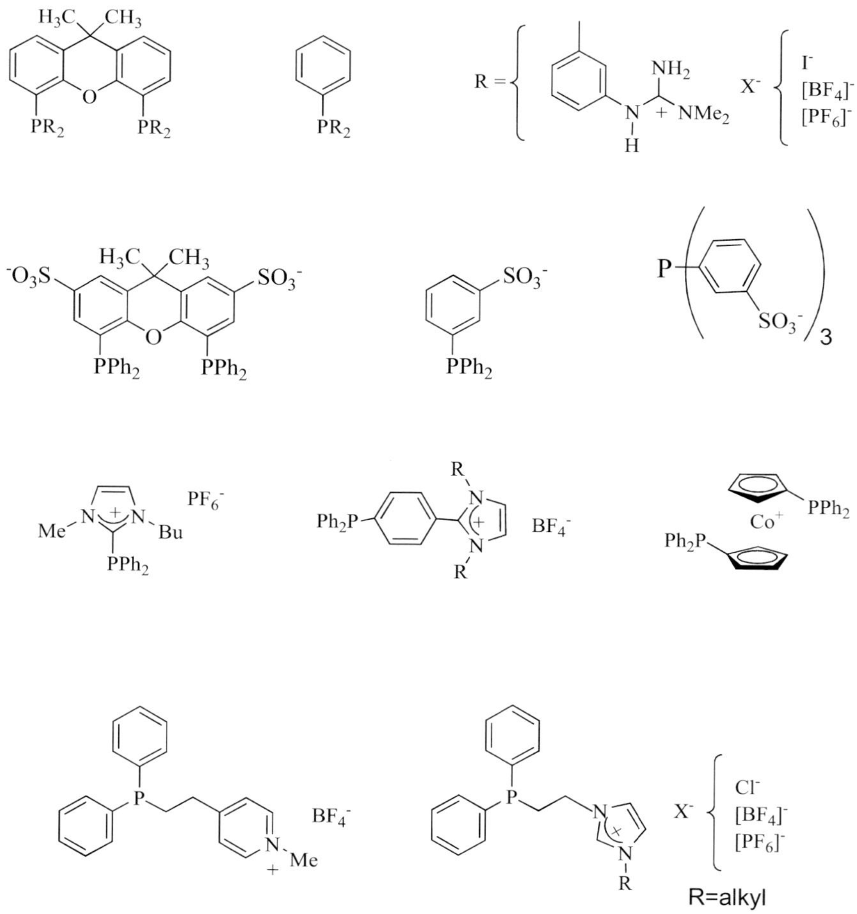

Figure 3.1 Ionic ligands to immobilize transition metal catalysts in ionic liquids. Adapted from [12].

3.2
Ionic Liquid–Catalyst Interaction

Depending on the coordinative properties of the anion and on the nature of the ionic liquid cation, ionic liquid–catalyst interactions can be very different. For green chemistry applications, it is important to understand which ionic liquid provides the ideal environment for the catalytic center to perform in the most selective and most active way with a very long lifetime.

Depending on the type of metal complex dissolved in the ionic liquid, the strategies to achieve this ambitious goal differ considerably. Two cases can be distinguished in general:

- cases in which the catalyst in the ionic liquid is catalytically active as such or in which it is self activating, e.g. by thermal abstraction of a protecting group;
- cases in which a chemical reaction between the ionic liquid and the catalyst precursor forms the active catalyst in the ionic liquid *in situ*.

In the first case, ionic liquids with weakly coordinating, inert anions, e.g. $[NTf_2]^-$, $[BF_4]^-$, $[SbF_6]^-$ and $[PF_3(C_2F_5)_3]^-$, and inert cations are preferred. In these cases, the role of the ionic liquid is to provide a moderately polar medium that does not compete with the substrate for free coordination sites. Moreover, the ionic liquid should be selected in order to provide sufficient feedstock solubility and efficient options for product isolation. As many ionic liquids offer a very special combination of polarity and nucleophilicity [13], they provide a superior reaction environment for many electrophilic and cationic transition metal complexes. For example, the success of ionic liquids in multiphasic oligomerization reactions with cationic Ni complexes [14] is based to a large extent on this often unique combination of solvation power for the cationic Ni complexes and the very weak coordination of the ionic liquid.

It is noteworthy, however, that the concentration of anions in ionic liquids is high (compared with, e.g., molecular solvents in which a salt is dissolved) so that even weakly coordinating anions coordinate to dissolved metal complexes and modify their reactivity [15]. Thus, even "innocent" ionic liquids have been shown to exhibit a distinctive influence on the reaction kinetics of ligand substitution reactions that can be rationalized using a linear solvation relationship based on the Kamlet–Taft solvent scale [16]. It is also very important to note that the "innocent" character of an ionic liquid is highly dependent on the ionic liquid's quality. Chloride impurities have been reported to act as strong catalyst poisons in many reactions (e.g. [17]), but also methylimidazole impurities have been reported to show a strong influence on the formation of the active transition metal complex [18].

In the second case, it is essential to understand and to control exactly the reaction that forms the active catalyst in the ionic liquid. To optimize the green chemistry performance of such an *in situ* catalytic system, it is crucial that 100% of the precious metal ends up after pre-formation in the most active, most selective and most stable catalyst complex.

Two processes for catalyst preformation in ionic liquids will be discussed in more detail in the following: activation of transition metal complexes by Lewis acidic ionic liquids and *in situ* formation of metal–carbene complexes. A third case is the *in situ* reduction of metal complexes to catalytic nanoparticles in ionic liquids. This topic will be highlighted in Section 3.5.

3.2.1
Activation of Transition Metal Complexes by Lewis Acidic Ionic Liquids

As chloroaluminate ionic liquids were among the first known and well-studied ionic liquids, many of the early examples of transition metal catalysis in ionic liquids used the Lewis acidity of these ionic liquids to abstract a chloride ligand from a catalyst precursor to create a cationic, electrophilic catalyst complex. The activation of Cp_2TiCl_2 [19] and $(ligand)_2NiCl_2$ [20] in acidic chloroaluminate melts and the activation of $(PR_3)_2PtCl_2$ in chlorostannate melts [21] are shown as examples in Equations 3.1–3.3:

$$Cp_2TiCl_2 + [cation][Al_2Cl_7] \rightleftharpoons [Cp_2TiCl][AlCl_4] + [cation][AlCl_4] \quad (3.1)$$

$$(ligand)_2NiCl_2 + [cation][Al_2Cl_7] + [cation][Al_2EtCl_6] \rightleftharpoons$$
$$[(ligand)Ni\text{-}CH_2\text{-}CH_3][AlCl_4] + 2[cation][AlCl_4] + AlCl_3\text{-ligand} \quad (3.2)$$

$$(PR_3)_2PtCl_2 + [cation][Sn_2Cl_5] \rightleftharpoons [(PR_3)_2PtCl][SnCl_3] + [cation][SnCl_3] \quad (3.3)$$

For example, van Eldik and co-workers studied in detail the kinetics of the formation of the active hydroformylation catalysts *cis*-[Pt(PPh$_3$)$_2$Cl(SnCl$_3$)] and *cis*-[Pt(PPh$_3$)$_2$(SnCl$_3$)$_2$] from the precursor *cis*-[Pt(PPh$_3$)$_2$Cl$_2$] in presence of SnCl$_2$ in different imidazolium-based chlorostannate ionic liquids [22].

It is obvious that in these cases the ionic liquid plays the role of a co-catalyst and therefore the nature of the ionic liquid becomes very important for the effective catalyst species that is formed. The potential to improve the catalytic performance and thus the greenness of the catalyst includes variation of the halide salt, the Lewis acid and the ratio of the two components forming the ionic liquid.

It is noteworthy, however, that the use of Lewis acidic ionic liquids may cause some restrictions concerning the reactivity of feedstock and products towards the Lewis acidic anion. The water content of the feedstock is also an important issue as water reacts with most Lewis acidic anions in an irreversible manner.

3.2.2
***In Situ* Carbene Complex Formation**

1,3-Dialkylimidazolium ions, which are most widely used as cations in ionic liquids, can be converted under certain circumstances to carbene ligands. As a consequence, catalytically active metal–carbene complexes can form *in situ* in these ionic liquids. Formation of metal–carbene complexes has been described by oxidative addition of the imidazolium cation [23] (see Scheme 3.1a) and by deprotonation of the imidazolium salt (based on the well-known acidity of the hydrogen in 2-position of the 1,3-dialkylimidazolium salt [24]) followed by coordination of the imidazolylidene on the metal center [25] (see Scheme 3.1b).

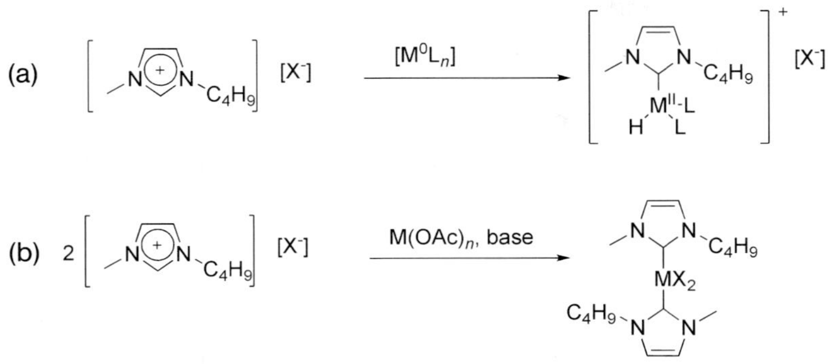

Scheme 3.1 Different routes for an *in situ* ligand formation from the ionic liquid's 1,3-dialkylimidazolium cation.

To avoid the formation of metal–carbene complexes by deprotonation of the imidazolium cation under basic conditions, the use of 2-methyl-substituted imidazolium is frequently suggested. However, it should be mentioned that strong bases can abstract a proton here also to form the vinylimidazolidene species which may also act as a strong ligand to electrophilic metal centers. Moreover, deprotonation of the protons on the 4- and 5-positions of the imidazolium has also been observed [26]. The following procedures can be recommended to check whether *in situ* carbene formation is of relevance for a given transition metal-catalyzed reaction in ionic liquids:

- variation of ligands in the catalytic system;
- application of independently prepared, defined metal–carbene complexes;
- investigation of the reaction in pyridinium-, tetraalkylammonium-, tetraalkylphosphonium- or 1,2,3-trialkylimidazolium-based ionic liquids.

It is important to note that the *in situ* formation of transition metal–carbene complexes does not automatically result in a more efficient catalyst or to greener transition metal catalysis. On the contrary, the enormous excess of pre-ligand in a 1,3-dialkylimidazolium ionic liquid can easily lead to a situation in which all available coordination centers at the metal are blocked with carbene ligands and consequently the catalyst loses all activity. Such behavior has been described by Basset and co-workers for the Pd-catalyzed telomerization of butadiene and methanol in 1,3-dialkylimidazolium ionic liquids [27]. Welton and co-workers described a very elegant approach to circumvent this problem by using a mixed ligand system in which a bulky phosphine ligand avoids complete coordination of a Pd center by carbene ligands. Consequently, a phosphine–carbene complex was formed *in situ* in the reaction mixture that showed good catalytic performance in the Suzuki cross-coupling [28]. A similar approach was later applied by de Souza and Antunes to obtain an active catalyst for the Heck reaction of isopropenyl acetate with aryl and benzyl halides by using $Pd(OAc)_2$–K_2CO_3–PPh_3 in [C_4mim]Br [29]. The groups of Cole-Hamilton and Chauvin prepared *in situ* N-heterocyclic carbene complexes of iridium and rhodium for catalytic hydrogenation reactions [30]. Scholten and Dupont provided evidence by NMR experiments that the formation of *in situ* carbene complexes plays a role even in Rh-catalyzed hydroformylation, but here the carbene ligands are easily displaced and therefore do not significantly change the catalytic performance [31]. Very recently, Kim *et al.* reported the *in situ* formation of an Ru–carbene complex from an Ru-carboxylate complex in the presence of [C_4mim]Br which showed high catalytic activity in ring-closing metathesis [32].

3.3
Distillative Product Isolation from Ionic Catalyst Solutions

The extremely low vapor pressure of ionic liquids permits efficient product isolation by distillative processes (see above). During these processes, the ionic liquid is able to stabilize the homogeneous catalyst by providing a non-volatile and stabilizing solvent environment [33].

Ionic liquid recycling from homogeneous monophasic solutions via distillation can be a particularly green approach to perform transition metal catalysis with ionic liquids. The monophasic reaction mode does not require extensive stirring and prevents all mass transfer problems during reaction. The energetic effort for the distillation step is usually not detrimental to the energy balance of the process as the products would need a distillative work-up anyway. Additionally, distillative product separation and recycling of the ionic catalyst solution from the distillation bottom do not require ligand modification and can be easily implemented in traditional reaction equipment. Thus, extra investment in equipment (e.g. phase separators, settlers) can be avoided. Of course, a prerequisite for this mode of operation is a thermally robust ionic catalyst solution and sufficient volatility of products and by-products.

Recently, a new concept to process transition metal catalysis in ionic liquids has been introduced that makes explicit use of the above-described concept but tries at the same time to reduce the amount of ionic liquid to an absolute minimum – the supported ionic liquid phase (SILP) concept. Here, the ionic catalyst solution is confined on the surface of a highly porous solid by various methods, such as physisorption, tethering or covalent anchoring of ionic liquid fragments [34]. This preparation results in a solid that appears macroscopically as a heterogeneous catalyst but acts chemically like a homogeneous catalyst due to the molecular defined, uniform ionic liquid environment provided to the dissolved metal complex by the supported ionic liquid film.

During catalytic reaction using a SILP catalyst, the feedstock molecules diffuse through the residual pores of the catalyst, dissolve in the ionic liquid catalyst phase and react at dissolved transition metal complexes within the thin liquid catalyst film which is dispersed on the walls of the support material. The products diffuse back into the void pore space and further out of the catalyst particle (see Figure 3.2).

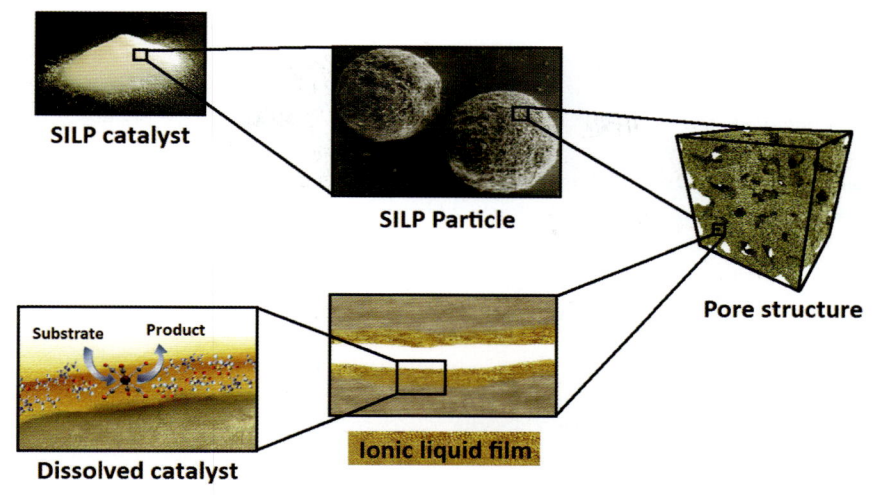

Figure 3.2 Schematic representation of a supported ionic liquid phase (SILP) catalyst.

SILP catalysis is a particularly green way to perform transition metal catalysis, especially if all the reactants are contacted with the SILP material in gaseous form. In this case, a completely solvent-free process can be realized that makes use of the very small amount of ionic liquid in a highly efficient manner. Moreover, the highly selective and active nature of the homogeneous catalyst can be combined with the very simple processing of classical heterogeneous catalysis in fixed-bed reactors. Note that no specific ligand design is necessary to immobilize the transition metal complex in the ionic liquid provided that the metal complex does not evaporate or sublime from the ionic liquid in gas-phase contact.

It is noteworthy in this context, that the underlying principles of SILP catalysis are not new at all. Earlier attempts were made to achieve the same goal of reducing the amount of solvent by a thin-film technique for continuous catalytic reactions using high-boiling organic liquid phases [35] or water [36] as the immobilized liquid phase. However, in continuous gas-phase contact, the volatile nature of these solvents resulted in too unstable systems for technical use. In contrast, the negligible vapor pressure of ionic liquids together with their often very suitable solvent nature for transition metal catalysis offer tremendous potential for continuous catalytic processes even under industrial conditions. Operational stabilities of more than 45 days have been demonstrated recently in our laboratories for a SILP hydroformylation catalyst in continuous gas contact [37].

Convincing examples of SILP transition metal catalysis in continuous gas-phase contact have been reported for hydroformylation [38] (see also Section 3.7), hydrogenation [39], hydroamination [40], methanol carbonylation [41] and water gas shift reactions [42].

Processing of SILP catalysts in liquid reaction phases (slurry mode reaction) is also possible but requires extremely low solubility of the ionic liquid film in the liquid reaction mixture. Moreover, there are special constraints on the wettability of the ionic liquid film on the support's surface to avoid convective, mechanical removal of the film. Another complication is that the slurry mode operation requires complex immobilization (e.g. by modified ligands) in the ionic liquid film to avoid catalyst leaching from the supported ionic liquid film into the reaction phase. Despite these additional requirements, a significant number of very promising applications of SILP-slurry reactions have been reported in recent times, including examples of hydrogenations [43], hydroaminations [44], carbonylations [45], Heck reactions [46], Suzuki cross-coupling reactions [47] and allylic substitution reactions [48]. For more details, the reader is referred to a number of recent excellent reviews on the topic of SILP transition metal catalysis [49].

3.4
New Opportunities for Biphasic Catalysis

Despite the obvious advantages of homogeneous over heterogeneous catalysis – such as specificity, variability through ligand design, utilization of every metal center – the proportion of homogeneous catalyzed reactions in industrial chemistry is still fairly low. The main reason for this is the difficulty in separating the homogeneously dissolved catalyst from the products and by-products after the reaction. Since the transition metal complexes used in homogeneous catalysis are usually valuable, complete catalyst recovery is crucial in the context of green chemistry.

Biphasic catalysis in a liquid–liquid system is a powerful approach for combining the advantages of both homogeneous and heterogeneous catalysis. Here, the catalyst is immobilized in one of two immiscible phases. Product separation works by decantation and the catalyst phase can be recycled without any further treatment, thermal stress or chemical transformation. To realize liquid–liquid biphasic catalysis successfully, a miscibility gap between the two phases, excellent catalyst immobilization without catalyst deactivation and sufficient feedstock solubility in the catalyst phase are the main prerequisites [50]. These demanding requirements have limited the number of industrially realized examples of liquid–liquid biphasic catalysis with aqueous or highly polar organic phases [51]. Many ionic liquids offer unique miscibility properties in combination with superior solvent properties for dissolved catalyst complexes and much potential for complex immobilization (see above). For these reasons, ionic liquids are considered as a new generation of solvents for green liquid–liquid biphasic catalysis.

In addition to catalyst immobilization and recycling, ionic liquid–organic biphasic catalysis offers the possibility of optimizing the reaction's activity and selectivity. This can be realized by *in situ* extraction of catalyst poisons or reaction intermediates from the catalytic layer. Figure 3.3 demonstrates this concept, exemplified for selectivity enhancement in a dimerization reaction. It is shown that the dimer selectivity of the reaction of compound A can be enhanced significantly if the reaction is carried out in biphasic mode using a catalyst solvent with high solubility for the monomer A and low solubility for the dimer A–A. Thus, the dimer A–A is readily extracted from the

organic layer: A/A-A = 1:1

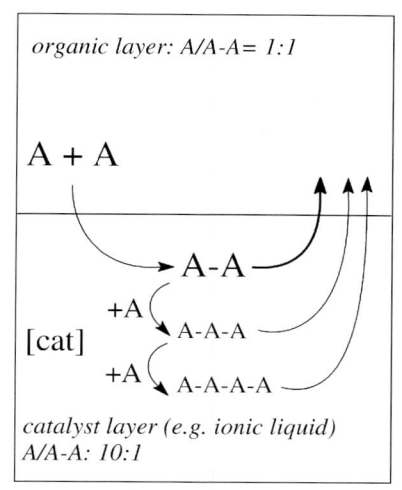

A + A

→ A-A

[cat] +A A-A-A

+A A-A-A-A

catalyst layer (e.g. ionic liquid)
A/A-A: 10:1

Figure 3.3 Enhanced dimer selectivity in the oligomerization of compound A due to a biphasic reaction mode with a catalyst solvent of high preferential solubility for A.

catalyst phase into the product layer, which reduces the chance for the formation of the undesired higher oligomers A–A–A and A–A–A–A.

In the context of green chemistry, there are three convincing arguments to apply ionic liquid-organic biphasic catalysis:

1. This reaction mode offers a simple means for catalyst recovery and recycling.

2. This reaction mode offers a simple means for ionic liquid recovery and recycling. In fact, the ionic catalyst solution will be recovered and recycled together as in heterogeneous catalysis where the support and the metal loading are filtered off the products in one step with separation.

3. Ionic liquid–organic biphasic catalysis allows specific ionic liquid effects to be achieved (e.g. special feedstock solubility or particular catalyst activation), even with a small ionic liquid to organic ratio, as an immiscible ionic catalyst solution retains its specific ionic liquid character even if the immiscible, organic reactant–product phase is used in large volumetric excess.

To demonstrate the green chemistry potential of ionic liquids based on liquid–liquid biphasic catalysis, it is instructive to look at a specific example, Pt-catalyzed hydrosilylation.

Hydrosilylation involves the metal-catalyzed anti-Markovnikov addition of an Si–H moiety to alkenes with formation of a new Si–C bond [52]. The reaction is of high preparative value and has found extensive use for the technical production of organosilane compounds and organo-modified oligosiloxanes [53]. However, attempts to realize liquid–liquid biphasic hydrosilylation failed in the past due to the reactivity of most Si–H compounds with water or polar organic solvents (such as e.g. diols). As a consequence, it is industrially applied practice in hydrosilylation to leave the Pt catalyst (usually in amounts of 5–50 ppm) in the products if the latter are too high boiling for a distillative work-up. Thus, platinum that has been extracted and refined from a Pt-rich ore ends up at ppm levels in a product that is later used, for example, as an additive to polymers where it finally ends up at the ppb level. In terms of sustainability, the result of this practice is that a part of our Pt resource is diluted and distributed in a manner that never allows its subsequent recycling, an irreversible waste of an important precious metal resource (even if the amount per ton of product is relatively low). In contrast, liquid–liquid biphasic catalysis with ionic liquid catalyst phases allows the Pt required for hydrosilylation to be immobilized very efficiently, as has been demonstrated by several industrial [54] and academic groups [55]. Even continuous operation in a loop reactor has been demonstrated over more than 50 h time on-stream [56] (see Figure 3.4).

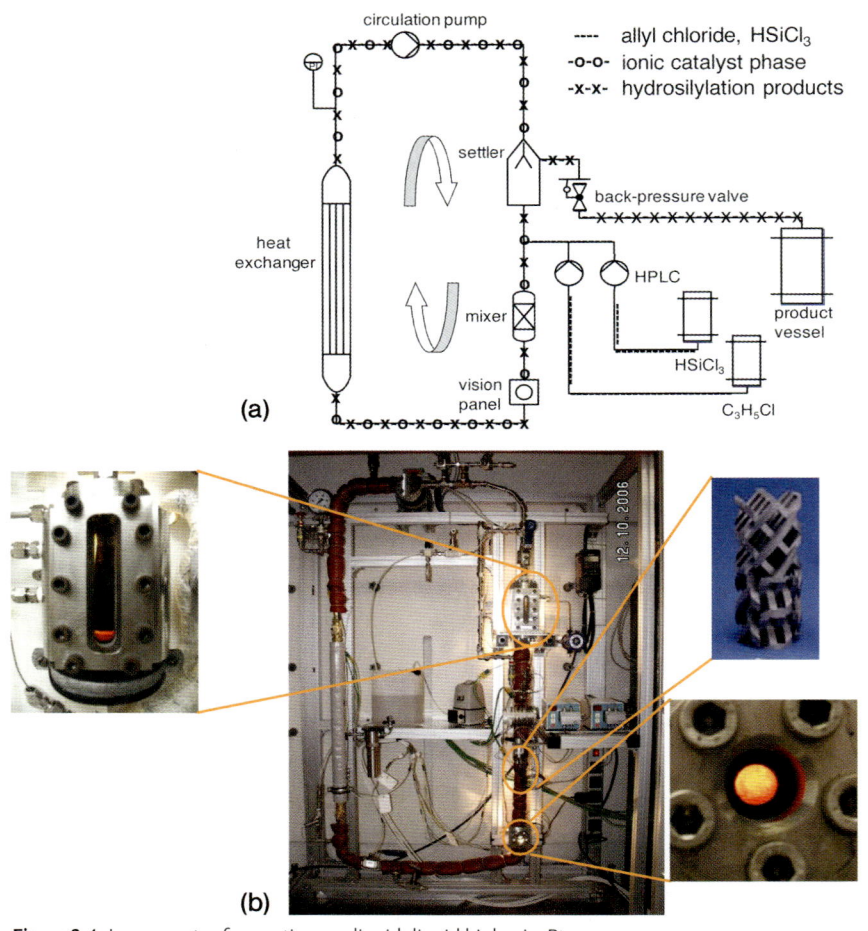

Figure 3.4 Loop reactor for continuous liquid–liquid biphasic, Pt-catalyzed hydrosilylation of allyl chloride and $HSiCl_3$ at the University of Erlangen-Nuremberg. (a) Schematic view; (b) reactor during continuous reaction.

Here Pt is present in relatively concentrated form and can be easily recycled even if the catalyst may lose some of its activity over time. In contrast to water or polar ionic liquids, many ionic liquids do not react with Si–H compounds and still form a miscibility gap with hydrosilylation products, which is a unique feature that has great potential to improve the greenness of future hydrosilylation processes.

At the end of this section, two interesting variations of multiphase catalysis including ionic liquids should be mentioned that promise significant potential with respect to green chemistry: catalysis in ionic liquid–scCO$_2$ biphasic systems and catalysis in ionic liquid–organic thermomorphic systems.

The combination of ionic liquids and compressed CO$_2$ – which are on the extreme ends of the volatility and polarity scales – offers great potential in the design of environmentally benign catalytic processes. In comparison with homogeneous catalysis in compressed CO$_2$ alone, the active catalyst complexes are much easier to dissolve and immobilize in ionic liquids. Furthermore, in the ionic liquid–scCO$_2$ biphasic system, product separation is possible without exposing the catalyst to variations in temperature, pressure or substrate concentration. Compared with the use of pure ionic liquid, the presence of compressed CO$_2$ greatly decreases the viscosity of the ionic catalyst solution and increases drastically the solubility of reactive gases (such as hydrogen, carbon monoxide and oxygen) in the ionic liquid.

Transition metal catalysis in thermomorphic ionic liquid systems provides the green advantage that the reaction system is monophasic under reaction conditions [57]. Thus, in contrast to permanent liquid–liquid biphasic catalysis, energy intake to create a large interfacial surface for interphase mass transport is not needed. However, the complete immobilization of the catalyst in one of the phases of the biphasic system that forms in the cold remains a challenge in this approach.

3.5
Green Aspects of Nanoparticle and Nanocluster Catalysis in Ionic Liquids

The scientific interest in catalysis by transition metal nanoparticles has seen a dramatic increase in recent years [58] and ionic liquids have been shown to be very suitable media for the generation and stabilization of metallic nanoparticles [59].

In this section, some green aspects of catalysis with nanoparticles or nanoclusters in ionic liquids are discussed. This approach has been mainly tested for selective hydrogenation [60], Heck reactions [61], hydrosilylation reactions [62], Suzuki coupling [63] and Stille coupling [64].

High catalytic activities and long catalytic lifetimes could be demonstrated for nanoparticles immobilized in ionic liquids in some selected cases. An impressive example is the hydrogenation of 1-hexene with Ru nanoparticles in [C$_4$mim][PF$_6$] described by Dupont and co-workers [60e]. Turnover numbers of 110 000 (based on total Ru) or 320 000 (based on Ru surface atoms) could be demonstrated and the nanoparticle–ionic liquid solution could be recycled over 17 times.

Despite this impressive performance, nanoparticles require in some cases (note that the reaction temperature is an important parameter here) additional stabilization in ionic liquids by additives to provide good lifetime for catalysis. For example, phenanthroline was used as a stabilizer for Pd nanoparticles by Han and co-workers [60a]. The ionic liquid suspension of the particles was applied to catalyze the hydrogenation of alkenes. When the phenanthroline was omitted, the system severely lost its reactivity on the first recycle, presumably due to particle agglomeration.

If nanoparticle catalysts are generated *in situ* and dispersed in ionic liquids, it should be noted that the solubility properties of the ionic liquid still determine the concentration of reactants at the catalytic surface. Solubility effects on reaction selectivity have been found by Huang *et al.* [60b] and particle stabilization against oxidation from air (due to the low oxygen solubility of the applied ionic liquid) has been reported by Dupont and co-workers [60e]. Moreover, the ionic liquid around a catalytic particle will not necessarily be inert, but in many cases chemical modification of the particle surface by the ionic liquid can be expected. An example of this case has been reported by Karimi and Enders [61g]. They studied the Heck-reaction of bromobenzene with methyl acrylate in a system in which the nanoparticles were obtained as a result of the covalent anchoring of an *N*-heterocyclic carbene palladium–ionic liquid matrix on a silica surface.

The preparation of nanoparticulate metal catalysts in porous supports with the help of ionic liquids was reported in 2003 by Hardacre and co-workers, who prepared a Pd-loaded monolith for hydrogenation and Heck coupling reactions by reduction of Pd acetate in [C_4mim][NTf$_2$] and incubation in (EtO)$_4$Si and formic acid [65]. Another convincing example of an SILP–nanoparticle catalyst has been reported by Han and co-workers [66]. They applied 1,1,3,3-tetramethylguanidinium (TGA)-exchanged montmorillonite (naturally occurring, negatively charged two-dimensional silicate sheets separated by interlayers of TGA ions) as supporting and stabilizing medium for Ru nanoparticles. The favorable synergistic effects of the applied guanidinium ions, the montmorillonite support and the Ru particles led to a catalytic material exhibiting high activity in benzene hydrogenation [turnover frequency (TOF) = 4000 h^{-1}] and excellent stability. Dupont and co-workers reported more recently a catalytic system for alkene hydrogenation that consisted of Ru nanoparticles dispersed in the ionic liquid [C_4mim][BF$_4$] immobilized on a silica network prepared in a sol–gel process [67]. A similar approach using an ionic liquid-modified xerogel has recently been demonstrated to be very effective also in Pd-catalyzed Suzuki and Heck coupling reactions [68].

SILP materials that contain catalytically active nanoparticles instead of dissolved homogeneous complexes in the ionic liquid film on a porous support are of high interest for green chemistry approaches. They have the potential to combine the interesting catalytic properties of nanoparticles with the stabilization of the ionic liquid (which is used in very small amounts) in a macroscopically solid catalyst material suitable for simple fixed-bed reactor technologies.

Finally, it should be noted that not in all cases in which nanoparticles are observed in ionic liquids are the latter the catalytic species responsible for the reactivity of the system. Dupont *et al.* interpreted their results in the Pd-catalyzed coupling of aryl halides with butyl acrylate by the Pd(0) nanoparticles serving as a reservoir for a homogeneous catalytically active species and not being the catalytically active species themselves [61f].

3.6
Green Aspects of Heterogeneous Catalysis in Ionic Liquids

In contrast to the large number of examples using homogeneously dissolved metal complexes or metal nanoparticles in ionic liquids for catalysis, literature reports on true heterogeneous catalysts suspended in ionic liquids or coated with ionic liquid layers are rather rare. Despite this fact, this approach is very attractive from a green chemistry perspective if the presence of the ionic liquid has a distinct positive effect compared with the application of the pure heterogeneous catalyst. The modification of an existing catalytic solid by contacting it with an ionic liquid is straightforward as it does not require any additional synthetic step.

In principle, the contact of the ionic liquid with the catalytic surface of a heterogeneous catalyst can have two major effects:

1. The ionic liquid's solubility properties can changes the relative concentrations of the reactants and products at the catalytic center, which in many cases has a huge effect on activity (e.g. in the case of product poisoning) and selectivity (e.g. in the case of consecutive reactions). Note that mass transfer restrictions caused by the relatively viscous ionic liquid play a role only for heterogeneous catalysts suspended in ionic liquids and not for thin film-coated heterogeneous catalysts. In the latter case, the time of diffusion is usually much shorter compared to the time of reaction due to the thin film of ionic liquid.

2. The ionic liquid can modify the chemical nature of the active site of the applied heterogeneous catalyst in a kind of ligand type interaction. In this case, the catalytic center may change completely its electronic and steric properties depending on the ionic liquid's chemical interaction potential.

Unfortunately, most reported examples of heterogeneous catalysis in contact with ionic liquids do not differentiate between these two influences, but it is obvious that both can easily act together.

Hardacre and co-workers were the first to explore suspensions of a classical Pd-on-support catalyst in ionic liquids for the hydrogenation of unsaturated aldehydes [69]. Another example was reported by Claus and co-workers, dealing with the hydrogenation of citral applying Rh–Sn/SiO_2 suspended in [C_4mim][NTf_2]) [70]. They found that treatment of a Pd-on-silica catalyst with dicyanamide ionic liquids resulted in a significant selectivity enhancement of the desired product, citronellal [71]. With 29 mass% of a pyridinium-based dicyanamide ionic liquid in combination with Pd/SiO_2, the quantitative synthesis of citronellal was achieved.

A new aspect has recently been added to the concept of heterogeneous catalysis in ionic liquids in a study by Obert et al. [72]. They demonstrated that ionic liquid–organic biphasic systems can offer a very efficient way to separate highly

Figure 3.5 Ru-on-C suspended in 1,2,4-trichlorobenzene (bottom phase) and dimethylcyclohexylammonium hydrogensulfate (top phase).

dispersed heterogeneous catalysts from liquid reaction products avoiding any tedious filtration process. Due to the difference in wettability, it was shown that an Ru-on-C catalyst remains selectively attached to the ionic liquid phase even in cases where the latter is the less dense phase (e.g. versus trichlorobenzene, as shown in Figure 3.5). The immobilization of Ru-on-C in the ionic liquid layer was used to recycle this ionic catalyst suspension after propionitrile hydrogenation by a simple decantation process.

Jess and co-workers developed the concept of adding ionic liquids to classical heterogeneous catalysts further by drastically reducing the amount of ionic liquid [73]. They prepared a thin ionic liquid coating on the top of a commercial heterogeneous Ni-based hydrogenation catalyst and applied the resulting material in cyclooctadiene hydrogenation. It was found that the presence of the ionic liquid film led to an increase in the selectivity to the desired reaction intermediate cyclooctene from 40 to 70% at comparable conversions. They demonstrated by particle size variation that the observed effect was not due to mass transfer restrictions. They coined the term "solid catalyst with ionic liquid layer" (SCILL) for their approach. More spectroscopic work adding to the same concept was recently presented by Lercher and co-workers [74], who investigated Pt/SiO_2 coated with a thin film of $[C_4C_1mim][TfO]$ by IR, inelastic neutron scattering and NMR spectroscopy. The coverage of the catalyst surface and the catalytic centers by the ionic liquid was confirmed by these methods and the resulting catalyst was still found to be active and stable for the hydrogenation of ethylene.

So far, this chapter has introduced and highlighted some important aspects of green chemistry with respect to transition metal catalysis in ionic liquids. Of course, the selection of topics is far from comprehensive and reflects the present author's personal perspective and expertise. The following section will focus on one specific reaction to illustrate many of the above-discussed aspects in more detail and in more practical terms. For this purpose, catalytic hydroformylation was chosen, a reaction that has been successfully carried out in ionic liquid media since 1972 [75]. Many of the recent developments in greener ionic liquids, greener recycling strategies and greener solvent combinations have been tested using hydroformylation, which makes this example particularly instructive. The author is fully aware that focusing on hydroformylation is a somewhat subjective choice. Many other transition metal-catalyzed reactions would have been equally worth treating as an example in more detail. The reader more interested in other examples of transition metal catalysis in ionic liquids is referred to a large number of excellent reviews that have appeared on the topic in the last 10 years. They all include many more examples for the above-described green chemistry aspects of transition metal catalysis in ionic liquids. Important examples originated from Wu *et al.* [76], Geldbach [77], Parvulescu and Hardacre [78], Wasserscheid and co-workers [79], Giernoth [80], Welton [81], MacFarlane and co-workers [82], Pozzi and Shepperson [83], Dupont *et al.* [84], Zhao *et al.* [85], Haag and co-workers [86], Dobbs and Kimberley [87], Olivier-Bourbigou and Magna [88], Sheldon [89], Gordon [90] and Holbrey and Seddon [91] on the same topic.

3.7
Green Chemistry Aspects of Hydroformylation Catalysis in Ionic Liquids

In hydroformylation, biphasic catalysis is a well-established, industrially realized [92] method for effective catalyst separation and recycling. Due to this industrial success of liquid–liquid biphasic hydroformylation, nearly all other concepts applying alternative solvents or catalyst immobilization strategies have been tested using hydroformylation. In a very interesting book edited by Cole-Hamilton and Tooze, these different approaches have been benchmarked against each other [93].

Concerning liquid–liquid biphasic, Rh-catalyzed hydroformylation in ionic liquids, the development started in 1995 with the pioneering work of Chauvin and co-workers [94]. In the following years, a huge amount of scientific activity has developed in this specific field and this was extensively reviewed recently by Haumann and Riisager [95]. Consequently, the scope of this section is restricted to a brief discussion of some important aspects that concern the greenness of hydroformylation catalysis with ionic liquids and illustrate further the arguments put forward in Sections 3.1–3.5.

3.7.1
Feedstock Solubility

Catalytic hydroformylation in liquid–liquid biphasic systems using pure water as the catalyst phase is restricted to the conversion of C_2-C_5 alkenes due to the low water solubility of higher alkenes. However, the hydroformylation of many higher alkenes is also of commercial interest. Ionic liquids can be structurally designed to dissolve large amounts of higher alkenes, and even completely alkene-miscible ionic liquids can be obtained with tetraalkylammonium or tetraalkylphosphonium salts [96]. As the feedstock concentration in the reaction phase largely determines the reaction rate, high and tunable feedstock solubility is a real green point which directly results in higher reaction rates, reduced reactor size and significantly lower capital investment. Concerning the reaction gases carbon monoxide and hydrogen, most ionic liquids show relatively low but sufficient gas solubilities to allow reasonable reaction rates [97].

3.7.2
Catalyst Solubility and Immobilization

The first hydroformylation study in ionic liquids by Chauvin *et al.* revealed the important point of immobilizing the dissolved hydroformylation catalyst by means of highly polar or ionic anchoring groups [94a]. Whereas in the hydroformylation of 1-pentene the neutral catalyst system [Rh(CO)$_2$(acac)]–triarylphosphine showed significant leaching from the [C$_4$mim][PF$_6$] ionic liquid, complete catalyst immobilization could be realized by the application of sulfonated triarylphosphine ligands. Later, a large number of phosphine ligands with ionic anchoring groups were designed and synthesized specifically for the immobilization of Rh hydroformylation catalysts in ionic liquids (see Figure 3.1). Scheme 3.2 shows the four-step synthesis to obtain a dicationic phenoxaphosphino-modified xantphos ligand for highly regioselective hydroformylation in ionic liquids to produce linear aldehydes [98]. From this scheme, it becomes clear that attaching immobilizing ionic groups to highly optimized (with respect to their electronic and steric properties) ligand structures is a challenging task and involves significant synthetic effort, including cost, waste production and energy consumption related to this.

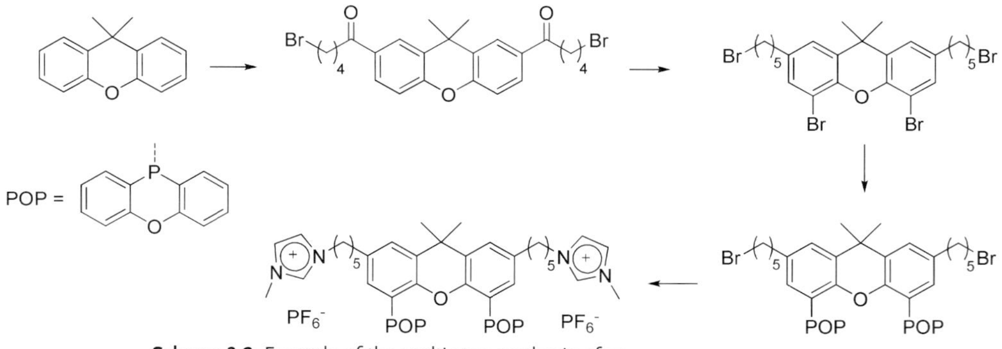

Scheme 3.2 Example of the multi-step synthesis of an imidazolium-tagged ligand for hydroformylation according to [98].

3.7.3
Use of Phosphite Ligands in Ionic Liquids

In addition to the traditional phosphine ligands, phosphite ligands are also very attractive in hydroformylation, for two reasons: they often lead to highly active hydroformylation systems and they allow hydroformylation and isomerization activity to be combined in one catalyst. The latter fact is particularly interesting for converting internal alkenes into linear aldehydes in one reactor. Unfortunately these attractive features come along with the disadvantage of a significant hydrolysis lability of phosphite ligands that prevents all immobilization and recycling concepts using an aqueous catalyst phase.

Here many ionic liquids have the important green advantage of being stable in the presence of phosphite ligands, which opens up a new approach to immobilizing Rh–phosphite complexes in liquid–liquid biphasic hydroformylation systems for efficient catalyst recycling. This concept was first demonstrated by Keim *et al.* in the regioselective hydroformylation of methyl-3-pentenoate using Rh complexes with bulky phosphite ligands in $[C_4mim][PF_6]$ [99]. Later, Olivier-Bourbigou's group demonstrated liquid–liquid biphasic, Rh-catalyzed hydroformylation of 1-hexene in various imidazolium and pyrrolidinium ionic liquids using phosphite ligands and better selectivities to the linear aldehyde product were found compared with the use of traditional phosphine ligand systems [100].

3.7.4
Halogen-containing Ionic Liquids Versus Halogen-free Ionic Liquids in Hydroformylation

All early examples of hydroformylation catalysis with ionic liquids were carried out in tetrafluoroborate and hexafluorophosphate salts. However, as we know very well today, these anions are sensitive to hydrolysis [101] and toxic HF is released when these anions react with water. In hydroformylation catalysis, water impurities may be introduced to the reaction system by wet reactants or by water formation in aldol condensation, which is one of the main consecutive reactions of the aldehyde products. In every case hydrolysis of tetrafluoroborate or hexafluorophospate anions has severe consequences for the green chemistry assessment of a hydroformylation reaction:

- Anion hydrolysis causes loss or partial loss of the ionic liquid solvent.
- Anion hydrolysis causes severe safety and corrosion problems related to the HF formed.
- Anion hydrolysis causes deactivation of the transition metal catalyst through irreversible Rh complexation by the released F^- ions.

Consequently, the application of tetrafluoroborate and hexafluorophosphate ionic liquids should be avoided in green hydroformylation catalysis.

However, the use of hydrolysis-stable fluorinated anions, such as $[NTf_2]^-$ salts [102], also is not ideal for green hydroformylation catalysis. This is due to the relatively high price of the anion, its problematic biodegradation and toxicity profile (see Chapter 4) and its complicated disposal by combustion (liberation of HF during incineration). These negative aspects cannot be overcompensated by the advantages of [NTf$_2$]$^-$ salts, such as low viscosity, high thermal stability and easy preparation in pure form, as long as suitable ionic liquid alternatives are available for the application, which is definitely the case for hydroformylation catalysis.

In 1998, Andersen and co-workers published a paper describing the use of phosphonium tosylates (melting points >70 °C) in the Rh-catalyzed hydroformylation of 1-hexene [103]. Later, imidazolium alkylsulfate and oligoether sulfate ionic liquids were introduced by our group as "even greener ionic liquids" for hydroformylation catalysis [104] and used as solvents in transition metal catalysis. For example, [C$_4$mim][n-C$_8$H$_{17}$SO$_4$] (m.p. = 35 °C) was successfully applied in the Rh-catalyzed hydroformylation of 1-octene. Excellent catalyst immobilization and attractive catalytic activities and selectivities could be demonstrated in these studies. A more recent example of Rh-catalyzed hydroformylation in 1,3-dialkylimidazolium tosylate ionic liquids was reported by Lin *et al.* for the conversion of higher alkenes [105]. In this study, higher rates were found with more lipophilic cations, reflecting the improved alkene feedstock solubility.

3.7.5
Hydroformylation in scCO$_2$–Ionic Liquid Multiphasic Systems

Hydroformylation catalysis was among the pioneering examples of ionic liquid–scCO$_2$ biphasic systems. Cole-Hamilton's group demonstrated the hydroformylation of 1-octene using [C$_4$mim][PF$_6$] as ionic liquid and [C$_3$mim]$_2$[PhP(C$_6$H$_4$SO$_3$)$_2$] as ligand [106]. During 33 h time on-stream in a continuous flow apparatus, no catalyst decomposition was observed and Rh leaching into the scCO$_2$–product stream was less than 1 ppm. During the continuous reaction, alkene, CO, H$_2$ and CO$_2$ were separately fed into the reactor, which contained the ionic liquid catalyst solution. The products and unconverted feedstock left the reactor dissolved in the mobile scCO$_2$ phase. After decompression, the liquid product was collected and analyzed. A schematic view of the apparatus used is shown in Figure 3.6.

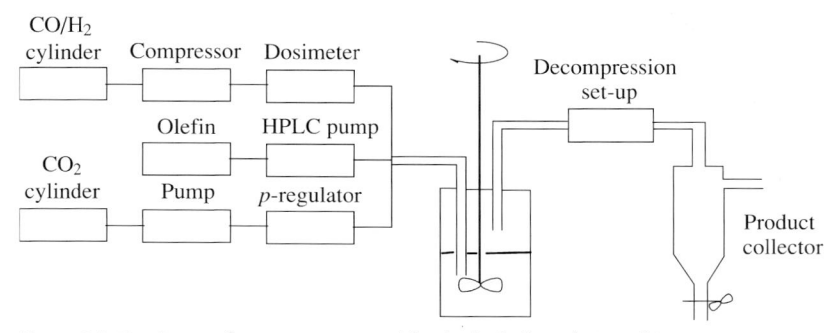

Figure 3.6 Continuous flow apparatus used for the hydroformylation of 1-octene in the biphasic system [C$_4$mim][PF$_6$]–scCO$_2$.

It is noteworthy that this first study demonstrated the importance of continuous operation with this special multiphasic system. Although in repetitive batch mode decreasing selectivities and significant catalyst leaching were observed over the recycling in nine repetitive runs – a behavior that could be clearly attributed to ligand oxidation during the batchwise recycling of the ionic catalyst solution – all these problems could be convincingly overcome by applying the above-described continuous operation mode. Obviously, the continuous flow ionic liquid–$scCO_2$ biphasic system provides a method for continuous flow homogeneous catalysis with integrated separation of the products. From a green chemistry perspective, it is a remarkable feature of this methodology that despite the relatively high boiling points of the nonanals and the aldol side-products, the use of any organic solvent is avoided. Compared with the use of pure ionic liquid, the presence of compressed CO_2 greatly decreases the viscosity of the ionic catalyst solution and increases drastically the hydrogen solubility – and thus the catalyst activity – in the ionic liquid. Moreover, the relatively high-boiling nonanal products and aldol by-products can be removed very elegantly by $scCO_2$ without the use of an additional organic solvent. In subsequent work by the same group, the methodology was also demonstrated for the hydroformylation of 1-dodecene [107].

Later, the same group published a concept for more efficient utilization of the ionic liquid in ionic liquid–$scCO_2$ catalysis [108]. In a "solventless" continuous flow hydroformylation process that promises to be particularly green, they applied P-functionalized imidazolium salts of the general type $[C_n mim][TPPMS]$ [TPPMS = (3-$C_6H_4SO_3)PPh_2$] to immobilize an Rh hydroformylation catalyst in a reaction system composed otherwise only of the 1-octene feedstock and the nonanal product. Through this liquid, syngas and 1-octene dissolved in supercritical CO_2 were passed. By optimizing the flows of these components in the optimum way, a TOF of $180\,h^{-1}$ could be realized with an Rh loss of only 100 ppb.

Very recently, a study by Scurto and co-workers investigated in more detail the underlying thermodynamic properties of the system $[C_6mim][Tf_2N]$–$scCO_2$ with respect to 1-octene hydroformylation catalysis [109]. Phase equilibria, volume expansion, viscosity effects and diffusion coefficients were measured as a function of the CO_2 pressure.

3.7.6
Reducing the Amount of Ionic liquid Necessary – the Supported Ionic Liquid Phase (SILP) Catalyst Technology in Hydroformylation

Reducing the amount of applied ionic liquid is not only an important green chemistry aspect in the context of ionic liquid–$scCO_2$ mixtures, it is also a main green driver for the further development of SILP technology (see Section 3.3 for details). In SILP catalysis, only a diffusion layer thick catalytically active phase is applied that is dispersed on a porous solid, which leads in total to the use of a much smaller amount of ionic liquid compared with corresponding liquid–liquid biphasic processing.

A first example of SILP hydroformylation catalysis using a SILP catalyst suspended in the liquid alkene feedstock in slurry mode was reported by Mehnert et al. in 2002 [110].

For better immobilization of the ionic liquid film, they applied a surface-modified silica gel containing covalently anchored ionic liquid fragments as support. The latter was impregnated with tetrafluoroborate and hexafluorophosphate ionic liquids containing dissolved [Rh(acetylacetonate)(CO)$_2$] and a trisulfonated triphenylphosphine ligand. The prepared catalysts were investigated for the hydroformylation of 1-hexene in a batchwise, slurry phase reaction, showing slightly enhanced activity compared with the corresponding liquid–liquid biphasic system. Unfortunately, at high aldehyde concentrations the applied ionic liquids were found to dissolve partially in the organic phase, and with that an unacceptably high rhodium loss of up to 2.1 mol% was observed.

A later approach using an MCM-41-supported Rh–TPPTS complex in the ionic liquid 1,1,3,3-tetramethylguanidinium lactate as catalyst for 1-hexene hydroformylation showed more promising results [111]. Here, the catalysts exhibited practically unchanged performance in 12 consecutive runs, providing about 50% conversion per run at *n:iso* ratios of about 2.5.

Another variation of slurry phase SILP hydroformylation catalysis was reported recently by Hamza and Blum [112]. They immobilized a mixture of [Rh(COD)Cl]$_2$ and Na[TPPMS] in a silica sol–gel matrix that was modified with 5% of a trimethoxysilyl-functionalized ionic liquid. The resulting ceramic catalyst material was applied to the hydroformylation of vinylarenes and the authors claimed that the catalyst could be recycled without leaching. Both the sol–gel component and the ionic liquid ingredient proved to be necessary to obtain such stable, active and recyclable catalytic material.

Hydroformylation with SILP catalysts in continuous gas-phase contact has the great advantage that ionic liquid or catalyst leaching is avoided due to the lack of solvation power of the gas phase, which makes ligand modification unnecessary [113]. Consequently, complex ligand structures known for their high activity and regioselectivity in hydroformylation can be directly applied to this concept.

The use of catalyst systems containing the bidentate phosphine ligand sulfoxantphos proved particular interesting. To obtain a stable catalytic system, a suitable thermal pretreatment of the silica support and a 10-fold excess of the ligand versus Rh dissolved in the [C$_4$mim][*n*-C$_8$H$_{17}$OSO$_3$] ionic liquid on a support proved to be necessary to obtain a catalytic system with stable activity and selectivity over 60 h time on-stream in propene hydroformylation [turnover number (TON) >2400; *n:iso* ratio 21–23] [113c]. In contrast to this early SILP system, a corresponding Rh–sulfoxantphos catalyst on silica without ionic liquid showed a sharp decrease in activity and selectivity in the first 10 h time on-stream, demonstrating the need for the ionic liquid film as a stabilizing medium for the active Rh catalyst on a support.

The concept of SILP hydroformylation catalysis was later extended to 1-butene hydroformylation using [C$_4$mim][C$_8$H$_{17}$OSO$_3$] as ionic liquid and a disulfonated xantphos ligand in a gradient-free Berty reactor [114]. In this work, a reaction first order in Rh was determined over a wide concentration range. Moreover, 1-butene hydroformylation was found to be more active than propene hydroformylation with similar catalyst systems and this difference could be attributed to the higher solubility of the higher alkene in the applied ionic liquid.

Recently, the combination of SILP catalysis with scCO$_2$ as mobile phase was reported by Cole-Hamilton and co-workers for the first time [115]. They converted 1-octene in a hydroformylation reaction with high catalytic activity (TOF up to 800 h^{-1}) for 40 h with minimal Rh leaching (0.5 ppm). This achievement is remarkable in the context of green chemistry as it demonstrates that the application of the green solvent scCO$_2$ can successfully broaden the applicability of SILP catalysis to much less volatile substrates and products while retaining the typical advantages of SILP systems in continuous gas-phase contact.

3.8
Conclusion

As we reach the end of this chapter, an innovative approach should be briefly mentioned that may have the potential to move hydroformylation catalysis with ionic liquids to even greener grounds in the future. Tominga described the hydroformylation of 1-hexene using CO$_2$ and hydrogen as further reactants [116]. This approach is relevant with respect to green chemistry in terms of both the absence of toxic CO starting material and the potential chemical utilization of CO$_2$ (note that this approach can only lead to a net CO$_2$ consumption if the hydrogen used for the reaction has been obtained from a non-fossil starting material). For the reaction, a mixture of [C$_4$mim]Cl and [C$_4$mim][NTf$_2$] was applied together with a Ru complex, the reaction products were isolated by distillation and the ionic catalyst solution was successfully recycled.

References

1 (a) Welton, T. (2007) in *Ionic Liquids in Synthesis* (eds P. Wasserscheid and T. Welton), Wiley-VCH Verlag GmbH, Weinheim, pp. 130–140; (b) Chiappe, C., Malvaldi, M. and Pomelli, C.S. (2009) *Pure and Applied Chemistry*, **81** (4), 767.

2 Chauvin, Y., Mussmann, L. and Olivier, H. (1995) *Angewandte Chemie (International Edition in English)*, **34**, 1149.

3 (a) Favre, F., Olivier-Bourbigou, H., Commereuc, D. and Saussine, L. (2001) *Chemical Communications*, 1360; (b) Wasserscheid, P., Waffenschmidt, H., Machnitzki, P., Kottsieper, K.W. and Stelzer, O. (2001) *Chemical Communications*, 451.

4 Audic, N., Clavier, H., Mauduit, M. and Guillemin, J.-C. (2003) *Journal of the American Chemical Society*, **125**, 9248–9249.

5 Brauer, D.J., Kottsieper, K.W., Liek, C., Stelzer, O., Waffenschmidt, H. and Wasserscheid, P. (2001) *Journal of Organometallic Chemistry*, **630**, 177.

6 Kottsieper, K.W., Stelzer, O. and Wasserscheid, P. (2001) *Journal of Molecular Catalysis*, **175**, 285.

7 Brasse, C.C., Englert, U., Salzer, A., Waffenschmidt, H. and Wasserscheid, P. (2000) *Organometallics*, **19**, 3818.

8 Dyson, P.J., Ellis, D.J., Parker, D.G. and Welton, T. (1999) *Chemical Communications*, 25.

9 Chauvin, Y., Einloft, S. and Olivier, H. (1995) *Industrial & Engineering Chemistry Research*, **34**, 1149.

10. Mathews, C.J., Smith, P.J. and Welton, T. (2000) *Chemical Communications*, 1249.
11. Sebesta, R. Kmentova, I. and Toma, S. (2008) *Green Chemistry*, **10** (5), 484.
12. Olivier-Bourbigou, H. and Favre, F. (2007) in *Ionic Liquids in Synthesis* (eds P. Wasserscheid and T. Welton), Wiley-VCH Verlag GmbH, Weinheim, pp. 464–488.
13. Wasserscheid, P., Gordon, C.M., Hilgers, C., Maldoon, M.J. and Dunkin, I.R. (2001) *Chemical Communications*, 1186.
14. Wasserscheid, P., Hilgers, C. and Keim, W. (2004) *Journal of Molecular Catalysis A-Chemical*, **214**, 83.
15. Welton, T. (2004) *Coordination Chemistry Reviews*, **248**, 2459.
16. (a) Correia, I. and Welton, T. (2009) *Dalton Transactions*, 4115; (b) Illner, P., Begel, S., Kern, S., Puchta, R. and van Eldik, R. (2009) *Inorganic Chemistry*, **48** (2), 588.
17. Stark, A., Ajam, M., Green, M., Raubenheimer, H.G., Ranwell, A. and Ondruschka, B. (2006) *Advanced Synthesis and Catalysis*, **348**, 1934.
18. Schmeisser, M. and van Eldik, R. (2009) *Inorganic Chemistry*, **48** (15), 7466.
19. Carlin, R.T. and Osteryoung, R.A. (1990) *Journal of Molecular Catalysis*, **63**, 125.
20. Chauvin, Y., Einloft, S. and Olivier, H. (1995) *Industrial & Engineering Chemistry Research*, **34**, 1149.
21. (a) Waffenschmidt, H. and Wasserscheid, P. (2001) *Journal of Molecular Catalysis*, **164**, 61–66; (b) Illner, P., Zahl, A., Puchta, R., van Eikema Hommes, N., Wasserscheid, P. and van Eldik, R. (2005) *Journal of Organometallic Chemistry*, **690**, 3567.
22. Illner, P., Zahl, A., Puchta, R., van Eikema Hommes, N., Wasserscheid, P. and van Eldik, R. (2005) *Journal of Organometallic Chemistry*, **690**, 3567.
23. (a) McGuinness, D.S., Cavell, K.J. and Yates, B.F. (2001) *Chemical Communications*, 355; (b) Hasan, M., Kozhevnikow, I.V., Siddiqui, M.R.H., Fermoni, C., Steiner, A. and Winterton, N. (2001) *Inorganic Chemistry*, **40** (4), 795; (c) Clement, N.D. and Cavell, K.J. (2004) *Angewandte Chemie (International Edition in English)*, **43**, 3845–3847; (d) Clement, N.D., Cavell, K.J., Jones, C. and Elsevier, C.J. (2004) *Angewandte Chemie (International Edition in English)*, **43**, 1277–1279.
24. (a) Arduengo, A.J., Harlow, R.L. and Kline, M. (1991) *Journal of the American Chemical Society*, **113**, 361; (b) Arduengo, A.J., Dias, H.V.R. and Harlow, R.L. (1992) *Journal of the American Chemical Society*, **114**, 5530; (c) Cheek, G.T. and Spencer, J.A. (1994) in *9th International Symposium on Molten Salts* (eds C.L. Hussey, D.S. Newman, G. Mamantov and Y. Ito), Electrochemical Society, New York, p. 426; (d) Herrmann, W.A., Elison, M., Fischer, J., Koecher, C. and Artus, G.R.J. (1995) *Angewandte Chemie (International Edition in English)*, **34**, 2371; (e) Bourissou, D., Guerret, O., Gabbaï, F.P. and Bertrand, G. (2000) *Chemical Reviews*, **100**, 39.
25. (a) Xu, L., Chen, W. and Xiao, J. (2000) *Organometallics*, **19**, 1123; (b) Mathews, C.J., Smith, P.J., Welton, T. and White, A.J.P. (2001) *Organometallics*, **20**, (18), 3848.
26. Bacciu, D., Cavell, K.J., Fallis, I.A. and Ooi, L. (2005) *Angewandte Chemie (International Edition in English)*, **44**, 5282.
27. Magna, L., Chauvin, Y., Niccolai, G.P. and Basset, J.-M. (2003) *Organometallics*, **22**, 4418.
28. McLachlan, F., Mathews, C.J., Smith, P.J. and Welton, T. (2003) *Organometallics*, **22**, 5350.
29. de Souza, R.O.M.A. and Antunes, O.A.C. (2008) *Catalysis Communications*, **9** (15), 2538.
30. Hintermair, U., Gutel, T., Slawin, A.M.Z., Cole-Hamilton, D.J., Santini, C.C. and Chauvin, Y. (2008) *Journal of Organometallic Chemistry*, **693** (14), 2407.
31. Scholten, J.D. and Dupont, J. (2008) *Organometallics*, **27** (17), 4439.
32. Kim, J.H., Boyoung, Y., Chen, S.-W. and Lee, S. (2009) *European Journal of Organic Chemistry*, **14**, 2239.

33 Keim, W., Vogt, D., Waffenschmidt, H. and Wasserscheid, P. (1999) *Journal of Catalysis*, **186**, 481.

34 (a) Mehnert, C.P. (2005) *Chemistry - A European Journal*, **11**, 50; (b) Riisager, A., Fehrmann, R., Haumann, M. and Wasserscheid, P. (2006) *European Journal of Inorganic Chemistry*, 695.

35 Hjortkjaer, J., Heinrich, B. and Capka, M. (1990) *Applied Organometallic Chemistry*, **4** (4), 369–374.

36 (a) Arhancet, J.P., Davis, M.E., Merola, J.S. and Hanson, B.E. (1989) *Nature*, **339**, 454–455; (b) Davis, M.E. (1992) *CHEMTECH*, 498–502; (c) Davis, M.E., Arhancet, J.P. and Hanson, B.E. (1991) US Patent 4 994 427, (d) Davis, M.E., Arhancet, J.P. and Hanson, B.E. (1990) US Patent 4 947 003.

37 Jakuttis, M. (2009) Dissertation, University of Erlangen-Nuremberg.

38 Riisager, A., Fehrmann, R., Haumann, M. and Wasserscheid, P. (2006) *Topics in Catalysis*, **40** (1–4), 91.

39 (a) Ruta, M., Yuranov, I., Dyson, P.J., Laurenczy, G. and Kiwi-Minsker, L. (2007) *Journal of Catalysis*, **247** (2), 269; Ruta, M., Laurenczy, G., Dyson, P.J. Kiwi-Minsker, L. and Lioubov (2008) *The Journal of Physical Chemistry. C*, **112** (46), 17814.

40 Breitenlechner, S., Fleck, M., Müller, T.E. and Suppan, A. (2004) *Journal of Molecular Catalysis A-Chemical*, **214**, 175–179.

41 Riisager, A., Jørgensen, B., Wasserscheid, P. and Fehrmann, R. (2006) *Chemical Communications*, 994.

42 Werner, S., Szesni, N., Fischer, R.W., Haumann, M. and Wasserscheid, P. (2009) *Physical Chemistry Chemical Physics*, 10817.

43 Lou, L.-L., Peng, X., Yu, K. and Liu, S. (2008) *Catalysis Communications* **9** (9), 1891.

44 (a) Jiminez, O., Müller, T.E., Sievers, C., Spirkl, A. and Lercher, J.A. (2006) *Chemical Communications*, **28**, 2974; (b) Sievers, C., Jiminez, O., Knapp, R., Li, X., Müller, T.E., Türler, A., Wierczinski, B. and Lercher, J.A. (2008) *Journal of Molecular Catalysis A-Chemical*, **279** (2), 187.

45 Shi, F., Zhang, Q., Li, D. and Deng, Y. (2005) *Chemistry - A European Journal*, **11** (18), 5279.

46 (a) Hagiwara, H., Sugawara, Y., Isobe, K., Hoshi, T. and Suzuki, T. (2004) *Organic Letters*, **6**, 2325; (b) Jung, J.-Y., Taher, A., Kim, H.-J., Ahn, W.-S. and Jin, M.-J. (2009) *Synlett*, **1**, 39.

47 (a) Zhong, C., Sasaki, T., Takehido, T., Mizuki, I. and Iwasawa, Y. (2006) *Journal of Catalysis*, **242** (2), 357; (b) Han, P., Zhang, H., Qiu, X., Ji, X. and Gao, L. (2008) *Journal of Molecular Catalysis A-Chemical*, **295** (1–2), 57.

48 Baudoux, J., Perrigaud, K., Madec, P.-J., Gaumont, A.-C. and Dez, I. (2007) *Green Chemistry* **9** (12), 1346.

49 (a) Riisager, A. and Fehrmann, R. (2007) in *Ionic Liquids in Synthesis* (eds P. Wasserscheid and T. Welton), Wiley-VCH Verlag GmbH, Weinheim, p. 130; (b) Gu, Y. and Li, G. (2009) *Advanced Synthesis and Catalysis*, **351** (6), 817.

50 Driessen-Hölscher, B., Wasserscheid, P. and Keim, W. (1998) *CATTECH*, June, 47.

51 (a) Prinz, T., Keim, W. and Driessen-Hölscher, B. (1996) *Angewandte Chemie (International Edition in English)*, **35**, 1708–1710; (b) Dobler, C., Mehltretter, G. and Beller, M. (1999) *Angewandte Chemie (International Edition in English)*, **38**, 3026.

52 Chalk, A.J. and Harrod, J.F. (1977) in *Organic Synthesis via Metal Carbonyls* (eds I. Wender and P. Pino), John Wiley & Sons Inc., pp. 673–704; (b) Marciniec, B. (1996) in *Applied Homogeneous Catalysis with Organometallic Compounds*, Vol. 1 (eds B. Cornils and W.A. Herrmann), Wiley-VCH Verlag GmbH, Weinheim, pp. 487–506; (c) Cornils, B. and Herrmann, W.A. (1996) in *Applied Homogeneous Catalysis with Organometallic Compounds*, Vol. 2 (eds B. Cornils and W.A. Herrmann), Wiley-VCH Verlag GmbH, New York, Weinheim pp. 575–601.

53 (a) Stadtmueller, S. (2002) *PolymerS & Polymer Composites*, **10**, 49–62; (b) Brook, M.A. (2000) *Silicon in Organic, Organometallic and Polymer Chemistry*, John Wiley & Sons Inc. (c) Jones, R.G., Ando, W. and Cojnowski, J. (eds) (2000) *Silicon-containing Polymers*, Kluwer Academic, New York, Dordrecht.

54 (a) Geldbach, T.J., Zhao, D., Castillo, N.C., Laurenczy, G., Weyershausen, B. and Dyson, P.J. (2006) *Journal of the American Chemical Society*, **128** (30), 9773; (b) Weyershausen, B., Hell, K. and Hesse, U. (2005) *Green Chemistry*, **7**, 283

55 (a) Behr, A. and Toslu, N. (2000) *Chemical Engineering & Technology*, **23**, 122; (b) van den Broeke, J., Winter, F., Deelman, B.-J. and van Koten, G. (2002) *Organic Letters*, **22** (4), 3851; (c) Marciniec, B., Maciejewski, A., Szubert, K. and Kurdykowska, M. (2006) *Monatshefte fur Chemie*, **137**, 605.

56 Hofmann, N., Bauer, A., Frey, T., Auer, M., Stanjek, V., Schulz, P.S. and Taccardi, N. (2008) *Advanced Synthesis and Catalysis*, **350** (16), 2599.

57 Tan, B., Jing, J., Wang, Y., Wie, L., Chen, D. and Jin, Z. (2008) *Applied Organometallic Chemistry*, **22** (11), 620.

58 (a) Moreno-Mañas, M. and Pleixats, R. (2003) *Accounts of Chemical Research*, **36**, 638; (b) Astruc, D., Lu, F. and Aranaez, J.R. (2005) *Angewandte Chemie (International Edition in English)*, **44**, 7852.

59 Antonietti, M., Kuang, D., Smarsly, B. and Zhou, Y. (2004) *Angewandte Chemie (International Edition in English)*, **43**, 2.

60 (a) Huang, J., Jiang, T., Han, B., Gao, H., Chang, Y., Zhao, G. and Wu, W. (2003) *Chemical Communications*, 1654; (b) Huang, J., Jiang, T., Gao, H., Han, B., Liu, Z., Wu, W., Chanf, Y. and Zhao, G. (2004) *Angewandte Chemie (International Edition in English)*, **43**, 1397; (c) Umpierre, A.P., Machado, G., Fecher, G.H., Morais, J. and Dupont, J. (2005) *Advanced Synthesis and Catalysis*, **347**, 1404; (d) Scheeren, W., Machado, G., Dupont, J., Fichtner, P.F.P. and Texeira, S.R. (2003) *Inorganic Chemistry*, **42**, 4738; (e) Rossi, L.M., Machado, G., Fichtner, P.F.P., Teixera, S.R. and Dupont, J. (2004) *Catalysis Letters*, **92** (3–4), 149; (f) Rossi, L.M., Dupont, J., Machado, G., Fichtner, P.F.P., Radtke, C., Baumvol, I.J.R. and Teixeira, S.R. (2004) *Journal of the Brazilian Chemical Society*, **15** (6), 904; (g) Sileira, E.T., Umpierre, A.P., Rossi, L.M., Machado, G., Morais, J., Soares, G.V., Baumvol, I.J.R., Teixera, S.R., Fichtner, P.F.P. and Dupont, J. (2004) *Chemistry - A European Journal*, **10**, 3734; (h) Miao, A., Liu, Z., Han, B., Huang, J., Sun, Z., Zhang, J. and Jiang, T. (2006) *Angewandte Chemie (International Edition in English)*, **45**, 266; (i) Huang, J., Jiang, T., Han, B., Wu, W., Liu, Z., Xie, Z. and Zhang, J. (2005) *Catalysis Letters*, **103** (1–2), 59; (j) Dupont, J., Fonseca, G.S., Umpierre, A.P., Fichtner, P.F.P. and Teixeira, S.R. (2002) *Journal of the American Chemical Society*, **124**, 4228; (k) Fonseca, G.S., Umpierre, A.P., Fichtner, P.F.P., Teixeira, S.R. and Dupont, J. (2003) *Chemistry - A European Journal*, **9**, 3263; (l) Mu, X., Meng, J., Li, Z. and Kou, Y. (2005) *Journal of the American Chemical Society*, **127**, 9694; (m) Anderson, K., Cortinas Fernandez, S., Hardacre, C. and Marr, P.C. (2004) *Inorganic Chemistry Communications*, **7**, 73–76; (n) Tatumi, R. and Fujihara, H. (2005) *Chemical Communications*, 83; (o) Mévellec, V., Leger, B., Mauduit, M. and Roucoux, A. (2005) *Chemical Communications*, 2838; (p) Fonseca, G.S., Silveira, E.T., Gelesky, M.A. and Dupont, J. (2005) *Advanced Synthesis and Catalysis*, **347**, 847; (q) Fonseca, G.S., Domingos, J.B., Nome, F. and Dupont, J. (2006) *Journal of Molecular Catalysis A-Chemical*, **248**, 10.

61 (a) Deshmaukh, R.R., Rajagopal, R. and Srinivasan, K.V. (2001) *Chemical Communications*, 1544; (b) Calò, V., Nacci, A., Monopoli, A., Laera, S. and Cioffi, N. (2003) *The Journal of Organic Chemistry*, **68**, 2929; (c) Calo, V., Nacci, A., Monopoli,

A., Detomaso, A. and Iliade, P. (2003) *Organometallics*, **22**, 4193; (d) Calo, V., Nacci, A. and Monopoli, A. (2004) *Journal of Molecular Catalysis A-Chemical*, **214**, 45; (e) Forsyth, S.A., Nimal Gunaratne, H.Q., Hardacre, C., Mc Keown, A., Rooney, D.W. and Seddon, K.R. (2005) *Journal of Molecular Catalysis A-Chemical*, **231**, 61; (f) Cassol, C.C., Umpierre, A.P., Machado, G., Wolke, S.I. and Dupont, J. (2005) *Journal of the American Chemical Society*, **127**, 3298; (g) Karimi, B. and Enders, D. (2006) *Organic Letters*, **8** (6), 1237.

62 Geldbach, T., Zhao, D., Castillo, N.C., Laurenczy, G., Weyershausen, B. and Dyson, P.J. (2006) *Journal of the American Chemical Society*, **128** (30), 9773.

63 Calo, V., Nacci, A., Monopoli, A. and Montingelli, F. (2005) *The Journal of Organic Chemistry*, **70** (15), 6040–6044.

64 Chiappe, C., Pieraccini, D., Zhao, D., Fei, Z. and Dyson, P.J. (2006) *Advanced Synthesis and Catalysis*, **348** (1–2), 68.

65 Anderson, K., Cortinas Fernandez, S., Hardacre, C. and Marr, P.C. (2003) *Inorganic Chemistry Communications*, **7** (1), 73.

66 Miao, A., Liu, Z., Han, B., Huang, J., Zhang, Z.Sun.J., and Jiang, T. (2006) *Angewandte Chemie (International Edition in English)*, **45**, 266.

67 Gelesky, M.A., Chiaro, S.S.X., Pavan, F.A., dos Santos, J.H.Z. and Dupont, J. (2007) *Dalton Transactions*, 5549.

68 Safavi, N., Maleki, N., Iranpoor, N., Firouzababadi, H., Banazadeh, A.R., Azadi, R. and Sedaghati, F. (2008) *Chemical Communications*, 6155.

69 Anderson, K., Goodrich, P., Hardacre, C. and Rooney, D. (2003) *Green Chemistry*, **5**, 448.

70 Steffan, M., Lucas, M., Brandner, A., Wollny, M., Oldenburg, N. and Claus, P. (2007) *Chemical Engineering & Technology*, **30** (4), 481.

71 Arras, J., Steffan, M., Shayeghi, Y., Ruppert, D. and Claus, P. (2009) *Green Chemistry*, **11** (5), 716.

72 Obert, K., Roth, D., Ehrig, M., Schönweiz, A., Assenbaum, D., Lange, H. and Wasserscheid, P. (2009) *Applied Catalysis A-General*, **356**, 43.

73 Kernchen, U., Etzold, B., Korth, W. and Jess, A. (2007) *Chemical Engineering & Technology*, **30** (8), 985.

74 Knapp, R., Jentys, A. and Lercher, J. (2009) *Green Chemistry*, **11** (5), 656.

75 Parshall, G.W. (1972) *Journal of the American Chemical Society*, **94**, 8716.

76 Wu, B., Wie, W., Zhang, Y.M. and Wang, H. (2009) *Chemistry - A European Journal*, **15** (8), 1804.

77 Geldbach, T. (2008) *Organometallic Chemistry*, **34**, 58.

78 Parvulescu, V.I. and Hardacre, C. (2007) *Chemical Reviews*, **107** (6), 2615.

79 (a) Wasserscheid, P. (2007) *Journal of Industrial and Engineering Chemistry*, **13** (3), 325; (b) Wasserscheid, P. and Schulz, P.S. (2007) in *Ionic Liquids in Synthesis* (eds P. Wasserscheid and T. Welton), Wiley-VCH Verlag GmbH, Weinheim, pp. 369–463; (c) Wasserscheid, P. and Keim, W. (2000) *Angewandte Chemie (International Edition in English)*, **39**, 3772.

80 Giernoth, R. (2007) *Topics in Current Chemistry*, **276**, 1.

81 (a) Welton, T. (2004) *Coordination Chemistry Reviews*, **248**, 2459; (b) Welton, T. (1999) *Chemical Reviews*, **99**, 2071–2083.

82 Forsyth, S.A., Pringle, J.M., and MacFarlane, D.R. (2004) *Australian Journal of Chemistry*, **57**, 113.

83 Pozzi, G. and Shepperson, I. (2003) *Coordination Chemistry Reviews*, **242**, 115.

84 Dupont, J., De Souza, R.F. and Suarez, P.A.Z. (2002) *Chemical Reviews*, **102**, 3667.

85 Zhao, D., Wu, M., Kou, Y. and Min, E. (2002) *Catalysis Today*, **74**, 157.

86 Tzschucke, C.C., Markert, C., Bannwarth, W., Roller, S., Hebel, A. and Haag, R. (2002) *Angewandte Chemie (International Edition in English)*, **41**, 3964.

87 Dobbs, A.P. and Kimberley, M.R.J. (2002) *Fluorine Chemistry*, **118**, 3.

88 Olivier-Bourbigou, H. and Magna, L. (2002) *Journal of Molecular Catalysis A-Chemical*, **2484**, 1.
89 Sheldon, R. (2001) *Chemical Communications*, **182–183**, 419.
90 Gordon, C.M. (2001) *Applied Catalysis A-General*, **222**, 101.
91 Holbrey, J.D. and Seddon, K.R. (1999) *Clean Products Processes*, **1**, 223.
92 (a) Kuntz, E.G. and Kuntz, E.(to Rhône-Poulenc SA) (1976) German Patent DE 2627354, (1977) *Chemical Abstracts*, **87**, 101944; (b) Kuntz, E.G. (1987) *CHEMTECH*, **17**, 570–575; (c) Cornils, B. and Herrmann, W.A. (1998) *Aqueous-phase Organometallic Catalysis*, Wiley-VCH Verlag GmbH, Weinheim.
93 Cole-Hamilton, D. and Tooze R. (eds) (2006) *Catalyst Separation, Recovery and Recycling*, Springer, Dordrecht.
94 (a) Chauvin, Y., Mussmann, L. and Olivier, H. (1995) *Angewandte Chemie (International Edition in English)*, **34**, 1149; (b) Chauvin, Y., Olivier, H. and Mussmann, L. (to Institut Français du Pétrole), (1997) European Patent EP 776880. (1997) *Chemical Abstracts*, **127**, 65507.
95 Haumann, M. and Riisager, A. (2008) *Chemical Reviews*, **108** (4), 1474.
96 Cocalia, V.A., Visser, A.E., Rogers, R.D. and Holbrey, J.D. (2007) in *Ionic Liquids in Synthesis* (eds P. Wasserscheid and T. Welton), Wiley-VCH Verlag GmbH, Weinheim, pp. 89–102.
97 (a) Dyson, J.P., Laurenczy, G., Ohlin, C.A., Vallance, J. and Welton, T. (2003) *Chemical Communications*, 2418; (b) Ohlin, C.A., Dyson, P.J. and Laurenczy, G. (1070) *Chemical Communications*, **2004**.
98 Bronger, R.P.J., Silva, S.M., Kamer, P.C.J. and van Leeuwen, P.W.N.M. (2002) *Chemical Communications*, 3044.
99 Keim, W., Vogt, D., Waffenschmidt, H. and Wasserscheid, P. (1999) *Journal of Catalysis*, **186**, 481.
100 Favre, F., Olivier-Bourbigou, H., Commereuc, D. and Saussine, L. (2001) *Chemical Communications*, 1360.
101 Swatloski, R.P., Holbrey, J.D. and Rogers, R.D. (2003) *Green Chemistry*, **5**, 361–363.
102 Bonhôte, P., Dias, A.-P., Papageorgiou, N., Kalyanasundaram, K. and Grätzel, M. (1996) *Inorganic Chemistry*, **35**, 1168–1178.
103 Karodia, N., Guise, S., Newlands, C. and Andersen, J.-A. (1998) *Chemical Communications*, 2341–2342.
104 (a) Wasserscheid, P., van Hal, R. and Bösmann, A. (2002) *Green Chemistry*, **4**, 400; (b) Wasserscheid, P., Himmler, S., Hörmann, S. and Schulz, P.S. (2006) *Green Chemistry*, **8** (10), 887–894; (c) Wasserscheid, P., van Hal, R., Roy, and Bösmann, A. (2002) *Proceedings of the Electrochemical Society*, **19**, 146–154.
105 Lin, Q., Jiang, W., Fu, H., Chen, H. and Li, Y. (2007) *Applied Catalysis A-General*, **328** (1), 83.
106 (a) Sellin, M.F., Webb, P.B. and Cole-Hamilton, D.J. (2001) *Chemical Communications*, 781; (b) Cole-Hamilton, D.J., Sellin, M.F. and Webb, P.B.(to the University of St. Andrews) (2002) World Patent WO 0202218. (2002) *Chemical Abstracts*, **136**, 104215.
107 Webb, P.B., Sellin, M.F., Kunene, T.E., Williamson, S., Slawin, A.M.Z. and Cole-Hamilton, D.J. (2003) *Journal of the American Chemical Society*, **125**, 15577.
108 Frisch, A.C., Webb, P.B., Zhao, G., Muldoon, M.J., Pogorzelec, P.J. and Cole-Hamilton, D.J. (2007) *Dalton Transactions*, 5531.
109 Ahosseini, A., Ren, W. and Scurto, A.M. (2009) *Industrial & Engineering Chemistry Research*, **48** (9), 4254.
110 Mehnert, C.P., Cook, R.A., Dispenziere, N.C. and Afeworki, M. (2002) *Journal of the American Chemical Society*, **124**, 12932.
111 (a) Yang, Y., Lin, H., Deng, C., She, J. and Yuan, Y. (2005) *Chemistry Letters*, **34**, 220–221; (b) Yang, Y., Deng, C. and Yuan, Y. (2005) *Journal of Catalysis*, **232** (1), 108.
112 Hamza, K. and Blum, J. (2007) *European Journal of Organic Chemistry*, **28**, 4706.
113 (a) Riisager, A., Wasserscheid, P., van Hal, R. and Fehrmann, R. (2003) *Journal of Catalysis*, **219**, 252; (b) Riisager, A.,

Eriksen, K.M., Wasserscheid, P. and Fehrmann, R. (2003) *Catalysis Letters*, **90**, 149; (c) Riisager, A., Fehrmann, R., Flicker, S., van Hal, R., Haumann, M. and Wasserscheid, P. (2005) *Angewandte Chemie (International Edition in English)*, **44**, 815.

114 Haumann, M., Dentler, K., Joni, J., Riisager, A. and Wasserscheid, P. (2007) *Advanced Synthesis and Catalysis*, **349** (3), 425.

115 Hintermair, U., Zhao, G., Santini, C.C., Muldoon, M.J. and Cole-Hamilton, D.J. (2007) *Chemical Communications*, 1462.

116 Tominaga, K. (2006) *Catalysis Today*, **115** (1–4), 70.

4
Ionic Liquids in the Manufacture of 5-Hydroxymethylfurfural from Saccharides. An Example of the Conversion of Renewable Resources to Platform Chemicals

Annegret Stark and Bernd Ondruschka

4.1
Introduction

In the course of the realization of depleting petrochemical feedstock, increasing global warming due to human interference into Nature's material flow and the globalization of markets, one of the most important current drivers for chemical research and development is sustainability, meaning economically profitable production at the lowest possible ecological consequence. Both the consumption of petrochemical feedstock and emission of CO_2 during the life cycle of a product can be reduced if products and processes become available which employ renewable resources for the production of chemicals.

In general, two strategies can be distinguished. The first strives for the production of established chemicals from renewable resources, so that existing product streams remain unaffected. An example is the production of ethanol by fermentation, which may be used in the same applications as ethanol produced from water and crude oil-based ethene, or, more generally, the gasification of biomass to yield synthesis gas, which can be used to build up chemicals along the Fischer–Tropsch or methanol routes to many traditional products of the chemical industry. Such an approach has the definite advantage that the product performance is well known and the products are already established in the market. Hence it may pave the way for a transition to renewable resources as feedstock, with a short- to medium-term perspective of integration.

The second strategy makes use of the functionalities incorporated into feeds produced by the unequalled, selective synthetic accomplishments of Nature. This requires the adaptation of catalysts, solvent systems and processing techniques to enable the conversion of highly functionalized feedstock. In addition to chemistry and engineering challenges, the resulting products do often not find structural equivalence in already established products, offer other properties and may target chemical product markets where the competition with cheap petrochemical-derived products is high (e.g. introduction of polylactic acid to the polymer market).

Therefore, establishing novel products will rather be a long-term goal and support by political and financial engagement may be needed in some cases.

It can be foreseen that the introduction of such products will occur most easily for upmarket products which in general possess very specific properties for a small area of application, with rather low tonnage and high profit margin. At the same time, production technologies will be established which will successively diffuse into bulk (platform) chemical production, finally leading to a bio-refinery-based tree of products.

From the point of view of an ionic liquid chemist, the production of platform chemicals deserves attention because of the large technical innovation potential, which might not be exploitable in already established and optimized, petrochemical processes. The utilization of ionic liquids, possibly in connection with other innovative tools such as supercritical fluid extraction, microwave-assisted synthesis or microreaction technology, may find entry into production in tandem with renewable resource utilization.

5-Hydroxymethylfurfural (HMF) is such a platform chemical which has recently found much attention in chemical research (Figure 4.1). As expected for a molecule bearing multiple functionalities (aldehyde, alcohol, ether, diene), the number of reactions that it undergoes is enormous (Scheme 4.1) and has been reviewed extensively [1–4].

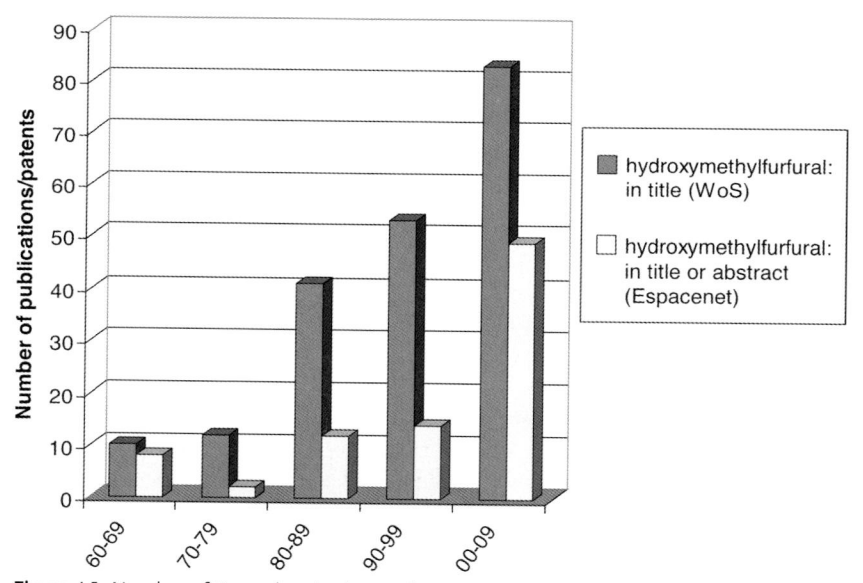

Figure 4.1 Number of times that the keyword "hydroxymethylfurfural" occurs in the title of an article (search engine: Web of Science) or in the title and/or abstract (search engine: Espacenet.org), displayed in decades; last updated May 2009.

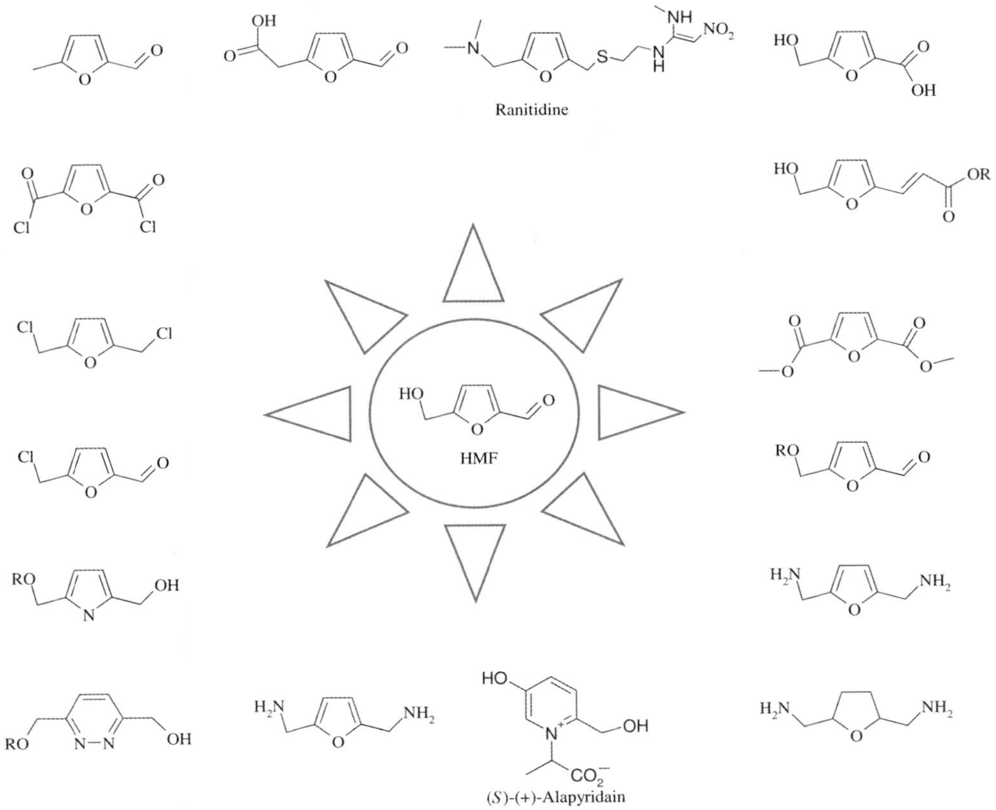

Scheme 4.1 Examples of HMF derivatives (for oxidized and reduced products, see Schemes 4.2 and 4.3).

4.1.1
Areas of Application for HMF and its Derivatives

Whereas the older literature dealt mostly with the production of monomers to substitute, e.g. terephthalic acid, and reveals much work of a rather qualitative nature, more recent studies have focused on the production of fuels and fuel additives, pharmaceuticals and fine chemicals. In the following, areas of application for HMF and its derivatives are summarized, with the focus on recent advancements.

4.1.1.1 Direct Uses of HMF
A US Patent [5] describes HMF as an ingredient for a beverage for weaning people off smoking or alcohol. A mood-boosting effect is ascribed to HMF. Another patent [6] mentions HMF in the context of treatment of skin diseases and improvement of hair growth. The use of HMF for the preparation of beverages for the cure and suppression of oxidative stress in humans and animals and performance enhancement has

also been claimed [7]. Two US Patents [8, 9] describe the use of HMF in a mixture with other compounds as an infusion, for oral or rectal administration or in cancer therapy due to its destructive effect on malignant tumors [10].

4.1.1.2 Derivatives of HMF

Esters such as acetoxymethylfurfural have a high energy content and can therefore be used as fuels or fuel additives [11]. Propionoxymethylfurfural has been described as an effective solvent for separating aromatics from naphthenics, diolefins from olefins and aromatics from aliphatics [12]. Monoesters of HMF with dicarboxylic acids may be useful as reactive diluents for adhesive, composite, coating and ink applications (heat-activated cross-linkers) [13].

Also, ethers have been explored as potential biofuel [14] or fuel additives [15]. Ethoxymethylfurfural possesses an energy density of $8.7\,\mathrm{kW\,h\,l^{-1}}$, which is substantially higher than that of ethanol ($6.1\,\mathrm{kW\,h\,l^{-1}}$) and comparable to that of gasoline ($8.8\,\mathrm{kW\,h\,l^{-1}}$) or diesel fuel ($9.7\,\mathrm{kW\,h\,l^{-1}}$) [16]. Furthermore, it can be used as a solvent and monomer and as a fine chemical or pharmaceutical intermediate [15].

Acetal derivatives have been grafted on polysaccharides by free radical polymerization to impart wet strength to paper (temporary wet strength additive), especially 5-(dimethoxymethyl)furfuryl-2-methyl acrylate and methacrylate [17]. Applications are found especially in the sector of paper towels, where the compounds are applied at 0.25–5.00 wt% either during the paper manufacturing process or by spray coating of the finished product. The acetal formed between HMF and glycerin and its derivatives may be used as antifreeze additives, monomers, solvents, resins, plasticizers, personal care additives, fabric softeners, lubricants, biofuel additives, coatings, adhesives and sealants [18].

Cyanovinyl-substituted furan derivatives find application in the electro-optical field, e.g. for the manufacture of optical data recording systems or organic electron-conducting materials or electron-transfer catalysts [19]. HMF can also be used as a starting material for the synthesis of nitrogen heterocycles [20]. In order to produce large libraries of compounds for pharmaceutical screening, HMF has been employed as a scaffold for the generation of highly functionalized furan-based libraries [21].

HMF derivatives have been claimed for the fluorescent detection of nucleic acids [22], organic luminophores [23] and pharmaceuticals exhibiting antifungal, antitumor [24] and antiviral activity [25–27]. A patent claims the use of HMF-based isothiocyanates in the control of plant-parasitic nematodes [28]. Several 5-substituted furfurals (esters, ethers) and 2,5-diformylfuran (DFF) have been patented for their antifungal and antibacterial action against a wide variety of viruses, including bacteriophages, rickettsiae and protozoa. Examples were given for the reduction of microbial spoilage of various foods, wood, textiles and leather [29]. Ranitidine (see Scheme 4.1) has been developed as an H2-antihistamine and has been used in pharmaceuticals against gastritis and ulcus. This medicine has been produced at a scale of about 500 t per year [3], highlighting the possibilities of large-scale HMF applications [30, 31]. (S)-(+)-Alapyridain (see Scheme 4.1) is naturally formed by heating mixtures of sugar and amino acids and can be found in, e.g., beef bouillon. It enhances taste, in particular sweet, umami and salty. The product can be produced as a racemic mixture by reacting HMF with L-alanine under alkaline conditions. The pure L- or D-forms are obtained by reductive amination of HMF with Raney nickel–dihydrogen in the presence of either L- or D-alanine, followed by mild oxidation [32–35].

Both oxidized and reduced products of HMF have attracted considerable interest in recent years. Scheme 4.2 shows the successive oxidation of HMF.

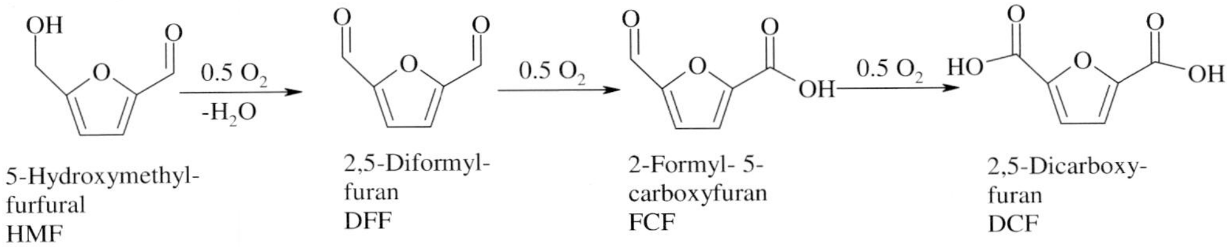

5-Hydroxymethyl-furfural
HMF

2,5-Diformyl-furan
DFF

2-Formyl- 5-carboxyfuran
FCF

2,5-Dicarboxy-furan
DCF

Scheme 4.2 Oxidation products of HMF.

DFF has been polymerized to form polypinacols and polyvinyls and used as a starting material for the synthesis of antifungal agents, drugs and ligands [26, 36, 37]. DFF can also be applied to the preparation of polymeric Schiff bases with good thermal stability and semiconducting response after doping [38]. Furthermore, the use of DFF in alkaline battery separators [39] has been described.

The Pacific Northwest National Laboratory (PNNL) selected 2,5-dicarboxyfuran (DCF) as a potential building block for polyurethanes (for molded plastics, liquid crystal displays, dyes, paper products and textiles) and polyhydroxy-polyesters (for paints, resins, siding, insulation, cements, coatings, varnishes, flame retardants, adhesives and carpeting), with a market size of commodity chemicals (high value–high volume chemical market) [40]. DCF can be converted to, e.g., 2,5-bis(aminomethyl)tetrahydrofuran, 2,5-bis(hydroxymethyl) tetrahydrofuran and 2,5-bis(hydroxymethyl)furan, which may have applications as poly(ethylene terephthalate) (PET) analogues, polyesters [41] and polyamides (nylons, fibers) with potentially new properties [40]. DCF is postulated as having a high potential to replace terephthalic acid, which is used in various polyesters, such as PET and poly(butylene terephthalate) (PBT). Market sizes of PBT, PET and nylon approaching 0.5×10^9, 2×10^9 and 4.5×10^9 kg per year, respectively, could be addressed, with product values between \$1.7 and $6.0\,\mathrm{kg}^{-1}$ [40].

Likewise, applications of the hydrogenated products of HMF shown in Scheme 4.3 have been reported.

2,5-Bis(hydroxymethyl)furan (DHMF) has been proposed as a fuel [42] and in the manufacture of polyesters and polyurethane foams [43] featuring excellent thermal insulating capabilities and fire resistance. In addition, these materials possess good dimensional stability and structural strength [44].

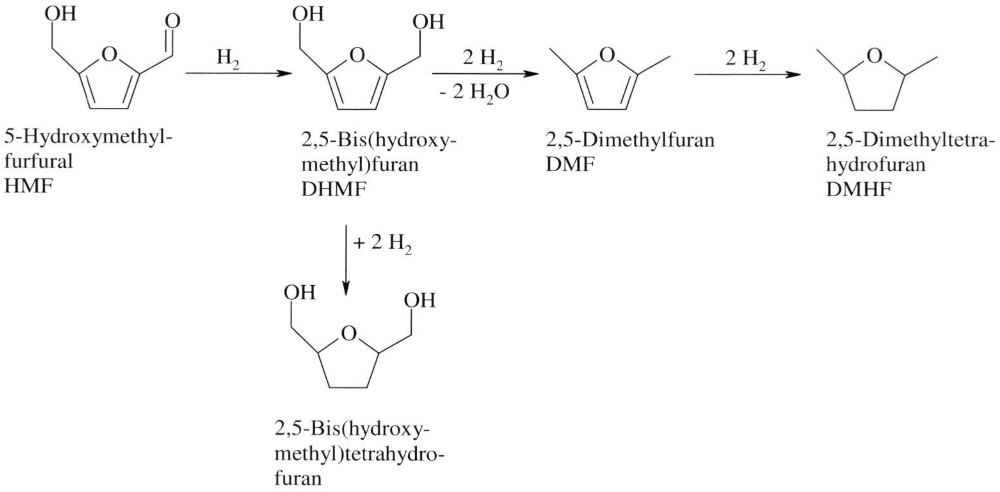

Scheme 4.3 Reduction products of HMF.

2,5-Dimethylfuran (DMF) has been described as a fuel, fuel additive (antiknock compound) and solvent with properties similar to those of tetrahydrofuran [45]. The energy content of DMF (8.7 kW h l^{-1}) is similar to that of gasoline and much greater than that of ethanol [46]. 2,5-Dimethyltetrahydrofuran (DMHF) can be used in the preparation of polyesters. The market for this compound is 8000 t per year at a price of around €7 kg^{-1} [4].

4.1.2
Summary: Application of HMF and Its Derivatives

Despite this great potential, it must be stated that relatively little is known about the specific properties of HMF-based products, in particular about the much-propagated polymers. None of the polymers exhibits properties (and monomer prices) which can directly substitute existing products. Hence HMF-based polymers will have to find niche markets before entering the bulk chemical sector.

For HMF and its derivatives, there is also a lack of up-to-date data on toxicological properties. Some toxicity studies conducted in the 1950s and 1960s reported a low acute toxicity, and a metabolite (5-sulfoxymethylfurfural) may form that is toxic/carcinogenic. The chronic toxicity is low (4 days to 11 months) and short-term carcinogenicity is moderate. *In vitro* genotoxicity tests were negative; effects were only observed at high dosage.

A technical report described the development of effective and selective dehydration and oxidation technology for sugar derivatives as one of the technical barriers [40].

For the advancement of the application of HMF derivatives, HMF production processes have to be established that give HMF in high space–time yield and sufficient (isolated) purity. Methods of preparation are reviewed in the following to give an overview of the state of the art.

4.2
HMF Manufacture

For the production of HMF, several renewable resources may serve as starting materials: in addition to monosaccharides (fructose, glucose), the disaccharide saccharose and various polymeric materials, such as inulin, starch and cellulose, can in principle be used. Several feeds, which may increasingly become accessible (bio-refinery) contain these compounds, e.g. invert sugar, high-fructose corn syrup (HFCS), mother liquors from fructose crystallization and hydrolyzed wood.

Several comprehensive summaries concerning HMF production and its derivatives have been published (e.g. [47–52]). Figure 4.1 shows that the interest in HMF is increasing rapidly, in both the patent and the open literature.

4.2.1
General Aspects of HMF Manufacture

The conversion of monosaccharides to HMF is formally an intramolecular condensation reaction, in which 3 mol of water are liberated per mole of monosaccharide converted. As with other condensation reactions, an acid catalyst is required in most cases. It is known that with very strong acid catalysts or at high reaction temperatures, several side-products occur: in addition to levulinic acid and formic acid, brown-blackish compounds occur, which have been collectively termed "humines" [53]. These lower the efficiency of the process considerably, since they make further product purification steps necessary. Interestingly, the formation of these side-products is dependent on the initial concentration of the monosaccharide solution: the higher it is, the higher also are their rates of formation. Additionally, in aqueous media, the side-product formation is enhanced due to rehydration reactions [54–56].

In general, glucose enolizes very slowly to fructose and the enolization is the rate-determining step in HMF formation. Kuster and van der Baan reported that the dehydration of D-glucose is 40 times slower than that of D-fructose [55]. Hence most work is concerned with the conversion of fructose, although isomerization and subsequent dehydration starting from glucose are a much more challenging task. From an economic point of view, however, the conversion of glucose (and glucose unit-containing feed) is more beneficial (Table 4.1).

Table 4.1 Global production and market prices of various potential starting materials for HMF production.

Resources	Global annual production (t)	Price (€/t)
Cellulose (processed)	320×10^6	500
Starch	55×10^6	250
Sugar (saccharose)	140×10^6	250
Fructose	8×10^6	500–700
Glucose	20×10^6	300
Inulin	110×10^3	1500–2000

Considering the "loss" of 3 equiv. of water per monosaccharide unit converted to HMF (Scheme 4.4), only 70% of the original molar mass ends up in the product. Based on molar mass and disregarding additional technological expenditure, the costs caused by fructose to produce 1 kg of HMF is between €0.7 and 1.0. This simplistic calculation shows that a long-term development of HMF production processes utilizing more cost-effective resources is required to achieve product prices comparable to current bulk chemical prices. These may in future be provided from bio-refinery feed streams (especially glucose from cellulose).

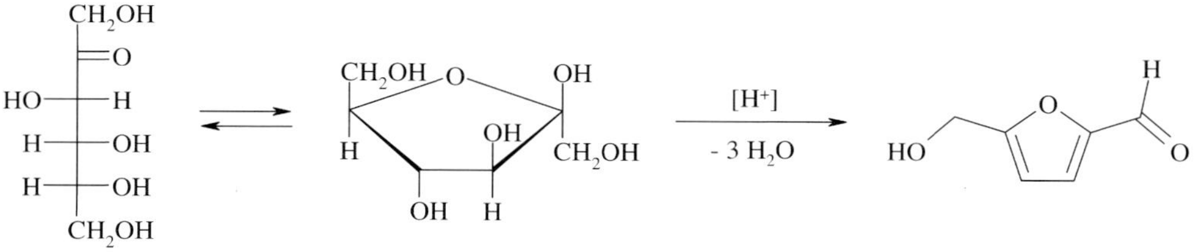

Scheme 4.4 Condensation of fructose to HMF.

4.2.2
Methods of Manufacture of HMF from Fructose

The synthesis of HMF from fructose is not *per se* problematic and has been extensively described in the literature. For example, the condensation can be achieved with various Brønsted acids [57–59] in sub- and supercritical water [53, 55, 60–66] or organic solvents [60, 61, 67, 68].

In addition to Brønsted acids, Lewis acids have also been investigated as catalysts, e.g. boron trifluoride etherate [69] and aluminum compounds [54, 70]. Several authors have reported the use of transition metals [63, 71–77], heterogeneous metal-containing catalysts such as cubic zirconium pyrophosphate, γ-titanium phosphate [78] or niobium phosphate [79] or zeolites and other ion exchangers [36, 56, 60, 80–93].

An interesting approach is the condensation of fructose to HMF in DMSO or DMF. Here, yields up to 90% can be achieved in the absence of any additional acidic catalyst and side-products are not formed [36, 49, 93–97]. Problems with this synthetic strategy are a low reproducibility [98], difficulties in the quantitative removal of these high-boiling solvents and the possibility of contamination with toxic sulfur-containing side-products.

Ionic liquids, in particular [C$_4$mim]Cl [99] and [C$_4$mim][BF$_4$] or [C$_4$mim][PF$_6$] [100], and their mixtures with DMSO, have been used as solvents for the condensation of fructose to HMF, catalyzed by acidic ion-exchange resins, Brønsted acids or tungsten salts [101]. Acido-basic media (nowadays a subgroup of ionic liquids), i.e. equimolar mixtures of organic bases (e.g. pyridine) with acids (e.g. HCl or *p*-toluenesulfonic acid), were developed to tune the acid strength and thus improve the selectivity of the system [102–105]. Yields of HMF between 70 and 90% can be obtained, completely avoiding the formation of levulinic acid or humines. Brønsted acidic ionic liquids, in some cases chemically immobilized on silica, have also been described [106].

From a chemical point of view, it is therefore obvious that HMF can be obtained from fructose with satisfactory yields under various conditions (Table 4.2). However, very few concepts have been proposed in which efficient product separation and purification are demonstrated, although these aspects contribute essentially to the production costs.

In the case that homogeneous Brønsted or Lewis acids are used as catalysts, neutralization by addition of a base often precedes the extraction (e.g. [53]). At least equimolar amounts of salts are produced, requiring a filtration step and disposal. From both ecological and economic points of view, this is nowadays not efficient. One concept of product separation is the use of heterogeneous catalysts, e.g. zeolites or other ion exchangers. Although these have been reported to give high yields, the catalysts often possess only a limited lifetime, especially at high processing temperatures due to their low thermal stability. A second concept, product separation by vacuum distillation, may prove to be difficult, since HMF has been described as being not thermally stable.

The most frequently applied concept is the use of biphasic systems, where product separation can be achieved by extraction [81, 107]. Examples are the combination of water with methyl isobutyl ketone (MIBK) [80, 82, 84, 86, 88, 89, 91, 108], water–butanol [109], pyridinium chloride–ethyl acetate [104], water–furfural [58], water with aromatic or halogenated organic solvents [110] and ionic liquids with tetrahydrofuran [101] or ethyl acetate [99]. Very few chemical engineering data are available to judge the extraction efficiency of these systems. However, it appears as if the determination of an efficient extractant with high selectivity and capacity for HMF was hampered by the fact that the monosaccharides, the catalyst acid and side-products possess solvation properties (polarity) similar to those of HMF itself and hence the extract is often impure.

One example in the open literature underlines the complexity of efficient product separation: in order to obtain high yields of HMF, fructose is converted using HCl as catalyst in a reaction phase consisting of aqueous DMSO, which is modified with poly (1-vinyl-2-pyrrolidone) (PVP) to improve selectivity. This process yields >80% HMF at 90% sugar conversion. The extractant is a solvent mixture consisting of MIBK and 2-butanol, both high-boiling solvents which have to be separated by distillation, lowering the energy efficiency of the overall process dramatically [56, 111, 112].

It has been shown [113] that an elaborate purification concept is required in many instances after extraction (concentration of solutions, column chromatography, subsequent extraction steps) to separate side-products. This aspect is also not given much attention in the literature, with the exception of using activated charcoal for HMF adsorption to improve selectivity and product recovery [114].

The majority of the above-described processes are conducted batchwise. With the exception of a continuous single-phase process described for poly(ethylene glycol) (PEG)-600 (featuring a rather low selectivity of 65%) [67], continuous processes are based on biphasic liquid–liquid solvent systems, e.g. water–MIBK. When solid catalysts are employed in a fixed bed, extraction with MIBK or other

Table 4.2 Overview of process parameters and performance of HMF production.

Starting material (concentration in solvent)	T, t, p	Solvent, catalyst	Conversion (%)	Selectivity (%)	Yield (%)	Ref.
Fructose	150 °C, 2 h	MIBK	–	–	75	[61, 67, 94]
Fructose (10 mol%)		DMSO	–	–	90	[95]
Fructose (10 wt%)	100 °C, 4–5 h	DMA-KI, H_2SO_4 (6 mol%)	–	–	90	[46]
Fructose (50 wt%)	180 °C, τ = 3 min	[Water–DMSO (8:2)]–PVP (7:3), 0.25 M HCl	92	77	70	[56]
Fructose (40 wt%)	80 °C, 0.5 h	Water, vanadyl phosphate, Fe-substituted	58	87	50	[75]
Fructose (0.05 M)	240 °C, τ = 120 s	Subcritical water, H_3PO_4	–	–	65	[118]
Fructose (10 g l^{-1})	180 °C, τ = 120 s 20 MPa	Sub-/supercritical acetone	99	77	75	[68]
Fructose (10 wt%)	165 °C, 1 h	Mordenite, water	60	–	70	[116]
Fructose	165 °C, 1 h	Water–MIBK (1:5), zeolite	76	91	70	[88]
Fructose (1.2 M)	88 °C, 15 h	Water–MIBK (1:9), macroporous highly acidic ion exchanger	–	–	66	[80]
Fructose (6 wt%)	100 °C, 1 h	Water, heterogeneous cubic zirconium pyrophosphate, γ-titanium phosphate	88	96	84	[78]
Fructose (9 g l^{-1})	80 °C, 24 h	Amberlyst, ionic liquids–DMSO (5:3)	70	85	60	[100]
Fructose (3 wt%)	100 °C, 4 min, MW	3-Allyl-1-(4-sulfobutyl)imidazolium trifluoromethanesulfonate, fructose–ionic liquid (4:1)	>95	–	80	[106]
Fructose (50 mol%)	120 °C, 30 min	[Hpy]Cl	–	–	>95	
Fructose (17 mol%)	90 °C, 45 min	[H-mim]Cl	–	–	70	[104]
Fructose (6 wt%)	80 °C, 1 h	Choline chloride–citric acid	98	93	92	[105]
Glucose (25 wt%)	200–230 °C, 20 min	Water–dioxane, [H-py][H_2PO_4]	–	–	91	[107]
					45	[102]

Substrate	Conditions	Solvent/catalyst			Yield (%)	Ref.
Glucose (10 wt%)	170 °C, $\tau = 3$ min	water–DMSO (4:6), 0.25 M HCl	43	53	23	[111]
Glucose (10 mol%)	100 °C, 3 h, 1 bar	[C$_2$mim]Cl, CrCl$_2$ (6 mol% rel. sugar)	95	74	70	[76]
Glucose (10 wt%)	100 °C, 6 h	[C$_4$mim]Cl, Cr(II) chloride, imidazolidene-modified	–	–	81	[119]
Glucose	100 °C, 5 h	DMA–NaBr, CrCl$_2$ (6 mol%)	–	–	80	[46]
Saccharose (160 g l^{-1})	100 °C, 1 h	DMSO, I$_2$	–	–	35	[93]
Cellulose (4 wt%)	140 °C, 2 h	DMA–LiCl–[C$_2$mim]Cl, CrCl$_2$ (10 mol%), HCl (10 mol%)	–	–	50	[46]

water-insoluble solvents can be used in counter-current mode [82, 108, 115–117]. Kuster and van der Stehen reported, however, that in many instances HMF decomposition may occur faster than extraction, thus lowering the efficiency of the process [115].

On a technical scale, following the procedures described in a patent [53] using a pressurized batch reactor, HMF has been produced in aqueous solution at pH 1.8 (H_2SO_4) in 2 h at 150 °C. After precipitation of the catalyst by neutralization with a base, filtration, removal of the solvent, followed by time- and energy-consuming ion-exchange chromatography and a crystallization step are required. At a conversion of 90%, yields between 40 and 50% are obtained [53]. Using this method, a large batch has been produced and distributed by Südzucker in the 1980s.

Table 4.2 gives a summary of the conditions and yields obtained from various starting materials (open literature).

4.2.3
Methods of Manufacture of HMF from Sugars Other Than Fructose

As pointed out above, the economic potential of HMF is closely connected with the type of bio-feedstock that is processable. Due to the availability of fructose, its cost and the fact that it is a foodstuff, hexose alternatives need to be exploited, even if these are converted less efficiently than fructose itself.

As early as 1962, Mednick described the synthesis of HMF in 45% yield from glucose (or starch) in acido-basic media as catalysts, consisting of, e.g., pyridine and phosphoric acid, in a mixture of water and dioxane as solvent at high temperatures (200–230 °C) [102].

The temperature can be significantly reduced by using metal-based catalysts. Hence a groundbreaking improvement in HMF manufacture was achieved by Zhao et al. with a system that allows for the single-stage isomerization of glucose to fructose prior to the condensation to HMF [76, 77]. Addition of a catalytic amount (0.6 mol% of $CrCl_2$) to [C_2mim]Cl gives a glucose conversion of 80–90% (yield ∼70%) at 100 °C. Although it is clear that from an environmental point of view the use of chromium catalysts is problematic, this study demonstrates that glucose isomerization is feasible using Lewis acid catalysts in ionic liquids. Further improvements in yields of HMF from both fructose and glucose were achieved by ligand design. Yong et al. showed that 1,3-substituted imidazolidene ligands on Cr(II) chloride (10 mol% catalyst concentration) gave yields of 96 and 81% for fructose and glucose, respectively, using 10 wt% sugar in [C_4mim]Cl [119].

N,N-Dimethylacetamide (DMA)–LiCl mixtures are known cellulose solvents (solvation capacity up to 15 wt%). However, strong ion-pair formation seems to inhibit catalytic interaction with fructose in the conversion to HMF. On the other hand, a softer halide (bromide) gave an excellent yield of 90%. Starting from glucose, DMA with $CrCl_2$ or $CrCl_3$ (modified with NaBr) resulted in HMF yields up to 80%. From both purified cellulose and untreated corn stover, HMF yields up to 50% were obtained in mixtures of DMA–LiCl, $CrCl_2$, HCl and [C_2mim]Cl at 4 wt% cellulose concentration [46].

Watanabe et al. have shown that glucose can be converted to HMF using TiO_2 (anatase) in sub- or supercritical water in 10 wt% solution [66]. A yield of ~20% was obtained.

4.2.4
Deficits in HMF Manufacture

5-Hydroxymethylfurfural is in principle an intermediate in the dehydration of hexoses to levulinic acid and formic acid. From a chemical point of view, the challenge is thus to stop the dehydration selectively at the appropriate time. This fact, in turn, defines the engineering challenge, i.e. the separation of HMF from the starting material, catalyst and side-products, which in many cases contributes a major part to the production costs.

Most publications on the production of HMF are concerned with the optimization of the reaction itself, i.e. the reaction temperature, pressure, catalyst and solvent system. Very few deal with the separation of the product from the solvent, catalyst and, if necessary, from the side-products to achieve the required product purities, leading to a deficit in engineering data on extraction efficiency.

Furthermore, alternative feeds to fructose have to be exploited in systematic studies, for which catalyst development is required. Especially where ionic liquids are involved, the purity of these materials has to be stated in order to yield reliable data. A lack of purity determination has led, in at least one case, to the formation of HMF from fructose, presumably in the absence of catalyst acid, while it appears likely that the $[C_4mim][BF_4]$ used hydrolyzed to HF, which then acted as a catalyst [100].

4.3
Goals of Study

It was therefore the goal to study the utility of ionic liquids in the conversion of various saccharides to HMF. Initial investigations focused primarily on the conversion of fructose, which is known to undergo the condensation reaction much more easily than glucose [55]. After understanding the reaction in terms of the effects of temperature, reaction time, fructose concentration, purity of the ionic liquid and choice of ionic liquid, several other saccharide-containing starting material were tested, leading to a continuous process for the production of HMF.

4.4
HMF Manufacture in Ionic Liquids – Results of Detailed Studies in the Jena Laboratories

Initially, the condensation of fructose to HMF was investigated in a large variety of ionic liquids in the absence of auxiliary acid, as it was thought that the ionic liquid's anion or cation may engage in the catalysis in a pseudo-organo-catalytic fashion.

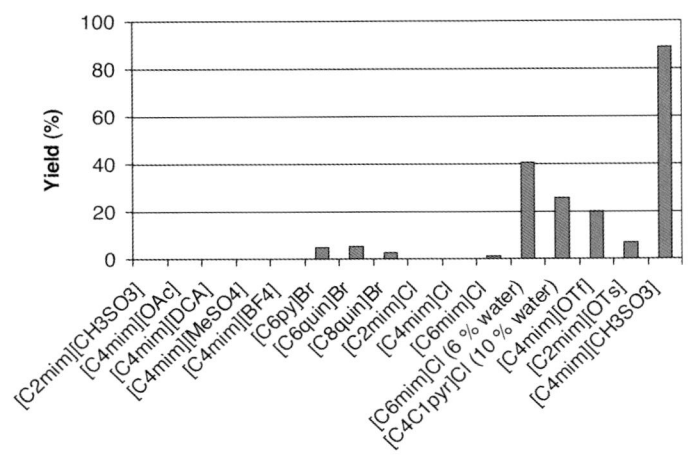

Figure 4.2 Yield of HMF obtained in various ionic liquids [10 g ionic liquid, 4.1 wt% fructose, 80 °C (microwave apparatus), 1 h].

Figure 4.2 shows the results of this preliminary study: in most cases, no or little HMF was obtained within 1 h, except with [C$_4$mim][CH$_3$SO$_3$], where a yield close to 90% was obtained.

Closer inspection of this batch (BASF technical grade, S2806, specified with ≥95% purity) showed that it contained acidic impurities (see below). As at that time it was one of the few ionic liquids commercially available at moderate cost on a kilogram scale, it was used in further studies detailed in the following.

4.4.1
Temperature

Using a 4.1 wt% fructose solution in [C$_4$mim][CH$_3$SO$_3$] (10 g scale, batch S2806), the effect of temperature was first tested. It was found that at 80 °C, an 89% yield of HMF was obtained after 1 h, whereas at 60 and 100 °C, the yields were lower (72 and 66%, respectively). Therefore, 80 °C was adopted as a constant reaction temperature in the remainder of the study. Several reactions were performed in a microwave (MW) apparatus (HPR 1000/10 from MLS GmbH, Leutkirch im Allgau, Germany, equipped with an internal temperature fiber sensor) and compared with experiments conducted in a Radleys carousel six-reactor station. The type of energy input did not affect either the yield nor the selectivity and both set-ups can therefore be used interchangeably.

4.4.2
Concentration and Time

The effect of the fructose concentration was investigated as a function of reaction time and the results are shown in Figure 4.3.

4.4 HMF Manufacture in Ionic Liquids – Results of Detailed Studies in the Jena Laboratories

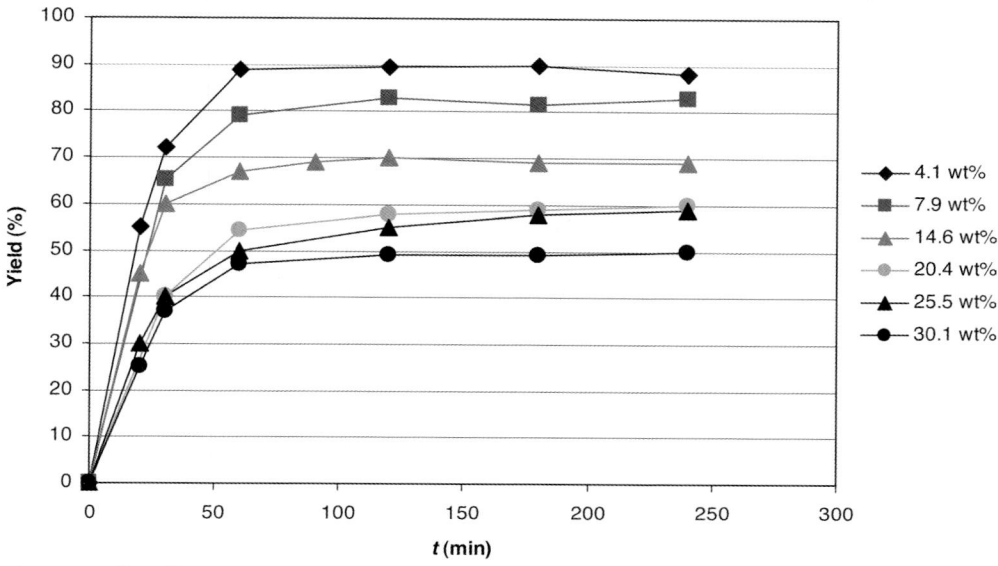

Figure 4.3 Effect of fructose concentration on yield [batch S2806, 10 g [C$_4$mim][CH$_3$SO$_3$], 80 °C, reactor station].

It was found that the reaction was complete within 1–2 h at any concentration used and that the yield decreases with increasing fructose concentration. In all instances, fructose was quantitatively (>95%) consumed after 2 h, indicating that the selectivity decreases with increasing fructose concentration. This is a well-known phenomenon which has been described also for water and organic solvents and has been attributed to the formation of humines and side-products from rehydration reactions [54–56]. Hence, for future efficient processing, a balance must be found between yield and the amount of ionic liquid used. While at low concentrations of fructose high yields are obtained, this also implies a large ionic liquid inventory, which will affect the production costs.

Figure 4.3 also demonstrates that HMF is stable under the reaction conditions and does not decompose after extended times, contrary to when the reaction is carried out in other solvents. Separate experiments on the stability of HMF in [C$_4$mim][CH$_3$SO$_3$] and in mixtures of [C$_4$mim][CH$_3$SO$_3$] and water (1 : 1 v/v) also showed that HMF does not decompose at 80 °C within 3 h.

Analyzing the information displayed in Figure 4.3 further, the productivity can be derived, which is defined in analogy with the turnover frequency in homogeneous catalysis as millimoles of HMF produced per mole of ionic liquid per hour. Figure 4.4 shows that both the yield and productivity are linear functions of the fructose concentration. As opposed to the yield, the productivity increases with increasing concentration, meaning that the rate of reaction is higher at higher fructose concentrations.

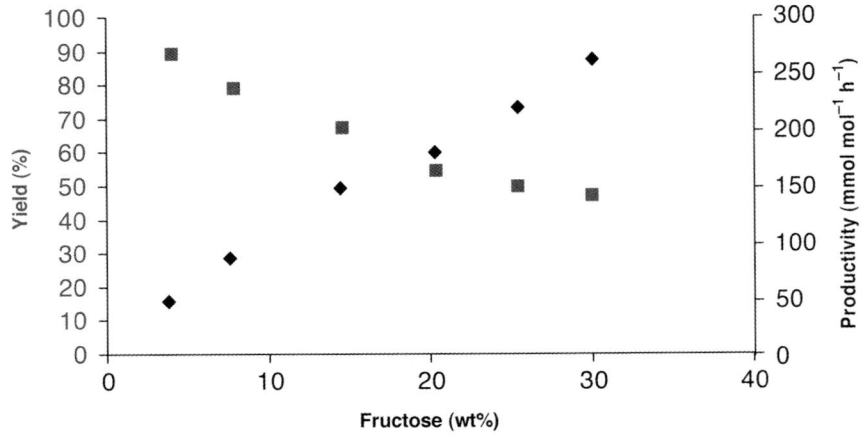

Figure 4.4 Effect of fructose concentration on yield (■) and productivity (◆) [batch S2806, 10 g [C$_4$mim][CH$_3$SO$_3$], 80 °C (reactor station), 1 h].

4.4.3
Effect of Water

Since the formation of HMF from saccharides is a condensation reaction, water is liberated (see Scheme 4.4). In view of the long-term use of the ionic liquid (IL), the question was raised of whether the reaction was influenced by the presence of water accumulating during recycling. Figure 4.5 shows the effect of the molar fraction X [IL/(IL + water)]. Clearly, two regimes exist: the yield of HMF is fairly stable for X (IL) > 0.5, with yields between 75 and 90%. If water is present in excess, i.e. X (IL) < 0.5, the yield decreases rapidly, with no conversion in water.

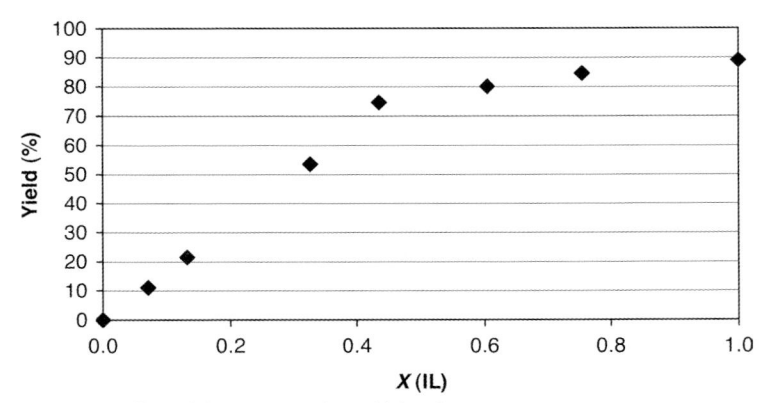

Figure 4.5 Effect of the water on the yield [batch S2806, 10 g ([C$_4$mim][CH$_3$SO$_3$] + water), 4.1 wt% fructose, 80 °C (reactor station), 1 h].

Hence, depending on the concentration of fructose used, water accumulating from the condensation must be removed occasionally before reuse, e.g. by evaporation.

It is interesting that two regimes seem to exist and the linear correlations between yield and water content intersect at a molar fraction of 0.5. Although not investigated in more detail, this feature may indicate a stoichiometric ionic liquid–water interaction.

4.4.4
Effect of Purity

Investigations into the purity of the ionic liquid were triggered by deviations in yield when changing the batches of ionic liquid used. All ionic liquid batches of [C_4mim][CH_3SO_3] were of technical grade (manufacturer's specification: $\geq$95% purity, manufactured by BASF). The purity was assessed with regard to the ionic liquid (>95%), water (<2.0%), halide (1.5%) and 1-methylimidazole content (<0.17%), making use of the analytical techniques described in [120], hence complying with the manufacturer's specification. Considering that residual halide was found, a synthetic strategy involving ion exchange between 1-butyl-3-methylimidazolium chloride and methanesulfonic acid during the preparation of the ionic liquid appears likely. As the pH value of the pure ionic liquid cannot be determined, 0.2 M aqueous solutions were prepared and measured. Of course, the pH values obtained (Table 4.3) do not represent the acidity of the pure ionic liquid. However, it is reasoned that any available Brønsted acid will contribute to the pH in the aqueous solution and therefore it should allow for the comparison between batches.

Table 4.3 also contains the results of various experiments conducted in these batches at different concentrations. First, it can be stated that in the ionic liquids of entries 1–3, the reaction gave lower yields at increased fructose concentrations, as already discussed above. Second, the pH values of 0.2 M aqueous solution are very low (between 1.6 and 3.6), indicating the presence of a strong acid, presumably methanesulfonic acid left from the ionic liquid's synthesis. Third, higher acidity

Table 4.3 pH values determined in 0.2 M aqueous solutions of different ionic liquid batches of [C_4mim][CH_3SO_3] used and comparison of the performance in the condensation of fructose to HMF [10 g [C_4mim][CH_3SO_3], 80 °C (reactor station), 1 h].

Entry	Batch	pH (0.2 M)	Yield (1 h) (%)					
			4.1 wt%[a]	7.9 wt%[a]	14.7 wt%[a]	20.4 wt%[a]	25.5 wt%[a]	30.1 wt%[a]
1	S2806	1.6	89.0	79.0	67.3	54.2	49.7	47.3
2	S40160	1.9	71.5	66.0	47.9	–	–	–
3	S31564	3.6	–	55.3	43.4	–	40.9	–
4	S56605	8.3	–	0	0	–	–	–
5	S60506	6.1	–	0	0	–	–	–

[a]wt% refers to proportion of fructose in the ionic liquid.

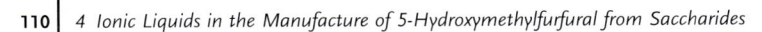

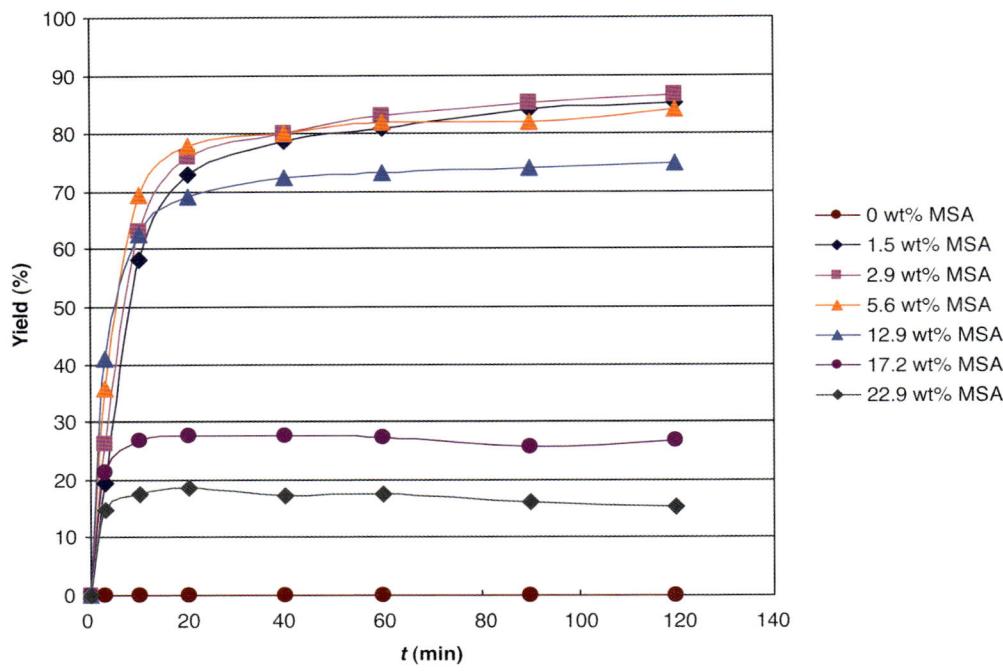

Figure 4.6 Effects of the incremental addition of MSA to pure [C_4mim][CH_3SO_3] [>99%, Solvent Innovation, 5 g ionic liquid, 14.7 wt% fructose, 80 °C (reactor station)].

leads to higher yields. Hence it is clear that the variation in the reaction performance is dependent on the acid content. In this particular case, the ionic liquid batches purchased more recently (entries 4 and 5) are not acidic (and do not yield HMF), probably due to improved manufacturing techniques at BASF.

For a final proof of this theory, a batch of high-purity ionic liquid (>99%, Solvent Innovation/Merck) was purchased, to which increments of methanesulfonic acid (MSA) were added to simulate this impurity. Figure 4.6 shows the effect of the MSA concentration on the yield of HMF as function of time.

Figure 4.6 shows that in the pure ionic liquid, HMF is not formed and the conversion of fructose is <2%. When small amounts of MSA (1.5–2.9 wt%, equaling 3.6–7.2 mol% relative to the ionic liquid) are added to the pure ionic liquid, the activity of the system increases tremendously, reaching 80% within 1 h. Larger amounts of MSA decrease the yield successively.

Parallel experiments measuring the pH value of 0.2 M aqueous solutions of [C_4mim][CH_3SO_3] containing no MSA (pH = 6.9) and increments between 2.0 and 65 wt% MSA (linear decrease in pH from 2.3 to 0.4) showed excellent agreement with the pH values of the acid in water in the absence of ionic liquid. Hence the ionic liquid does not act as a buffer and the decrease in yield at high MSA concentrations must be related to increasing rates of reactions producing side-products, since in all cases where MSA was added, the conversion of fructose was quantitative (>95%) within 1 h. These side-products are very likely not formed by successive reactions from HMF, as HMF is stable over time even at relatively high MSA concentrations, but alternative reaction pathways starting from fructose must be the source of side-product formation.

Batches S60506 and S56605 are too pure to give HMF. Indeed, when ~5.0 wt% MSA was added, yields of 51 and 55% were obtained within 1 h, respectively, using a 7.9 wt% fructose concentration. This compares reasonably well with the yields specified in Table 4.3, entries 2 and 3.

For acid-catalyzed reactions, the use of technical-grade batches may be beneficial from both cost and chemical performance points of view. However, for studies in which the effect of the ionic liquid structure on a reaction outcome is investigated, such impurities, even if present in only small quantities, will hamper the assignment of effects observed to structural characteristics.

4.4.5
Effect of the Choice of Ionic Liquid

Triggered by the availability of large amounts of "unusual" ionic liquids from continuous manufacture in a microreactor [121] at purities >95% (^{1}H NMR, HPLC) and at much lower chemical cost (€70 kg^{-1}) than the technical grade [C$_4$mim][CH$_3$SO$_3$] (€250 kg^{-1}, June 2009), the reaction of fructose was tested in 1,3-dibutyl-imidazolium chloride ([C$_4$C$_4$im]Cl). This homo-substituted dialkylimidazolium salt can be produced directly via a modified Radziszewski reaction from glyoxal, butylamine, formaldehyde and HCl, where a molar ratio of 1 : 2 : 1 : 1.2 has been shown to be optimal. The resulting ionic liquid is known to contain traces of HCl (pH in 0.2 M aqueous solution = 2.0), because the acid has to be used in excess to obtain good yields. However, since acid is required for the fructose condensation to proceed, this aspect is not relevant, as will be shown in the following.

Furthermore, a water-insoluble ionic liquid [C$_6$mim][OTf] (Merck, for synthesis) was also tested. Using a water-insoluble ionic liquid would allow for the extraction of HMF with water.

The application of [C$_4$C$_4$im]Cl as solvent for the condensation of fructose gives yields similar to those achieved with [C$_4$mim][CH$_3$SO$_3$] (batches S40160 and S31564; see Table 4.3). As with [C$_4$mim][CH$_3$SO$_3$], the yield of HMF of decreases in [C$_4$C$_4$im]Cl with increasing fructose concentration (see Table 4.4, entries 1–3). Incremental addition of MSA to [C$_4$C$_4$im]Cl, which already possesses a pH of 2.0 in 0.2 M aqueous solution, also decreases the yield (entries 1 and 4–6).

In pure [C$_6$mim][OTf], the reaction does not proceed. Addition of small amounts of MSA first increases the yield to ~43%, then it decreases again. Compared with the experiments conducted with [C$_4$mim][CH$_3$SO$_3$] and [C$_4$C$_4$im]Cl at the same fructose concentration, the water-insoluble [C$_6$mim][OTf] gives somewhat lower yields. This may be attributed to the water content of the saturated [C$_6$mim][OTf] used [X(IL) = 0.15], which was added to facilitate the dissolution of fructose.

Table 4.4 Yields obtained in various ionic liquids[a].

Entry	Ionic liquid	Fructose (wt%)	Yield (1 h) (%)
1	$[C_4C_4im]Cl$	7.9	61.0
2	$[C_4C_4im]Cl$	9.3	54.7
3	$[C_4C_4im]Cl$	14.7	51.7
4	$[C_4C_4im]Cl$ + 1.0 wt% MSA	7.9	53.2
5	$[C_4C_4im]Cl$ + 4.8 wt% MSA	7.9	41.3
6	$[C_4C_4im]Cl$ + 9.1 wt% MSA	7.9	33.8
7	$[C_6mim][OTf]$[b]	7.9	3.7
8	$[C_6mim][OTf]$[b] + 1.0 wt% MSA	7.9	39.9
9	$[C_6mim][OTf]$[b] + 4.8 wt% MSA	7.9	43.0
10	$[C_6mim][OTf]$[b] + 9.1 wt% MSA	7.9	30.0

[a] 5 g ionic liquid, 80 °C (reactor station), 1 h.
[b] Water saturated to facilitate the dissolution of fructose.

Hence ionic liquids other than $[C_4mim][CH_3SO_3]$ can be used in the condensation of fructose to HMF, if adequate amounts of an acid catalyst (e.g. HCl or MSA) are added.

The fact that the ionic liquid serves only as a solvent, i.e. it does not structurally influence the reaction, allows for its design with respect to cost efficiency and efficiency of recycling methods. Hence hydrophobic ionic liquids such as $[C_6mim][OTf]$ can also be used, from which HMF may be extracted with water. The extraction of HMF from water-soluble ionic liquids can be achieved with non-polar organic solvents, such as MIBK or diethyl ether, which are not very efficient, however.

4.4.6
Other Saccharides

Due to the relatively high price of fructose and the higher natural availability of glucose (and glucose-containing feedstock), other saccharides were tested in the reaction to HMF.

None of the saccharides (except fructose and inulin) gave satisfactory results (Table 4.5). With saccharose, only about a 50% yield was obtained, irrespective of the reaction conditions chosen, indicating that only its fructose moiety is consumed. Interestingly, maltose, a disaccharide consisting of two glucose units, yielded ~10% HMF within 3 h. This is surprising as glucose itself cannot be activated under these conditions. Inulin, a fructose polymer, gave an almost quantitative yield. However, for technical HMF production, inulin is not an alternative feedstock owing to its limited availability and high cost. Nevertheless, these results clearly show that the glycosidic cleavage occurs under the reaction conditions, but the isomerization of glucose to fructose is not effective.

The reduced reactivity of glucose as opposed to fructose is well documented in the literature [55]. The effect of temperature (60–120 °C) was again checked for both glucose and saccharose, with samples being taken after 1, 3 and 5 h. This study

Table 4.5 Yield of HMF using various saccharides at 4.1 wt% concentration in [C$_4$mim][CH$_3$SO$_3$] (S40160) at 80 °C (reactor station) within 1 and 3 h.

Saccharide	Yield (%)	
	1 h	3 h
Fructose	71.5	73.4
Lactose	4.3	5.6
Glucose	0	0
Saccharose	49.0	51.0
Starch	2.8	2.7
Maltose	5.5	10.1
Mannose	4.5	5.6
Inulin	85.0	95.0

revealed for glucose that whereas the yields were below 2% at temperatures below 100 °C, at 100 and 120 °C ~10 and ~20% yields (at quantitative glucose conversion) were obtained within 3 h. However, the reaction mixture turned black and viscous, indicating the formation of a number of side-products. For saccharose, the yield was ~50% for these temperatures within 30 min and remained constant thereafter, indicating complete consumption of the fructose moiety of saccharose.

Investigations were then carried out using an industrially available high-fructose corn syrup [HFCS 50, C*TruSweet 01750 (Cargill, Lot-Nummer 03120392)], which contains 29.1% water and 70.9% dry weight, of which 42.4% is fructose, 53.4% glucose and 4.2% other saccharides. Selected results are displayed in Figure 4.7 for two (total) sugar concentrations, 14.7 and 34.7 wt%, corresponding to fructose contents of 6.3 and 14.9 wt%, respectively. Yields obtained are given with respect

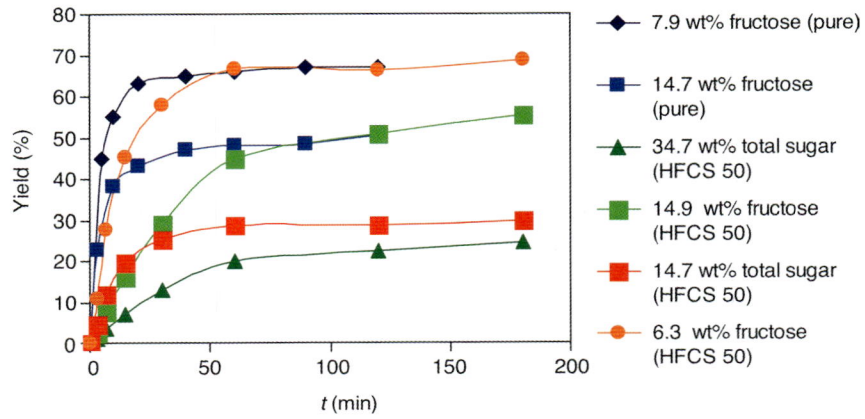

Figure 4.7 Comparison of the yields of HMF with HFCS 50 in [C$_4$mim][CH$_3$SO$_3$] [S40160, 5 g ionic liquid, 80 °C (reactor station)].

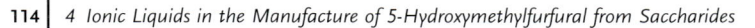

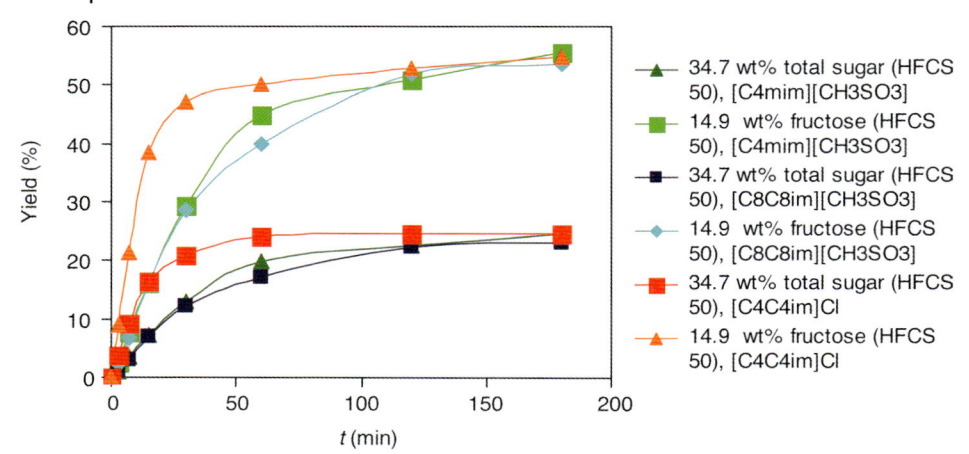

Figure 4.8 Comparison of the yields of HMF with HFCS 50 in different ionic liquids: $[C_4mim][CH_3SO_3]$ batch S40160, $[C_4C_4im]$Cl and $[C_8C_8im][CH_3SO_3]$ [5 g ionic liquid, 80 °C (reactor station)].

to the total sugar content and the fructose content and compared with results obtained in the same batch of ionic liquid with pure fructose. As with pure fructose, the yield decreases with increasing saccharide concentration. Moreover, the same yield of HMF regarding the fructose content is obtained as when pure fructose is converted. Therefore, HFCS 50 can be used in a future process and it can be predicted that HFCS 90, which is also commercially available and contains ~90% fructose, will perform even better.

Finally, the conversion of HFCS 50 was additionally tested in two alternative ionic liquids made by the modified Radziszewski reaction [121], i.e. $[C_8C_8im][CH_3SO_3]$ and $[C_4C_4im]$Cl. The reaction profiles shown in Figure 4.8 demonstrate that all three ionic liquids give similar performances and that the conversion of saccharides to HMF is not a direct function of the ionic liquid moieties.

Hence the ionic liquid's anion and cation can be chosen for optimal performance in the subsequent product separation and recycling steps. In particular, and as opposed to other polar (organic) solvents that are often used for the condensation of saccharides to HMF (see above), ionic liquids offer the unique advantage and opportunity to design a system which on the one hand provides an acidic medium for good reaction performance, i.e. high selectivity and conversion. On the other hand, the extraction may be achieved by combination with an organic solvent with high selectivity and capacity for HMF, leading to the development of a continuous extraction from the reactive mixture. This approach is detailed in the following.

4.4.7
Continuous Processing of HMF

A continuous process to manufacture HMF from hexoses, in particular fructose, using the concept of reactive extraction was developed based on the above re-

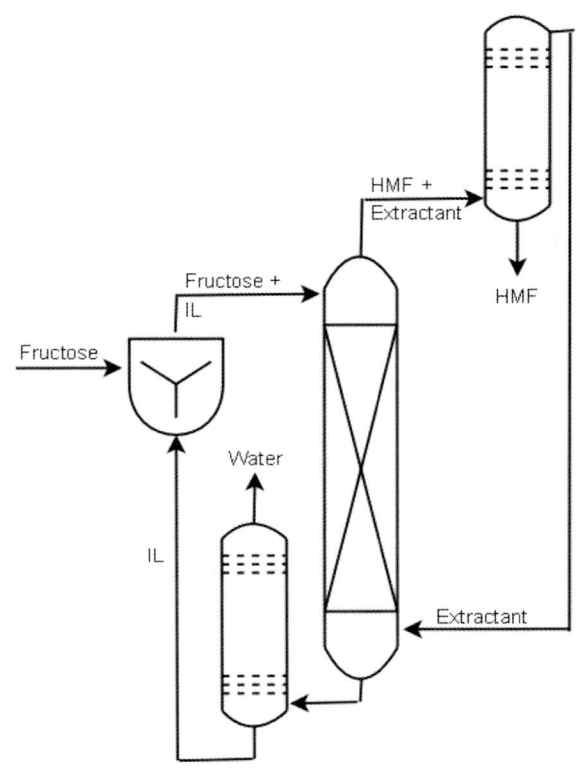

4.4 HMF Manufacture in Ionic Liquids – Results of Detailed Studies in the Jena Laboratories

Figure 4.9 Schematic diagram of experimental set-up to produce HMF continuously from fructose using ionic liquids [122].

sults [122]. An ionic liquid, in which the sugar is dissolved, is fed into a heated column and during its residence time, the conversion to HMF takes place. An organic solvent, which is not soluble in the ionic liquid, is fed into the reactor in counter-current mode. Hence HMF is extracted continuously, reducing its residence time in the heated reaction phase to a minimum. The extractant is removed *in vacuo*, yielding HMF in a purity >95% (Figure 4.9). In order to achieve high selectivities (>90%) and yields (>90%) in the reactor, it is advantageous to run the process at low spatio-temporal sugar concentrations (<10 wt%).

A cost assessment [123] was carried out based on the currently used 2 l laboratory set-up, run for 340 days per year at 12 h per day, assessed for costs relevant to Germany, in 2009. In total, 390 kg of HMF can be produced.

With regard to ionic liquid consumption, a recycling rate of 99.0% was taken into account. Due to the low sugar concentrations, this translates to about 0.15 kg of ionic liquid loss per kilogram of HMF produced. As for ionic liquid costs, both the actual market price (in kilogram quantities) and a reduced price (resulting from the optimization of ionic liquid manufacture) have been taken into account. The latter is achieved by using an alternative synthetic strategy using cheap reactants (aliphatic primary amines, glyoxal, formaldehyde and acid) to produce 1,3-dialkylimidazolium-based ionic liquids directly [121]. The resulting costs (€70 kg^{-1} ionic liquid) are in line with estimations by colleagues in the ionic liquids-producing industry, according to whom on a large scale ionic liquids will cost about €50 kg^{-1}. For the extractant, the assumed recycling rate was also estimated to be 99.0%.

The energy requirement of the reactive extraction (pump, stirrer, heating) and for the solvent recycling (distillation) were determined using an energy monitoring socket. Personnel and personnel-dependent (5%) costs are based on the salary of technicians working 40 h per month. Capital and capital-dependent costs (25%) are based on the cost of the extractive reactor, the pump and the distillation set-up, taking into account an amortization of 10 years. Table 4.6 shows that under these circumstances, HMF can be produced at a price between €125 and 150 kg^{-1}.

For the cost estimation of the production of 10 000 t per year, a reduction of ionic liquid price to €25 kg^{-1} and an improvement of the recycling efficiency to 99.3%, a reduction of extractant price and an improvement of the extraction efficiency (overall factor 4) and an improvement in energy efficiency (factor 2) were taken into account.

Table 4.6 Summary of costs (€) occurring in the production of HMF, as a function of reactor size.

Type of cost	Cost^a (390 kg per year)	Cost^a (390 kg per year)	Cost^b (10 000 t per year)
Materials:			
Fructose (700 €/t)	430	430	11 180 000
Ionic liquid	13 750^c	3 850^d	25 000 000
Extractant (1700 €/m^3)	2380	2380	15 470 000
Energies: electricity (0.3€/kW h)	860^e	860^e	11 180 000
Personnel/personnel-dependent costs	36 250	36 250	108 750
Capital/capital-dependent costs	3125	3125	2 825 000
Other costs (3% of above)^f	1700	1405	1 973 000
Total	58 500	48 300	67 736 750
	→ 150€/kg	→ 125€/kg	→ 7€/kg

^{a}Assumptions: continuous production at 1 bar, 80 °C in 2 l apparatus, Germany, 2009. 1 kg ionic liquid (10 wt% fructose), 90 mol% yield. Ionic liquid recycling: 0.15 kg ionic liquid loss per kg HMF (99.0% recycling). Recyclability extractant: 99.0%. Energies (pump, reactor, solvent recycling): 2.860 kW h per year. Personnel and personnel-dependent costs: 1.5 technicians, 8 h per day, 340 days per year. Capital and capital-dependent costs: reactor, pump, rotary evaporator. Amortization: 10 years.
^{b}Assumptions: as for laboratory scale, except ionic liquid recycling: 0.10 kg ionic liquid loss per kg HMF (99.3% recycling). Personnel and personnel-dependent costs: 4 shifts, 3 technicians, 24 h per day, 250 days per year. Extraction efficiency doubled, costs halved. Improvement in energy efficiency by a factor of 2, e.g. by energy recirculation. Capital and capital-dependent costs were calculated with $P_1/P_2 = (C_1/C_2)^m$, where P_1 and P_2 are the prices of apparatus 1 and 2 with capacity C 1 and 2 and $m = 2/3$.
^{c}Actual ionic liquid price on kg scale (€250 kg^{-1}).
^{d}€70 kg^{-1} [121].
^{e}Heat loss and energy recirculation not included.
^{f}Waste disposal not included.

Table 4.6 shows that under these conditions, an HMF production price of approximately €7 kg^{-1} is obtained.

4.5
Conclusion

The above study has shown that pure ionic liquids do not catalyze the reaction, but that traces of acidic impurities (e.g. from their manufacture) are responsible for the activity. It is therefore important not only to specify the purities of the materials used, but also to determine the effect of added acid catalyst. Small amounts of acid increase the rate of reaction to HMF tremendously, but too high concentrations lead to increasing rates of side-product(s) formation. It is interesting that the experimental evidence suggests that the side-product formation does not proceed via HMF, as HMF is stable under any of the conditions investigated. Instead, other pathways starting from fructose must occur. This is contrary to reports on side-product formation in conventional solvents [54–56]. In addition to the acid concentration, similar (negative) effects on the yield and selectivity have been observed on increasing the fructose concentration. It is therefore advantageous to run the reaction at the lowest possible acid and fructose concentrations.

Of course, this implies a relatively large ionic liquid inventory, which will greatly affect the processing costs. Therefore, a balance between chemical and engineering performances will have to be established.

As the ionic liquid's structural moiety does not seem to have an impact on the reaction *per se*, the opportunity arises to control the cost of the ionic liquid by using alternative cation structures. One such possibility has been demonstrated: Using homo-substituted 1,3-dialkylimidazolium-based ionic liquids made by a modified Radziszewski reaction, starting from glyoxal, alkylamine, formaldehyde and an acid, the ionic liquid price can be lowered to about one-third of the current ionic liquid prices. Further improvements regarding the choice of ionic liquid seem feasible.

Since fructose is a rather costly starting material, several other hexose-based feedstocks were tested. It was found that the glycosidic cleavage of dimeric (saccharose) and polymeric (inulin) saccharides was achieved, but that the isomerization of glucose did not proceed selectively, although under harsher conditions yields up to 20% (from glucose, at 120 °C, 3 h) were obtained. This, on the one hand, indicates future optimization potential with regard to the feed source.

On the other hand, a commercially available high-fructose corn syrup (HFCS 50), which is manufactured primarily from starch, has been employed with success in the manufacture of HMF. Although only the fructose content was converted, this is a first step to using bio-refinery feed streams, and higher fructose content syrups (HFCS 90) are also available.

The performance of the continuous laboratory-scale process has shown that the technology permits the production of HMF at a cost of €125 kg^{-1}, which is much too high to compete with any bulk chemical products or fuels on the market. The preliminary scale-up calculation indicated a production cost of €7 kg^{-1} on a 10 000 t per year scale, which would certainly allow for the introduction of HMF-based compounds into upmarket products. Future research is required to determine specific properties of these compounds to push market entry for innovative products.

References

1 Faury, A., Gaset, A. and Gorrichon, J.P. (1981) *Informations Chimique*, **214**, 203–209.
2 Musau, R.M. and Munavu, R.M. (1990) *Biomass*, **23**, 275–287.
3 Gandini, A. and Belgacem, M.N. (1997) *Progress in Polymer Science*, **22**, 1203–1379.
4 Moreau, C., Belgacem, M.N. and Gandini, A. (2004) *Topics in Catalysis*, **27**, 11–30.
5 Groke, K., Kager, E. and Bucher, C. (2005) US Patent 2005274391.
6 Gardovic, M. (2003) World Patent WO 2004000305.
7 Moser, P.M., Greilberger, J., Maier, A., Juan, H., Buecherl-Harder, C. and Kager, E.(to CYL Pharmaceutika GmbH) (2007) European Patent EP 1842536.
8 Groke, K., Miggitsch, H., Musil, H. and Polzer, J. (to Leopold & Co. Chem. Pharm. Fabrik AG.) (1989) US Patent 5 006 551.
9 Groke, K., Groke, I., Groke, V., Groke, P., Herwig, R. and Ferdinand, P. (2003) US Patent 20060292218.
10 Michail, K., Matzi, V., Maier, A., Herwig, R., Greilberger, J., Juan, H., Kunert, O. and Wintersteiger, R. (2007) *Analytical and Bioanalytical Chemistry*, **387**, 2801–2814.
11 Gruter, G.J.M., Dautzenberg, F. and Purmov, J.(to Avantium International BV) (2006) World Patent WO 2007104515.
12 Merck & Co. Inc., British Patent 925812, (1959).
13 Benecke, H.P., King, J.L. II, Kawczak, A.W., Zehnder, D.W. and Hirschl, E.E. (to Battelle Memorial Institute) (2007) World Patent WO 2007092569.
14 Mascal, M. and Nikitin, E.B. (2008) *Angewandte Chemie-International Edition*, **47**, 7924–7926.
15 Gruter, G.J.M.(to Furanix Technologies) (2009) World Patent WO 2009030510.
16 Gruter, G.J.M. and Dauzenberg, F.(to Avantium International BV) (2006) World Patent WO 2007104514.
17 Tsai, J.J.-H., Jobe, P.G. and Billmers, R.L.(to National Starch and Chemical Corporation) (1988) European Patent EP 0283824.
18 Bonsignore, P.V.(to Stefan Company) (2006) World Patent WO 2008022287.
19 Daup, J., Rapp, K.M., Seitz, P., Wild, R. and Salbeck, J.(to Sueddeutsche Zucker-Aktiengesellschaft) (1992) US Patent 5 091 538.
20 Lichtenthaler, F.W., Brust, A. and Cuny, E. (2001) *Green Chemistry*, **3**, 201–209.
21 Gupta, P., Singh, S.K., Pathak, A. and Kundu, B. (2002) *Tetrahedron*, **58**, 10469–10474.
22 Dykstra, C.C., Tidwell, R.R., Boykin, D.W., Wilson, D.W., Spychala, J., Das, B. and Kumar, A.(to The University of North Carolina at Chapel Hill, Georgia State University Research Foundation, Inc.) (2006) World Patent WO 9530901.
23 Newinomyssk, J.N., Tichonowitsch, P.F., Michajlowitsch, S.A., Isaakowitsch, K.M. and Borisowitsch, S.M.(to Rostowskij ordena Trudowowo Krasnowo Snameni Gosudarstwenny Uniwersitet) (1975) German Patent 2 349 803.
24 Dykstra, C.C., Perfect, J.R., Boykin, D.W., Wilson, W.D. and Tidwell, R.R.(to The University of North Carolina at Chapel Hill, Duke University, Georgia State University Research Foundation, Inc.) (1996) World Patent WO 9640145.
25 Dykstra, C.C., Boykin, D. and Tidwell, R.R.(to The University of North Carolina

at Chapel Hill, Auburn University, Georgia State University Research Foundation, Inc.) (1998) World Patent WO 9838170.
26. Poeta, M.D., Schell, W.A., Dykstra, C.C., Jones, S., Tidwell, R.R., Czarny, A., Bajic, M., Bajic, M., Kumar, A., Boykin, D. and Perfect, J.R. (1998) *Antimicrobial Agents and Chemotherapy*, **42**, 2495–2502.
27. Hopkins, K.T., Wilson, W.D., Bender, B.C., McCurdy, D.R., Hall, J.E., Tidwell, R.R., Kumar, A., Bajic, M. and Boykin, D.W. (1998) *Journal of Medicinal Chemistry*, **41**, 3872–3878.
28. Rodriguez-Kabana, R.(to Illovo Sugar Limited) (2000) World Patent WO 0067577.
29. Constantin, J.M., William, H.T., Lange, H.B., Shew, D. and Wagner, J.R.(to Merck & Co Inc.) (1963) US Patent 3 080 279.
30. Glushkov, R.G., Adamskaya, E.V., Vosyakova, T.I. and Oleinik, A.F. (1990) *Journal of Pharmacological Chemistry*, **24**, 369–373.
31. Aasen, A.J. and Skramstad, J. (1998) *Archives of Pharmacology*, **331**, 228–229.
32. Soldo, T., Blank, I. and Hofmann, T. (2003) *Chemical Senses*, **28**, 371–379.
33. Ottinger, H., Soldo, T. and Hofmann, T. (2003) *Journal of Agricultural and Food Chemistry*, **51**, 1035–1041.
34. Villard, R., Robert, F., Blank, I., Bernardinelli, G. and Soldo, T. (2003) *Journal of Agricultural and Food Chemistry*, **51**, 4040–4045.
35. Boghani, N., Gebreselassie, P. and Hargreaves, C.A.(to Cadbury Adams USA LLC) (2006) World Patent WO 2006127936.
36. Grushin, V., Herron, N. and Halliday, G.A.,(to DuPont) (2003) World Patent WO 03024947.
37. Chmielewski, P.J., Latos-Grazynski, L., Olmstead, M.M. and Balch, A.L. (1997) *Chemistry - A European Journal*, **3**, 268–278.
38. Hui, Z. and Gandini, A. (1992) *European Polymer Journal*, **28**, 1461–1469.
39. Sheibley, D.W., Manzo, M.A. and Gonzalez-Sanabria, O.D. (1983) *Journal of the Electrochemical Society: Electrochemical Science Technology*, **130**, 255–259.
40. Werpy, T. and Peterson, G. (eds.) (2004) *Top Value Added Chemicals from Biomass: Vol. 1 – Results of Screening for Potential Candidates from Sugars and Synthesis Gas*, National Renewable Energy Laboratory, Golden, CO.
41. Partenheimer, W. and Grushin, V.V. (2001) *Advanced Synthesis and Catalysis*, **343**, 102–111.
42. Correia, P. (2008) World Patent WO 2008053284.
43. Pentz, W.J.(to Quaker Oats Company) (1984) US Patent 4 426 460.
44. Pentz, W.J.(to Quaker Oats Company) (1981) US Patent 4 426 460.
45. Román-Leshkov, Y., Barett, C.J., Liu, Z.Y. and Dumesic, J.A. (2007) *Nature*, **447**, 982–985.
46. Binder, J.B. and Raines, R.T. (2009) *Journal of the American Chemical Society*, **131**, 1979–1985.
47. Gaset, G., Gorrichon, J.P. and Truchot, E. (1981) *Informations Chimie*, **212**, 179–184.
48. Kulkarni, A.D., Modak, H.M. and Jadhav, S.J. (1988) *Journal of Scientific & Industrial Research India*, **47**, 335–339.
49. Kuster, B.F.M. (1990) *Starch-Stärcke*, **8**, 314–321.
50. Cottier, L. and Descotes, G. (1991) *Trends Heterocyclic Chemistry*, **2**, 233–248.
51. Lewkowski, J. (2001) *Arkivoc*, 17–54.
52. Corma, A., Iborra, S. and Velty, A. (2007) *Chemical Reviews*, **107**, 2411–2502.
53. Rapp, K.M.(to Sueddeutsche Zucker AG) (1987) European Patent EP 0230250.
54. US Patent 3071599, R. A. Hales, J. W. Le Maistre, G. O. Orth, to Atlas Chemical Industries, Inc., 1963–01–01.
55. Kuster, B.F.M. and van der Baan, H.S. (1977) *Carbohydrate Research*, **54**, 165–176.
56. Román-Leshkov, Y., Chheda, J.N. and Dumesic, J.A. (2006) *Science*, **312**, 1933–1937.
57. Haworth, W.N. and Jones, W. (1944) *Journal of the Chemical Society A: Inorganic, Physical, Theoretical*, 667–670.

58 Garber, J.D. and Jones, R.E.(to Merck & Co Inc.) (1960) US Patent 2 929 823.
59 Asghari, F.S. and Yoshida, H. (2006) *Industrial & Engineering Chemistry Research*, **45**, 2163–2173.
60 The Colonial Sugar Refining Company Limited (1962) British Patent 1 019 512.
61 Kuster, B.F.M. (1977) *Carbohydrate Research*, **54**, 177–183.
62 Takenubo, K., Hirohide, M., Hiroshi, M., Hitoshi, H. and Katsuhiro, M.(to Canon Inc.) (2005) Japanese Patent 2005232116.
63 Katsuhiro, M., Takenobu, K. and Hitoshi, H.(to Canon Inc.) (2005) Japanese Patent 2005200321.
64 F.H. Snyder(to Entrol Inc.) (1958) US Patent 2 851 468.
65 Faber, M.(to Hydrocarbon Research Inc.) (1983) US Patent 4 400 468.
66 Watanabe, M., Aizawa, Y., Iida, T., Aida, T.M., Levy, C., Sue, K. and Inomata, H. (2005) *Carbohydrate Research*, **340**, 1925–1930.
67 Kuster, B.F.M. and Laurens, J. (1977) *Starch-Stärke*, **29**, 172–176.
68 Bicker, M., Hirth, J. and Vogel, H. (2003) *Green Chemistry*, **3**, 280–284.
69 Szmant, H.H. and Chundury, D.D. (1981) *Journal of Chemical Technology and Biotechnology*, **31**, 135–145.
70 Garber, J.D. and Jones, R.E.(to Merck & Co., Inc.) (1969) US Patent 3 483 228.
71 Krupenskii, V.I.(to Ukhta Industrial Institute) (1983) Russian Patent 1 054 349.
72 Krupenskii, V.I. and Potapov, G.P.(to Ukhta Industrial Institute) (1987) Russian Patent 1 351 932.
73 Ishida, H. and Seri, K.-I. (1996) *Journal of Molecular Catalysis A-Chemical*, **112**, L163–L165.
74 Seri, K.-I., Inoue, Y. and Ishida, H. (2000) *Chemistry Letters*, **29**, 22–23.
75 Carlini, C., Patrono, P., Galletti, A.M.R. and Sbrana, G. (2004) *Applied Catalysis A-General*, **275**, 111–118.
76 Zhao, H., Holladay, J.E., Brown, H. and Zhang, Z.C. (2007) *Science*, **316**, 1597–1600.
77 Zhao, H., Holladay, J.E. and Zhang, Z.C.(to Battelle Memorial Institute) (2008) World Patent WO 2008019219.
78 Benvenuti, F., Carlini, C., Patrono, P., Galletti, A.M.R., Sbrana, G., Massucci, M.A. and Galli, P. (2000) *Applied Catalysis A-General*, **193**, 147–153.
79 Armaroli, T., Busca, G., Carlini, C., Giuttari, M., Galletti, A.M.R. and Sbrana, G. (2000) *Journal of Molecular Catalysis A-Chemical*, **151**, 233–243.
80 Rigal, L., Gaset, A. and Gorrichon, J.-P. (1981) *Industrial & Engineering Chemistry Product Research and Development*, **20**, 719–721.
81 Hamada, K., Yoshihara, H. and Suzukamo, G. (1982) *Chemistry Letters*, 617–618.
82 Fléche, G., Gaset, A., Gorrichon, J.-P., Truchot, E. and Sicard, P.(to Roquette Frères) (1982) US Patent 4 339 387.
83 Rigal, L., Gorrichon, J.-P., Gaset, A. and Heughebaert, J.-C. (1985) *Biomass*, **7**, 27–45.
84 Gaset, A., Rigal, L., Paillassa, G., Salmé, J.P. and Flèche, G.(to Roquette Frères) (1986) US Patent 4 590 283.
85 Neyrete, C., Cottier, L., Nigay, H. and Descores, G.(to Beghin-Say) (1990) French Patent 2 664 273.
86 Avignon, G., Durand, R., Faugeras, P., Geneste, P., Moreau, C., Rivalier, P. and Ros, P.(to Commissariat de l'Energie Atomique) (1990) French Patent 2 670 209.
87 Hiroshi, T. and Takashi, S.(to Kao Corporation) (1991) Japanese Patent 03099072.
88 Moreau, C., Durand, R., Pourcheron, C. and Razigade, S. (1994) *Industrial Crops and Products*, **3**, 85–90.
89 Durand, G.-R., Faugeras, P., Laporte, F., Moreau, C., Neau, M.-C., Roux, G. and Trousselier, S.(to Agrichimie) (1996) World Patent WO 9617837.
90 T. Martin(to Episucres SA) (1996) German Patent 19619075.
91 Moreau, C., Durand, R., Razigade, S., Duhamet, J., Faugeras, P., Rivalier, P.,

Ros, P. and Avignon, G. (1996) *Applied Catalysis A-General*, **145**, 211–224.
92 Moreau, C., Durand, R., Duhamet, J. and Rivalier, P. (1997) *Journal of Carbohydrate Chemistry*, **16**, 709–714.
93 Bonner, T.G., Bourne, E.J. and Ruszkiewicz, M. (1960) *Journal of the Chemical Society A: Inorganic, Physical, Theoretical*, 787–791.
94 van Dam, H.E., Kieboom, A.P.G. and van Bekkum, H. (1986) *Starch-Stärke*, **38**, 95–101.
95 Musau, R.M. and Munavu, R.M. (1987) *Biomass*, **13**, 67–74.
96 M'Bazoa, C., Raymond, F., Riga, L. and Gaset, A.(to Furchim SRL) (1992) French Patent 2 669 635.
97 Antal, M.J.J., Mok, W.S.L. and Richards, G.N. (1990) *Carbohydrate Research*, **199**, 91–109.
98 Halliday, G.A., Young, R.J. and Grushin, V.V. (2003) *Organic Letters*, **5**, 2003–2005.
99 Qi, X., Watanabe, M., Aida, T.M. and Smith, R.L. (2009) *Green Chemistry*, **11**, 1327–1331.
100 Lansalot-Matras, C. and Moreau, C. (2003) *Catalysis Communications*, **4**, 517–520.
101 Chan, J.Y.G. and Zhang, Y. (2009) *ChemSusChem*, **2**, 731–734.
102 Mednick, M.L. (1962) *The Journal of Organic Chemistry*, **27**, 398–403.
103 Smith, N.H.(to Rayonier, Inc.) (1964) US Patent 3 118 912.
104 Fayet, C. and Gelas, J. (1983) *Carbohydrate Research*, **122**, 59–68.
105 Moreau, C., Finiels, A. and Vanoye, L. (2006) *Journal of Molecular Catalysis A-Chemical*, **253**, 165–169.
106 Bao, Q., Qiao, K., Tomida, D. and Yokoyama, C. (2008) *Catalysis Communications*, **9**, 1383–1388.
107 Hu, S., Zhan, Z., Zhou, Y., Han, B., Fan, H., Li, W., Song, J. and Xie, Y. (2008) *Green Chemistry*, **10**, 1280–1283.
108 Cope, A.C. (1959) US Patent 2 917 520.
109 Peniston, Q.P.(to Chemical and Research Laboratories) (1956) US Patent 2 750 394.
110 Lightner, G.E. (2002) US Patent 6 441 202.
111 Chheda, J.N., Román-Leshkov, Y. and Dumesic, J.A. (2007) *Green Chemistry*, **9**, 342–350.
112 Dumesic, J.A., Roman-Leshkov, Y. and Chheda, J.N.(to Wisconsin Alumni Research Foundation) (2007) World Patent WO 2007146636.
113 Atlas Powder Co. (1961) British Patent 876 463.
114 Vinke, P. and van Bekkum, H. (1992) *Starch-Stärke*, **44**, 90–96.
115 Kuster, B.F.M. and van der Stehen, H.J.C. (1977) *Starch-Stärke*, **29**, 99–103.
116 Rivalier, P., Duhamet, J., Moreau, C. and Durand, R. (1995) *Catalysis Today*, **24**, 165–171.
117 Laurent, F., Seguinaud, A.-F. and Faugeras, P.(to Sucreries et Raffineries d'Erstein) (2005) World Patent WO 2005018799.
118 Asghari, F.S. and Yoshida, H. (2006) *Industrial and Engineering Chemical Research*, **45**, 2163–2173.
119 Yong, G., Zhang, Y. and Ying, J.Y. (2008) *Angewandte Chemie-International Edition*, **47**, 9345–9348.
120 Stark, A., Behrend, P., Braun, O., Muller, A., Ranke, J., Ondruschka, B. and Jastorff, B. (2008) *Green Chemistry*, **10**, 1152–1161.
121 Zimmermann, J., Ondruschka, B. and Stark, A. (2009) *Green Chemistry*, submitted for publication.
122 Stark, A., Lifka, J. and Ondruschka, B.(to Friedrich-Schiller-Universität Jena) (2008) German Patent 2008009933.
123 Baerns, M., Behr, A., Brehm, A. Gmehling, J., Hoffmann, H., Onken, U. and Renken, A. (2006) in *Lehrbuch Technische Chemie*, Wiley-VCH Verlag GmbH, Weinheim., pp. 463–478.

5
Cellulose Dissolution and Processing with Ionic Liquids
Uwe Vagt

5.1
General Aspects

Occurring at a bulk of some 700×10^9 t, cellulose is Earth's most widespread natural organic chemical and therefore our most important bio-renewable resource. Together with more or less lignin, cellulose occurs as the main part of every lignocellulosic biomass. It is produced by biosynthesis on land (wood, grasses) and in the sea (algae) as component of plant cells in quantities of approximately 75×10^9 t per year (see Figure 5.1) [1].

However, out of these 75×10^9 t that Nature renews every year, only 0.2×10^9 t are used as feedstock for further processing – mainly for the pulp and paper industry. As feedstock for chemical processes, only a very small amount of 5×10^6 t per year (i.e. 0.007%) is utilized. This is surprising and also in a way disappointing, as the properties of the cellulose biopolymer given by Nature should be a very good basis for manufacturing polymeric materials based on renewable feedstock such as lignocellulosic biomass (Figure 5.2) [1, 2].

More intensive exploitation of cellulose as a bio-renewable feedstock has so far been prevented by a strong focus on petrochemical raw materials since the 1940s, but also by the lack of suitable solvents for processing both lignocellulosic biomass and cellulose itself.

Large-scale processing requiring the dissolution of cellulosic raw material is impeded by high prices due to the high manufacturing costs of pulping processes. The main challenge is to separate lignin and cellulose from each other, which appear as strongly bonded conglomerates in the lignocellulosic biomass. Suitable solvents allowing for the design of more cost-efficient and environmentally more friendly pulping processes would therefore be very helpful. They may in future provide cellulose as a raw material for chemical processes at costs comparable to those of materials from petrochemical feedstocks.

Today, however, any further processing of cellulose also suffers from the lack of suitable solvents. This is true for any reshaping process leading to fibers (viscose

Handbook of Green Chemistry, Volume 6: Ionic Liquids
Edited by Peter Wasserscheid and Annegret Stark
Copyright © 2010 WILEY-VCH Verlag GmbH & Co. KGaA, Weinheim
ISBN: 978-3-527-32592-4

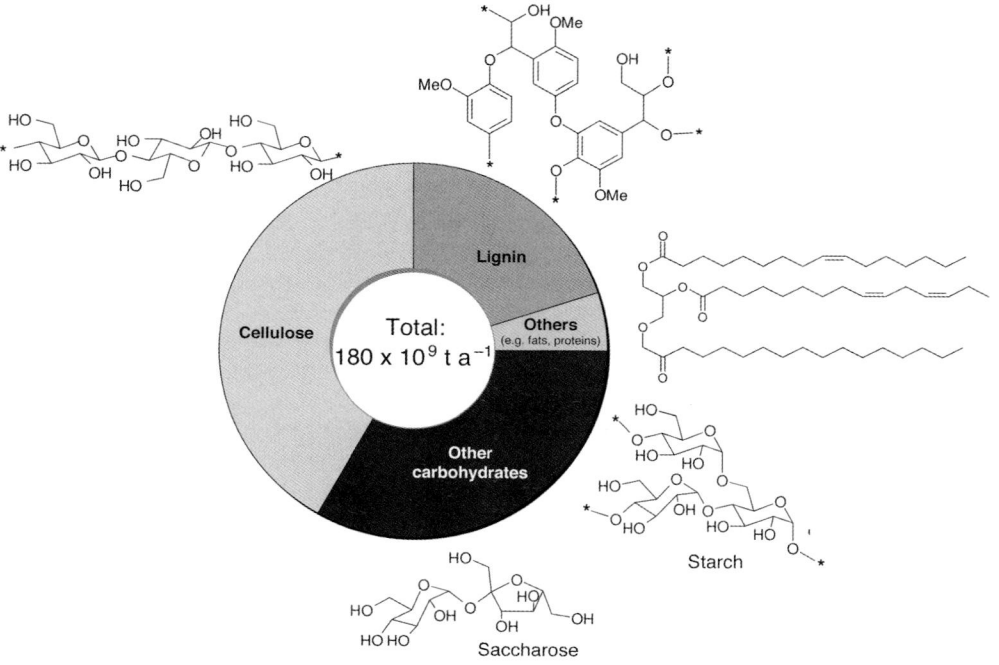

Figure 5.1 Cellulose – the most widespread raw material source on Earth.

fibers) or films (Cellophane) and also for cellulose derivatives, such as cellulose acetate and carboxymethylcellulose. Furthermore, processing dissolved cellulose in an inert solvent may open the way to new derivatives not accessible today and to totally new materials with superior properties.

The oldest method to dissolve cellulose, which was discovered in 1857, is the dissolution in a mixture of copper(II) salts, ammonia and sodium hydroxide. The solutions are called Schweitzer's reagent and can be used, for example, to spin fibers

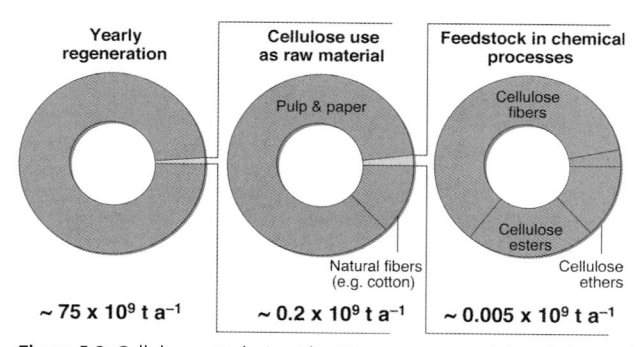

Figure 5.2 Cellulose – today's utilization as raw material and chemical feedstock.

5.1 General Aspects

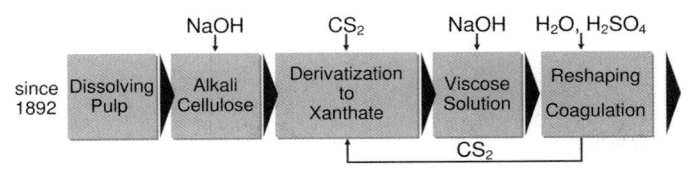

Figure 5.3 Viscose process.

(cupro fibers). Although coagulation processes using this reagent perform fairly well, the challenge in this process is the necessity to recycle copper and ammonia from the dilute aqueous solutions of the coagulation bath. Therefore, this dissolution process was never realized on a larger scale. Nevertheless, the cupro fibers are still a valuable niche product for textile applications due to the superior properties of the fibers obtained [2].

The majority of cellulose today is processed by using carbon disulfide (CS_2) as a solvent, mainly leading to the well-known viscose fibers produced worldwide in amounts of 2.5×10^6 t per year (viscose process; see Figure 5.3). The principle disadvantage of the CS_2 process is that this process is not a physical dissolution process. Cellulose is dissolved after derivatization as sodium xanthate and the major drawback results from the need for equimolar quantities of auxiliaries such as sodium hydroxide and sulfuric acid, leading to significant amounts of waste and efforts in waste treatment. This process requires large amounts of water: 450–850 l of water for each kilogram of fiber. The waste water contains large amounts of sodium sulfate, which has to be recovered in some countries. Additionally, off-gases of viscose plants require subsequent removal of hydrogen sulfide and carbon disulfide, which adds significant costs to the process in order to fulfill strict legislations. Investments in new viscose plants therefore become more and more expensive and new processes with alternative solvents with closed recycle are increasingly attractive. The fibers obtained by this process can be further optimized by adding auxiliaries to the spinning dope or the coagulation bath (for example, amines, polyether glycols, zinc salts).

Most of the disadvantages of the viscose process can be overcome by a spinning process based on cellulose carbamate, starting with alkali cellulose and urea as raw materials. This process leads to cellulosic fibers – partially containing carbamate functions – with properties equal to those obtained in the xanthate process. Advantages of the carbamate process are lower overall costs and much lower environmental impact. Therefore, the further development of this process might be an attractive option for cellulosic fiber production in the future [3].

Since 1892, when the CS_2 process was introduced, there have been numerous attempts to overcome this disadvantage by finding better solvents. Most of the solvent systems developed include combinations of conventional organic solvents, for example dimethylacetamide (DMA) or dimethylformamide (DMF) with lithium chloride (LiCl). The combination of a strongly hydrated, polarizing cation with a rather voluminous, weakly hydrated, easily polarizable anion leads to the formation of soluble oxonium compounds with the alcoholic hydroxyl groups in the cellulose. Reshaping processes for cellulose using these alternative solvents turned out to be as

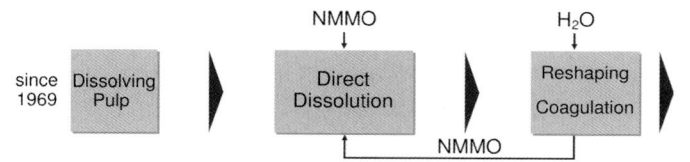

Figure 5.4 NMMO process.

problematic as the century-old viscose process and therefore were never commercialized on a larger scale [2].

Only one new solvent development made it to the commercial scale: today, the newest cellulose fiber production technology uses N-methylmorpholine N-oxide (NMMO) as a solvent (NMMO process; see Figure 5.4). This technology is commonly referred to as the Lyocell process. This process was developed at Eastman Kodak and American Enka and was commercialized by Courtaulds and Lenzing. With the realignment of fiber manufacturing companies, today Lenzing is the only company manufacturing cellulosic fibers by this process. The NMMO process is a direct dissolution process without the need to employ any auxiliaries and the only direct dissolution process commercialized on a large scale. However, the limitations are in market penetration due to the fibrillation of the fiber.

With regard to environmental effects, the Lyocell process is a major step in the right direction, but still faces significant engineering challenges with regard to solvent stability. Especially the autocatalytic decomposition of NMMO limits the opportunities of this process. In addition, the Lyocell fiber itself is fibrillating, which limits broader market penetration [4].

Another concept – aimed at dissolving cellulose in molten organic salts without any molecular organic solvent - was first introduced by Charles Graenacher in 1934 [5], who was able to dissolve cellulose in N-alkyl- or N-arylpyridinium chlorides in the presence of nitrogen-containing bases. At this time, the invention was treated probably as a novelty with little practical value, as molten salts were seen as somewhat "esoteric" and not readily available on a larger scale. Another drawback might have been that the concentrations of cellulose obtained in these molten salts were rather low. Hence this idea was not followed up and therefore not realized on a commercial scale [5].

It was Robin Rogers with his research team at the University of Alabama who in 2002 applied ionic liquids to the dissolution of cellulose. In particular, by using 1-butyl-3-methylimidazolium chloride ([C_4mim]Cl) they were the first to be able to dissolve cellulose in technically useful concentrations by physical dissolution in an inert solvent without using any auxiliaries [6]. From today's point of view, this has to be seen as a major breakthrough in opening up opportunities for a broader utilization of cellulose from natural feedstocks.

In 2005, BASF licensed the exclusive use of various intellectual property rights from the University of Alabama. These rights broadly cover "dissolution and further processing of cellulose in ionic liquids". Based on this, BASF started to explore the opportunities of this technology in more detail with a view to market and customer needs. This work was carried out in cooperation with Frank Hermanutz, Frank Gähr and their team at ITCF in Denkendorf, Germany, Birgit Kosan, Frank Meister and their colleagues at TITK in Rudolstadt, Germany, the teams of Rudolf Patt at the Department of Wood Science at the University of Hamburg, Germany, and Werner Mormann at the University of Siegen, Germany, and also Robin Rogers and his co-workers at the University of Alabama, USA. The following review gives a brief overview of what has been achieved so far and what can be expected in the future.

5.2
Dissolution of Cellulose in Ionic Liquids

Starting with [C_4mim]Cl, BASF set up a broad screening for the dissolution of cellulose in ionic liquids. The dissolution of cellulose (5–25 wt%) was investigated by including a broad range of cellulosic raw materials from different sources, for example cotton linters, eucalyptus pulp, peach pulp and alkali cellulose, varying the average degree of polymerization (DP) from 300 to 2000 and also applying two different dissolution techniques:

1. Cellulose pulp is first cut into small pieces of 1×1 cm and then ground to 1 mm particles in a Retsch cutting mill. This finely ground pulp is then added while stirring at 100 °C to the preheated ionic liquid and subsequently stirred at 100 °C for 20 min. These mix–store cycles are repeated over a period of 2 h or more if needed (dissolution method developed at the ITCF, Denkendorf).

2. The second dissolution method was derived from the NMMO spinning process and adapted to using ionic liquids as solvents. After cutting the cellulose pulp in small pieces of 1×1 cm, the pulp is suspended in water (roughly 5 wt% pulp) by using a Ultra-Turrax stirrer. The majority of the water is then removed by using a sieve press and the swollen cellulose obtained is added to a 70 wt% solution of the ionic liquid in water. This pulp suspension is then heated to 100 °C in a kneader and kneaded for about 2 h under reduced pressure (starting from 700 mbar and reducing to 5 mbar) in order to remove water and to achieve full cellulose dissolution (dissolution method developed at the TITK, Rudolstadt) [7].

It has been found that a given ionic liquid dissolves either all types of pulps or none at all, regardless of the method of dissolution. The dissolution process seems to be driven mainly by the anion of the ionic liquid. Anions such as halides, carboxylates and phosphates are able to break very effectively intermolecular hydrogen bonds within the cellulose structures as they are not hydrated and are strong hydrogen bond acceptors. The presence of water decreases the solubility though competitive hydrogen bonding processes. Ionic liquids with anions such as ethylsulfate, methanesulfonate, tetrafluoroborate and bis(trifluoromethanesulfonyl)amide showed no ability to dissolve cellulose.

The impact of the cation on the dissolution process is less important, but nevertheless has to be taken into consideration. Cations with cyclic structures such as pyridinium, pyrazolium, the protonated diazabicycloundecane (HDBU) and the most frequently used imidazolium cation gave the best results – leading to the suggestion that cations with a flatter molecular structure may support dissolution. The ability to dissolve cellulose decreases with increasing length of the alkyl chains on the cation. Overall, 1,3-dialkyllimidazolium salts with no alkyl substitution in the 2-position are preferred as they show lower viscosities and allow cellulose concentrations as high as 20 wt% and more.

These results are in line with the findings of other research groups, for example, those of J. Zhang [8], who preferably used 1-allyl-3-methylimidazolium chloride ([Allylmim]Cl) and H. Zhang [9] and Ohno [10], who described 1,3-dialkylimidazolium phosphonates as the preferred solvents – allowing cellulose to be dissolved at low temperatures or even at room temperature.

Stark and co-workers at the University of Jena correlated in ^{1}H NMR spectroscopic experiments the hydrogen bond interaction of ionic liquids with the difference in chemical shifts between the ethanol-**OH** and the ethanol-**CH$_3$** ("OH shift") [11]. These NMR experiments led to the following order of decreasing hydrogen bond interactions: [acetate] > [diethylphosphate] > [chloride] > [methanesulfonate] > [ethylsulfate] > [tetrafluoroborate] > [bis(trifluoromethanesulfonyl)imide], which correlates with the ability to dissolve cellulose described above. On the cation side, the NMR experiments showed a preference for those cations with a more aromatic character. Furthermore, the removal of acid donors in the cations – such as the 2H in the imidazolium ring, also resulted in reduced hydrogen bond interactions, resulting in a lower cellulose dissolution ability.

By using protonated acidic ionic liquids such as 1-H-3-methylimidazolium chloride ([H-mim]Cl) or acid-containing ionic liquids (e.g. [C$_4$mim]Cl with trifluoromethanesulfonic acid), cellulose is also dissolved, but rapidly shows a decrease in the degree of polymerization (DP) caused by the cleavage of the 1,4-glycosidic acetal bonds within the cellulosic polymer structure. Therefore, these types of ionic liquids are not suitable in those cases where the polymeric structure of the cellulose has to be retained for further processing, but they may offer interesting opportunities in fractionation processes of lignocellulosic biomass in bio-refinery concepts [12].

[C$_2$mim][OAc] turned out as the most preferred solvent for cellulose dissolution and processing as it is liquid at room temperature, offers relatively low viscosity (93 mPa s at 25 °C, 10 mPa s at 80 °C) and high dissolving power – even in the presence of up to 10 wt% of water. Pulps with DPs of up to 2000 dissolved – higher DPs led to increased dissolution times and higher viscosities of the solutions obtained. Concentrations of up to 25 wt% cellulose were achieved using [C$_2$mim][OAc].

Furthermore, [C$_2$mim][OAc] is not acutely toxic, shows no corrosion of stainless steel and is highly miscible with water. The only limitation to using [C$_2$mim][OAc] is the limited thermal stability of this ionic liquid. During processing of [C$_2$mim][OAc], temperatures below 150 °C should be applied – otherwise the decomposition of the imidazolium salt will lead to significant material loss.

5.3
Rheological Behavior of Cellulose Solutions in Ionic Liquids

With a view to further processing of the solutions of cellulose obtained in ionic liquids, their rheological behavior was investigated in detail. The knowledge of the rheological properties of cellulose solutions is an absolutely essential prerequisite for any further processing.

The viscosity of cellulose solutions without any shear – the so-called *zero shear viscosity* – shows strong dependences on the cellulose concentration, the DP of the cellulose, the temperature and the ionic liquid chosen. For example, a 10 wt% solution of cotton linters (DP 765) in [C_2mim][OAc] gave a zero shear viscosity of 10 000 Pa s at 25 °C, whereas a 16 wt% solution resulted in a viscosity of 70 000 Pa s at 25 °C, as shown in Figure 5.5. Very important for further processing, both solutions showed a significant decrease in viscosity with increasing temperature, leading to viscosities of 100 and 1500 Pa s at 85 °C, respectively. This opens up a broad range of opportunities to adapt the viscosity of the solutions easily to the requirements for further processing.

Even more important with regard to processing is the rheological behavior of the solutions as a function of the shear rate, as shown in Figures 5.6 and 5.7. These data were determined by oscillating measurements using a "plate-to-plate-viscometer". Figure 5.6 gives the viscosities for 8–16 wt% solutions of cotton linters at 95 °C as a function of the shear rate ("complex viscosity"). All solutions show typical non-Newtonian liquid behavior of decreasing viscosity with increasing shear rate ("shear thinning behavior"). Figure 5.7 gives the full set of rheological data for a 12 wt%

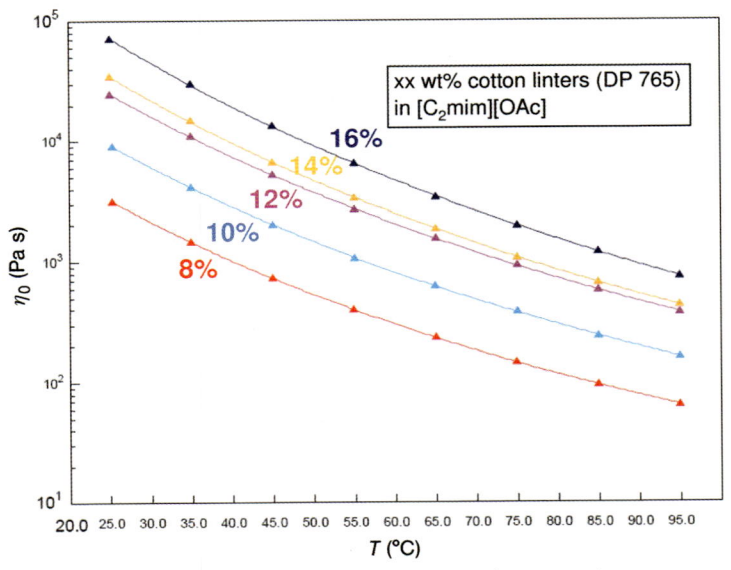

Figure 5.5 Zero shear viscosity versus temperature and concentration.

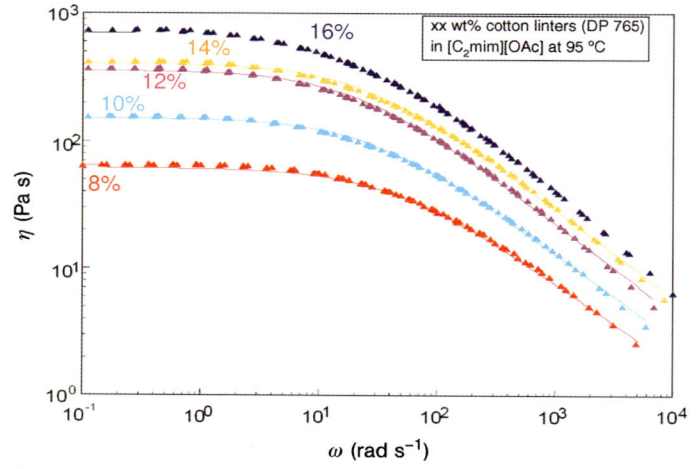

Figure 5.6 Viscosity versus shear rate and concentration.

solution of cotton linters (DP 765) in [C$_2$mim][OAc] at 25 °C. Both the storage modulus G' and the loss modulus G'' show typical viscoelastic behavior, which is representative of polymer solutions or melts to be used in spinning processes. Figure 5.7 also shows that cellulose solutions in [C$_2$mim][OAc] are stable even when stored at elevated temperatures for a longer period. All rheological data as a function of the shear rate remain unchanged after having stored the solution at 80 °C for 1 week. This clearly shows that the solutions are stable with only a very slight decrease in DP.

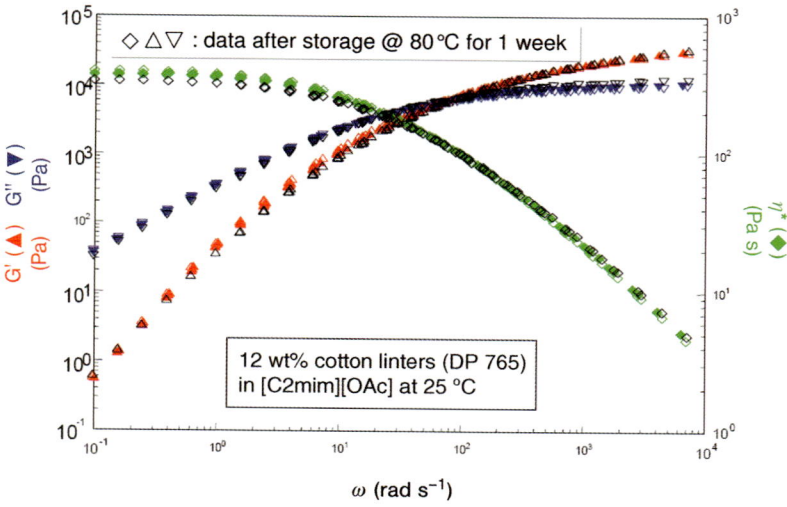

Figure 5.7 Rheological behavior of cellulose solution in [C$_2$mim][OAc].

Furthermore, none of the results on the rheological behavior of cellulose solutions in [C$_2$mim][OAc] gave any hint of micro-gel structures or phase separations. These solutions are expected to be homogeneous and stable and are therefore preferred for any kind of further processing, especially in spinning processes.

Before any further processing, the cellulose solution first needs to be degassed and filtered in order to remove any gases and small particles, which may later interrupt the spinning process by blocking the very small spinning nozzles. Ionic liquids offer significant advantages in this process step also, due to their very low solubility for gases, which leads to a very efficient removal procedures for gases. For the degassing, the spinning dope is heated to 80 °C, kept at a reduced pressure of approximately 50 mbar and subsequently passed through a filter.

5.4
Regeneration of the Cellulose and Recycling of the Ionic Liquid

By adding water or any other solvent miscible with the ionic liquid, such as methanol, ethanol or acetone, the dissolved cellulose is coagulated and regenerated quantitatively, e.g. during spinning. The regenerated cellulose has almost the same DP as the initial pulp, but the morphology changes significantly. The degree of crystallinity can be manipulated by exerting more or less stress on the regenerated material. Without any stress, cellulose is obtained as amorphous polymer – giving the opportunity to manufacture amorphous cellulosic materials.

During washing of the product with water, any remains of ionic liquid can easily be removed due to the very high affinity of this ionic liquid to water. Depending on the size of the particles, the content of ionic liquid in the cellulose material is between 0.1 and 1% – determined by measuring the nitrogen contents by elemental analysis.

After separation of the cellulose from the spin bath, a solution of the ionic liquid in water (or another solvent) is obtained. Both water and solvent can be removed, for example by evaporation under reduced pressure, allowing for the regeneration of the ionic liquid, which can then be reused for the dissolution step. Additional purification steps will be necessary after several regeneration cycles in order to remove impurities that were introduced into the process with the pulp, e.g. inorganic salts. These can be removed by filtration or, if necessary, by ion exchange.

The recently discovered volatility of ionic liquids offers an additional opportunity for further purification of the ionic liquid in the recycling step. At temperatures of 100–300 °C and under reduced pressure, the ionic liquid can be extensively purified [13].

5.5
Cellulosic Fibers

The primary goal in any cellulosic fiber spinning process is to orient and to align the cellulose molecules in the direction of the fiber axis in order to achieve the best

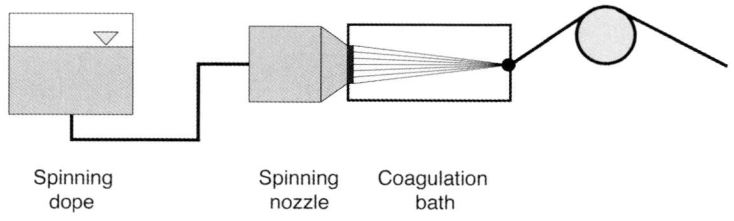

Figure 5.8 Wet/wet spinning process.

mechanical fiber properties. This is either carried out in wet/wet or dry/wet spinning processes. The key to optimizing the properties of the fibers is in separately controlling the rates of coagulation and regeneration and to use the differences in these two processes to maximize the stretching of the fibers thus formed ("draw ratio").

Although the ionic liquid-based spinning process for cellulosic fibers looks schematically very similar to the NMMO process, there is one important difference: for the NMMO process, only a dry/wet spinning process can be used, and wet/wet spinning does not lead to a stable fiber spinning process. Ionic liquids offer more opportunities and therefore the spinning process with ionic liquids could be investigated by using both spinning technologies:

1. Wet/wet spinning (ITCF, Denkendorf) (Figure 5.8)

    The spinning dope (up to 15 wt% cellulose) is spun at temperatures of 60–100 °C through a spinning nozzle directly into the aqueous spinning bath, which is kept at temperatures of 20–80 °C. Immediately after entering the spinning bath, the cellulose coagulates to fibers. By adjustment of the drawing conditions, the degree of crystallization and thereby the tenacities can be controlled.

2. Dry/wet spinning ("air gap spinning"; TITK, Rudolstadt) (Figure 5.9)

    The spinning dope (up to 20 wt% cellulose) is heated to temperatures of 80–120 °C and then spun through a spinning nozzle and through an air gap

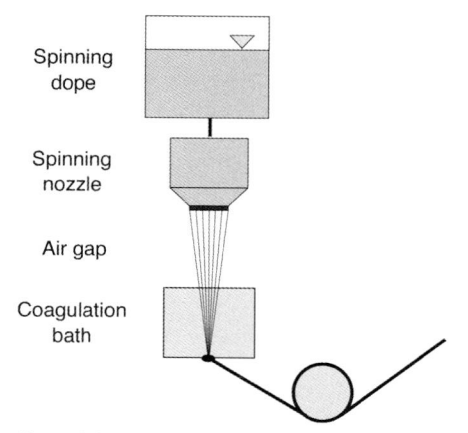

Figure 5.9 Dry/wet spinning process.

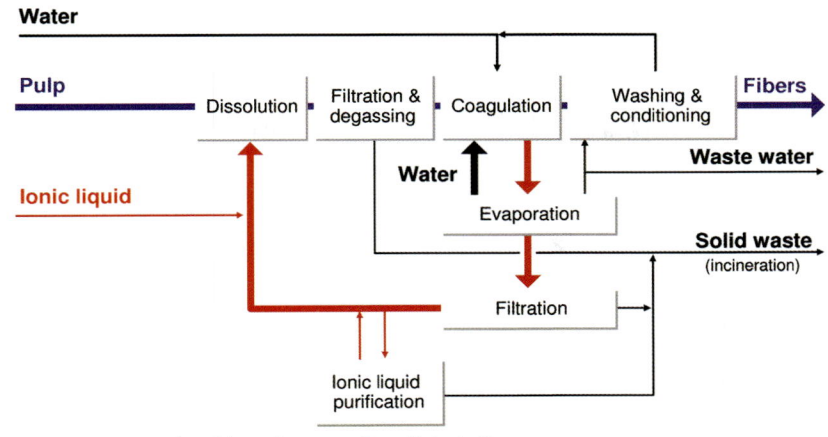

Figure 5.10 Ionic liquid-based process for cellulosic fibers.

(maximum 15 cm) into the aqueous coagulation bath at temperatures of 20–80 °C. Also in this case, the fiber formed immediately is drawn out in order to obtain optimum results regarding the fiber properties.

In both processes, the ionic liquid is subsequently extracted with water as the fibers pass through different washing steps, and is recovered as described above. Figure 5.10 gives an overview of how an ionic liquid-based process for manufacturing cellulosic fibers is set up, including the recycling of the ionic liquid.

Due to the high orientation already obtained after the air gap, the dry/wet spinning process is less flexible regarding variations of the spinning parameters, but leads to fibers with very high tenacities and allows for higher productivities than the wet/wet spinning process. The principle disadvantage of the dry/wet spinning process is a higher tendency of the fibers to fibrillate. Table 5.1 gives an overview of the fiber properties obtained with both processes in comparison with commercially available cellulosic fibers.

Table 5.1 Cellulosic fiber properties obtained by different spinning processes.

	Tenacity (dry) (cN tex^{-1})	Tenacity (wet) (cN tex^{-1})	Elongation (%)	Grade of fibrillation
Cotton	22–36	26–40	7–9	2
Rayon	24–26	12–13	18–23	1
Lyocell	34–36	28–30	15–17	4–5
IL (wet/wet)	15–25	8–14	10–20	1
IL (dry/wet)	35–55	28–42	10–15	4–5

5.6
Cellulose Derivatives

Homogeneous derivatization of cellulose has always been an important objective in polymer research. The advantages are more options in introducing functional groups and better control of the degree of polymerization – resulting in more opportunities to design new products, such as thermoplastic cellulosic materials. Additional opportunities are given by combining cellulosic with synthetic polymers in blends or composites.

Heinze *et al.*, for example, investigated the acylation and carbanilation of cellulose in ionic liquids [14]. The homogeneous reaction path allows for mild reaction conditions, low excesses of reagents, short reaction times and easy control of the degree of substitution (DS). These aspects result in cellulose derivatives with high purities. For example, with cellulose pulp (Avicel with DP 286) in [C_4mim][OAc] at 80 °C with a molar ratio of 3.0 mol of acetyl chloride per mole of anhydroglucose unit (AGU), a cellulose acetate with a DS of 1.87 is obtained after 2 h, whereas under the same reaction conditions with a higher molar ratio of 5.0 mol of acetyl chloride per mole of AGU, the result is a quantitatively functionalized cellulose acetate.

Another opportunity for carrying out the derivatization of cellulose in homogeneous reactions is the high chemical stability of these ionic liquids, which allows the use of highly active reagents such as ketenes or diketenes [15].

5.7
Fractionation of Biomass with Ionic Liquids

Most of the cellulose in Nature is part of lignocellulosic biomass with three major constituents: cellulose, hemicelluloses and lignin. The first step of any utilization of cellulose as raw material or chemical feedstock, therefore, is the separation of cellulose from the other components and especially from lignin, leading to chemical grade pulp or dissolving pulp which is used as feedstock for all artificial cellulosics. Sources for lignocellulosic biomass are both softwood and hardwood species. Among the delignification and purification processes used today, the acid bisulfite and the prehydrolysis Kraft pulping processes are the most important. Alternative pulping processes with lower levels of environmental pollution and energy consumption are under development, but have not yet been commercialized.

Ionic liquids offer new opportunities here, as they are able to dissolve lignocellulosic biomass partially or fully [16]. Futhermore, by subsequently adding solvents and water, lignin and cellulose/hemicelluloses can be obtained as separate fractions. As described above, the cellulose is obtained as more or less amorphous material, which shows much higher reactivity, for example in enzymatic hydrolysis to glucose [17]. After having removed the water and the solvent, the ionic liquid can be recovered and reused.

An alternative concept in utilizing ionic liquids in biomass fractionation might be to dissolve biomass as it is and to carry out a specific degradation of cellulose to the DP required for further processing or even to mono-sugars. After separation from the mixture, cellulose can be further processed by reshaping, derivatization or blending, whereas the mono-sugars could be processed to (cellulosic) bioethanol. In any of these processes, ionic liquids can be expected as valuable tools due to their ability to dissolve cellulose.

5.8 Conclusion and Outlook

Cellulosic polymers stand a good chance of a further increase in production volume in the future – especially if long-term prices of petrochemical feedstock increase further. Ionic liquids might be valuable tools to achieve this as they allow for the combination of the advantages of natural fibers with processing and structural modification leading to new, even more advantageous materials.

References

1 Peters, D. (2006) *Chemie Ingenieur Technik*, **78**, 229.
2 Klemm, D., Phillipp, B., Heinze, T. and Wagenknecht, H. (1998) *Comprehensive Cellulose Chemistry*, Wiley-VCH Verlag GmbH, Weinheim.Bredereck, K. and Hermanutz, F. (2005) *Rev. Prog. Color.*, **35**, 59; Klemm, D., Heublein, B., Fink, H.-P. and Bohn, A. (2005) *Angewandte Chemie*, **117**, 3422.
3 Turunen, T., Fors, J. and Huttunen, J. (1985) *Lenzinger Ber.* **59**, 111; Hermanutz, F. and Oppermann, W. (1998) German Patent 19635707 Gähr, F. and Hermanutz, F. (2002) *Melliand Textilber.* **83**, 149.
4 Rosenau, T., Potthast, A., Sixta, H. and Kosma, P. (2001) *Progress in Polymer Science*, **26**, 1763.
5 Graenacher, C. (1934) Cellulose solution, US Patent 1 943 176.
6 Swatloski, R., Spear, S., Holbrey, J. and Rogers, R. (2002) *Journal of the American Chemical Society*, **124**, 4974; Swatloski, R., Rogers, R. and Holbrey, J. (2003) World Patent WO 03029329.
7 Michels, C., Kosan, B. and Meister, F. (2005) German Patent 102004031025.
8 Xie, H., Li, S. and Zhang, S. (2005) *Green Chemistry*, **7**, 606; Xie, H., Zhang, S., and Li, S. (2006) *Green Chemistry*, **8**, 630.
9 Zhang, H., Wu, J., Zhang, J. and He, J.S. (2005) *Macromolecules* **38**, 8272; Wu, J., Zhang, J., He, J.S., Ren, Q. and Guo, M.L. (2004) *Biomacromolecules* **5**, 266.
10 Fukaya, Y., Heyashi, K., Wada, M. and Ohno, H. (2008) *Green Chemistry*, **10**, 44.
11 Sellin, M., Ondruschka, B. and Stark, A. Hydrogen bond acceptor properties of ionic liquids and their effect on cellulose solubility, in ACS National Meeting, New Orleans, 6–10 April 2008. *ACS Symposium Series* (eds T. Liebert, Th. Heinze and K. Edgar), American Chemical Society, Washington, DC.
12 Massonne, K., D'Andola, G., Stegmann, V., Mormann, W., Wezstein, M. and Leng, W. (2007) World Patent WO 2007101811.
13 Maase, M. (2005) World Patent WO 2005068404; Earle, M., Esperanca, J., Gilea, M., Lopes, J., Rebelo, L. Magee, J. Seddon, K. and Widegren, J. (2006) *Nature* **439**, 831.
14 Heinze, T., Schwikal, K. and Barthel, S. (2005) *Macromolecules Bioscience* **5**, 520.

15 Stegmann, V., Massonne, K., D'Andola, G., Mormann, W., Wezstein, M. and Leng, W. (2007) World Patent WO 2007144282.
16 Diego, A., Fort, R.C., Remsing, R.P., Swatloski, R.P., Moyna, P., Moyna, G. and Rogers, R.D. (2007) *Green Chemistry*, **9**, 63.
17 Balensiefer, T., Brodersen, J., D'Andola, G., Massonne, K., Freyer, S. and Stegmann, V. (2008) World Patent WO 2008090155; Balensiefer, T., Brodersen, J., D'Andola, G., Massonne, K., Freyer, S. and Stegmann, V. (2008) World Patent WO 090156; Edye, L. and Doherty, W. (2008) World Patent WO 2008095252.

Part III
Ionic Liquids in Green Engineering

6
Green Separation Processes with Ionic Liquids

Wytze (G. W.) Meindersma, Ferdy (S. A. F.) Onink, and André B. de Haan

6.1
Introduction

Since ionic liquids have the reputation of being "green", they are of interest in many fields, including separations. A low or negligible vapor pressure, non-flammability and non-toxicity are often equated with greenness. However, some ionic liquids can be distilled, some can be flammable or at least their decomposition products are and some are certainly toxic, as reported elsewhere in this book. Therefore, the application of ionic liquids in environmentally benign separation processes is not straightforward and is assessed in this chapter.

The designer character of ionic liquids is claimed to be an advantage, which is certainly true, but conventional solvents or their mixtures can also be designed and affinity extractants are molecularly tailored. Therefore, the designer character is not unique to ionic liquids and certainly not easier to understand.

However, ionic liquids do possess unique separation properties, enabling novel separation concepts and new products to be developed, and offer the opportunity for novel intensified processes. Ionic liquids can be used in separations because of their stability, non-volatility, adjustable miscibility and polarity. Ionic liquids can be hydrophilic or hydrophobic depending on the structure of their cations and anions. The anion seems more important in determining the water miscibility of ionic liquids [1]. For example, common 1,3-dialkylimidazolium salts with halide, acetate, nitrate and trifluoroacetate anions are completely miscible with water, whereas hexafluorophosphate- and bis(trifluoromethanesulfonyl)amide-based 1,3-dialkylimidazolium salts are water immiscible and tetrafluoroborate- and trifluoromethanesulfonate-based 1,3-dialkylimidazolium salts can be totally miscible or immiscible, depending on the substituents on the cation [2]. Since the ionic liquids with $[PF_6]^-$ and $[Tf_2N]^-$ anions are normally water immiscible, they are the ionic liquids of choice for a large number of investigations involving the formation of biphasic systems required for most ionic liquid extraction applications. However, $[PF_6]^-$-containing

Handbook of Green Chemistry, Volume 6: Ionic Liquids
Edited by Peter Wasserscheid and Annegret Stark
Copyright © 2010 WILEY-VCH Verlag GmbH & Co. KGaA, Weinheim
ISBN: 978-3-527-32592-4

ionic liquids cannot be considered as green: they are water sensitive and will form HF at higher temperatures [3], and ionic liquids with the $[Tf_2N]^-$ anion are usually expensive.

Green separation processes should possess one or more of the following properties, compared with conventional separation processes:

- reduced amount of waste/emissions, direct or indirect, into the environment (atmosphere/water)
- lower use of feedstock
- more efficient production/fewer separation steps: higher selectivity
- higher capacity
- higher energy efficiency
- lower investments in the separation process
- novel separation processes
- broader operational windows
- re-use of the ionic liquids in the process.

The advantages of ionic liquids in separation processes are:

- negligible vapor pressure: no emissions into the atmosphere
- wide temperature range
- tunable physico-chemical properties
- easy regeneration by evaporating or stripping of the compounds separated.

The disadvantages are:

- often high viscosity
- usually higher density
- higher price compared with standard organic solvents
- some ionic liquids toxic in aquatic environments.

In order to replace a conventional separation process by a process that utilizes ionic liquids, this process must be more economic, meaning a lower energy demand, lower use of feedstock, less waste, etc. In general, it is difficult to change existing processes, except when the savings in the investments and running costs are substantial and/or if the novel ionic liquid process can be carried out in existing equipment, with no or small changes. To apply a novel ionic liquid separation process is, in principle, easier than to retrofit, but the technology must be proven on a reasonable scale and during a prolonged period.

The main challenges in applying ionic liquids in separations, and other processes for that matter, are the stability of the ionic liquid in the end, especially at higher temperatures, and regeneration or re-use of the ionic liquid in the process. Moreover, for some candidates that may perform well in the separation task under investigation, the reduction of their toxicity in the aquatic environment remains an important issue. In addition, the physical properties of ionic liquids must be determined in order to predict their properties better in different applications and for modeling of (separation) processes.

6.2
Liquid Separations

6.2.1
Extraction

Liquid–liquid extraction is based on the partial miscibility of liquids and is used to separate a dissolved component from its solvent by its transfer into a second solvent. The equilibrium in extraction may be characterized by the distribution coefficient or partition coefficient, D_i, which is defined by the ratio of the concentrations of solute i in the extract phase (E) and in the raffinate phase (R), according to:

$$D_1 = C_1^E/C_1^R \text{ and } D_2 = C_2^E/C_2^R \tag{6.1}$$

The selectivity, $S_{1/2}$, of species 1 over species 2 is defined as the ratio of the distribution coefficients of species 1 and species 2:

$$S_{1/2} = D_1/D_2 = (C_1^E/C_1^R)/(C_2^E/C_2^R) \tag{6.2}$$

6.2.1.1 Metal Extraction

Conventional Process Conventional processes to recover metal oxides are based on dissolution in mineral acids and bases, followed by extraction with a wide variety of organic solvents or their mixtures. In traditional solvent extraction technologies, adding extractants that reside quantitatively in the extracting phase increases the metal ion partitioning to the more hydrophobic phase. The added extractant molecules dehydrate the metal ions and provide a more hydrophobic environment, allowing their transport to the extracting phase. The use of solvent combinations with specific extractants or chelating agents or ligands improves their combinatorial extractant performance on separation factors of a particular species. Commercial extractants, such as Alamine 336 and Aliquat 336, whose major components are ternary aliphatic amines and quaternary ammonium salts, respectively, have been used in solvent extraction of metals to a large extent [4]. In addition, crown ethers (cyclic polyether molecules) have been intensively investigated for metal extraction in liquid–liquid separations [5].

Extraction with Ionic Liquids Due to the hydrophobic character of some ionic liquids, it is possible to extract hydrophobic compounds in biphasic separations. In order to separate metal ions from the aqueous phase into hydrophobic ionic liquids, again extractants are required to form complexes to increase the hydrophobicity of the metal compounds [6–8]. The extractants used are neutral compounds, such as crown ethers [6, 7, 9–11], calix[4]arenes [11–16] and tri-*n*-butyl phosphate [17], and acidic or anionic extractants, such as organophosphorus acids and pseudohalides. Furthermore, task-specific ionic liquids, where the ionic liquid is both solvent and extractant, have recently been developed [18].

Dai et al. [6] stated that, from a thermodynamic perspective, the solvation of ionic species, such as crown ether complexes, in ionic liquids should be much more favored than in conventional solvent extractions. This is one of the key advantages of using ionic liquids in separations involving ionic species. Dai et al. used the crown ether dicyclohexyl-18-crown-6 (DC18C6) for extraction of Sr^{2+} from an aqueous solution, because DC18C6 is known to form strong complexes with Sr^{2+}. Without the use of an extractant, the D_{Sr} values are below 1, varying from not measurable with toluene or chloroform to 0.89 for the ionic liquid [C$_4$mim][PF$_6$]. The distribution coefficients of Sr^{2+} with 1,3-dialkylimidazolium ionic liquids with [PF$_6$]$^-$ and [Tf$_2$N]$^-$ as anions with the crown ether were in the range from 24 to as high as 1.1×10^4, which is exceptionally high compared with the distribution coefficients in organic solvents with this crown ether, such as chloroform ($D = 0.77$) and toluene ($D = 0.76$).

To achieve optimal extraction efficiency, there are several controlling factors [19]:

- Varying the structure of an ionic liquid (especially the side chain of its cation) to change its hydrophobicity can improve the distribution coefficients of metal ions [7, 20–22].
- The type of extractant (such as crown ethers) can be modified to achieve optimal selectivity for a specific application [23].
- The extraction efficiency of metal complexes can also be controlled by the pH of the system [24].

There are a variety of extraction mechanisms possible in metal extraction with ionic liquids, i.e. solvent ion-pair extraction, ion exchange and combinations of these [25, 26].

Partitioning of sodium ions between aqueous nitrate media and [C$_n$mim][Tf$_2$N] in the presence of DC18C6 takes place via as many as three pathways: conventional nitrato complex extraction and/or one or two ion-exchange processes, the relative importance of which is determined by the acidity of the aqueous phase and the hydrophobicity of the ionic liquid cation. Increasing the alkyl chain length of the ionic liquid cation is not always sufficient to eliminate the possibility of ion exchange as a mode of metal ion partitioning between the two phases, with negative implications for the utility of ionic liquids as environmentally benign extraction solvents [27]. Figure 6.1 shows the extraction of Na^+ from an aqueous phase.

The extraction of radioactive metals (lanthanides and actinides) has particular industrial significance in metal extraction for the handling of nuclear materials [19].

Octyl(phenyl)-N,N-diisobutylcarbamoylmethylphosphine oxide (CMPO) dissolved in [C$_4$mim][PF$_6$] showed extremely high extraction ability and selectivity of metal ions as compared with the ordinary diluent n-dodecane. Partitioning of the metal cations into the ionic liquid with CMPO appears to proceed via a cation-exchange mechanism, which is different from that of the conventional solvent extraction system [28]. The metal was successfully stripped from the ionic liquid.

Selective extraction–separation of yttrium(III) from heavy lanthanides into [C$_8$mim][PF$_6$] containing Cyanex 923 was achieved by adding a water-soluble complexing agent [ethylenediaminetetraacetic acid (EDTA)] to the aqueous phase. The

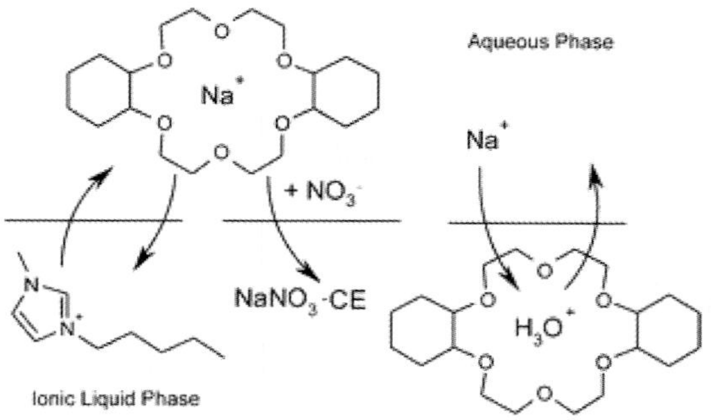

Figure 6.1 Extraction of Na$^+$ with DC18C6 [27]. CE = crown ether.

selectivity of the [C$_n$mim][PF$_6$]/[Tf$_2$N]-based extraction system was enhanced to about 4.3 without increasing the loss of [C$_n$mim]$^+$ [29].

Dietz et al. have shown that the mode of metal ion extraction by crown ethers can be shifted from cation exchange to extraction of a neutral ion-pair complex by increasing the lipophilicity of the ionic liquid cation [22, 23, 30, 31]. Formation of such neutral metal–crown ether–anion complexes is the conventional transport mechanism with crown ether extractants in organic solvents. In order to minimize the loss of ionic liquid, the ionic liquid cation can be added to the initial metal-containing solutions before extraction. Moreover, solutions with the ionic liquid anion are used for stripping. Several recycles were carried out for the extraction of Ce(IV) [32].

The mutual solubility of ionic liquids and water (and/or aqueous acid solutions) determines the suitability of ionic liquids as replacements for conventional molecular solvents in applications for the removal of metal ions. The solubility of water in ionic liquids can be increased by the presence of a neutral ligand, such as a crown ether, capable of extracting significant quantities of acid. The primary problem with cation- and anion-exchange equilibria in ionic liquid-based systems is that ionic liquid components are transferred to the aqueous phase. This process pollutes the water phase and this can make it difficult to subsequently recover the extracted metal ion [33]. Both cations and anions of the ionic liquids can be exchanged into the aqueous phase [34]. An extraction system in which metal ion partitioning is accompanied by increased dissolution of the ionic liquid in the aqueous phase cannot be regarded as "green" [35]. The process involved in the transfer of the metal ion into an ionic liquid in the presence of a crown ether is surprisingly complex. In principle, such complexity can provide a large number of opportunities for the design of ionic liquid-based extraction systems with improved efficiency and selectivity. However, the full potential of ionic liquids as environmentally benign extraction solvents will clearly not be realized until methods are devised to reduce or eliminate the aqueous dissolution of the ionic liquid that can accompany metal ion partitioning [27]. These are currently major limitations for using ionic liquids in metal ion separation.

Extraction with Task-specific Ionic Liquids (TSILs) Task-specific ionic liquids are designed to have targeted functionality by attaching a metal ion-coordinating group directly on to the ionic liquid cation, which makes the extractant an integral part of the hydrophobic phase and thereby reduces the chance of ionic liquid loss to the aqueous phase. Therefore, TSILs act as hydrophobic solvents and extractants at the same time. However, the cost of TSILs is generally high, as their synthesis usually requires multi-step organic syntheses. In order to eliminate this drawback, TSILs may be added to mixtures of less expensive ionic liquids.

Visser et al. [36] designed and synthesized several ionic liquids by appending different functional groups (namely thioether, urea and thiourea) to imidazolium cations to remove Hg^{2+} and Cd^{2+} from contaminated water. The synthesized TSIL cations were combined with the $[PF_6]^-$ anion and used alone or in a mixture with $[C_4mim][PF_6]$. The distribution coefficients for Hg^{2+} and Cd^{2+} in liquid–liquid separations with the thioether TSIL were high: 220 for Hg^{2+} and 330–375 for Cd^{2+}, depending on the pH of the aqueous phase. With the thiourea-appended TSIL, distribution coefficients of 345 for Hg^{2+} and 20 for Cd^{2+} were obtained. The separation of the metal ions from the TSILs was not reported.

Ammonium $[Tf_2N]$ salts with carboxylic acid functionalities show the ability to dissolve large quantities of metal oxides. This metal-solubilizing power is selective. The oxides of the trivalent rare earths, and also uranium(VI) oxide, zinc(II) oxide, cadmium(II) oxide, mercury(II) oxide, nickel(II) oxide, copper(II) oxide, palladium(II) oxide, lead(II) oxide, manganese(II) oxide and silver(I) oxide, are soluble in the ionic liquid. Iron(III), manganese(IV) and cobalt oxides, and also aluminum oxide and silicon dioxide, are insoluble or very poorly soluble. The metals can be stripped from the ionic liquid by treatment of the ionic liquid with an acidic aqueous solution. After transfer of the metal ions to the aqueous phase, the ionic liquid can be recycled for re-use [37].

Chelate extraction can be used for separating several metal ions from aqueous solutions [38–40]. 1,3-Dialkylimidazolium salts with appended ethylaminodiacetic acid moieties (employed as the di-tert-butyl ester) can be used for the formation of metal chelates with Cu(II), Ni(II) and Co(II) in aqueous solutions [40]. Figure 6.2 shows that an increase in alkyl chain length improves the hydrophobicity. Complex **d** is insoluble in water and precipitates after the chelate formation is completed. The efficiency of the extraction is 93%. However, the recovery of the metal from the complex and the regeneration of the complex have not been reported.

The separation of the noble organometallic Wilkinson's and Jacobsen's catalysts from a homogeneous organic phase is possible by extraction with functionalized ionic liquids [41], whereas with non-functionalized $[C_2mim][BF_4]$, no extraction of the catalysts was observed. The best functionalized ionic liquids were those with an amino acid-based anion (glycine or methionate): $[C_4mim][Gly]$ ($D_W = 1.7$, $D_J = 71$), $[C_6mim][Gly]$ ($D_W = 8.8$, $D_J = 889$) and [allyl-mim][Met] ($D_W = 76$, $D_J = 66.4$). The recovery of the catalysts from the ionic liquid solution is still under investigation.

Conclusions The removal of metal ions from aqueous solutions with ionic liquids is not straightforward. Ion exchange is not a suitable process, as either the ionic liquid cation or the ionic liquid anion will end up in the aqueous phase. The aqueous dissolution of the ionic liquid that can accompany metal ion partitioning must be

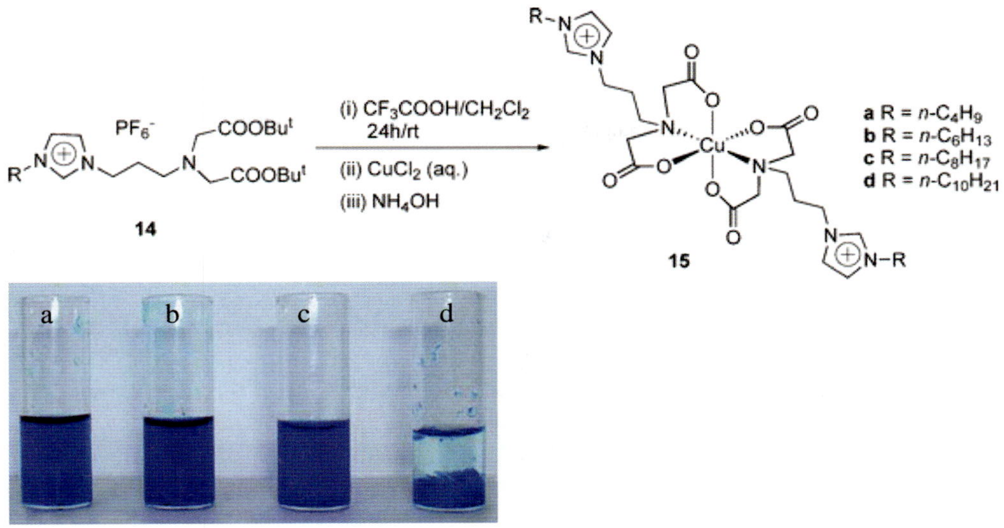

Figure 6.2 Formation of Cu^{2+} complexes [40].

avoided or largely eliminated. Therefore, further improvements of the ionic liquid process for the separation of metals are required. One option is to design task-specific ionic liquids. Additionally, many authors did not report the recovery of the metals from the ionic liquid solution and/or the regeneration of the ionic liquids. However, these steps are essential parts of the whole separation process.

6.2.1.2 Extraction of Aromatic Hydrocarbons

Conventional Process for Aromatic–Aliphatic Separation The separation of aromatic hydrocarbons (benzene, toluene, ethylbenzene and xylenes) from C_4–C_{10} aliphatic hydrocarbon mixtures is challenging, since these hydrocarbons have boiling points in a close range and several combinations form azeotropes. The conventional processes for the separation of these aromatic and aliphatic hydrocarbon mixtures are liquid extraction, suitable for the range 20–65 wt% aromatic content, extractive distillation for the range 65–90 wt% aromatics and azeotropic distillation for high aromatic content, >90 wt% [42]. Typical solvents used are polar components such as sulfolane [43–48], N-methylpyrrolidone (NMP) [46], N-formylmorpholine (NFM), ethylene glycol [47, 49, 50] and propylene carbonate [51]. This implies additional distillation steps to separate the extraction solvent from both the extract and raffinate phases and to purify the solvent, which leads consequently to additional investments and higher energy consumption. The costs of the regeneration of sulfolane are high, since sulfolane, which has a boiling point of 287.3 °C, is in the current aromatic–aliphatic separation process taken overhead from the regenerator and returned to the bottom of the aromatics stripper as a vapor [52]. Overviews of the use of extraction and extractive distillation for the separation of aromatic hydrocarbons from aliphatic hydrocarbons can be found elsewhere [53–56].

Extraction with Ionic Liquids The application of ionic liquids for extraction processes is promising because of their negligibly low vapor pressure [57]. This facilitates solvent recovery using techniques as simple as flash distillation or stripping. Thus, extraction of aromatics from mixed aromatic–aliphatic streams with ionic liquids is expected to require fewer process steps and a lower energy consumption than extraction with conventional solvents. The solvent sulfolane is used as a benchmark for the separation of aromatic and aliphatic hydrocarbons, because it is one of the most widely used solvents for this separation in industry.

In Table 6.1, aromatic distribution coefficients and aromatic/aliphatic selectivities for toluene–n-heptane and some other aromatic–aliphatic systems, determined by either extraction, solubility and/or by activity coefficients at infinite dilution, are shown. Only those references that report an aromatics distribution coefficient and/or an aromatic/aliphatic selectivity about equal to or larger than that of sulfolane are included in this table. If the distribution coefficient or the selectivity of the ionic liquids is lower than those for sulfolane, replacement of this solvent is not considered feasible.

From the data in Table 6.1, it is apparent that the values of the distribution coefficients and selectivities measured with activity coefficients at infinite dilution or extraction are not always consistent with each other. Only a few ionic liquids show both a higher aromatic distribution coefficient and a higher aromatic/aliphatic selectivity than sulfolane: [C_4mim][BF_4], [C_4mim][SCN], [C_4mim][DCA], 4-methyl-1-butylpyridinium tetrafluoroborate ([1-C_4,4-C_1py][BF_4]), [1-C_4,4-C_1py][CH_3SO_4], [1-C_4,3-C_1py][DCA], [C_1mim][Tf_2N], [C_4mim]Cl–2.0$AlCl_3$, [Me_3NH]Cl–$AlCl_3$ (1 : 2), [C_4mim]Cl–$AlCl_3$ (1 : 1), [C_2mim]Cl–$AlCl_3$ (1 : 1), [C_2mim][I_3] and [C_4mim][I_3]. However, the last two ionic liquids are not suitable due to extreme corrosion [58].

Figure 6.3 shows the ionic liquids suitable for aromatic–aliphatic hydrocarbon separation. The aromatic/aliphatic selectivity is shown as a function of the distribution coefficient of the aromatic compound.

Although the $AlCl_3$-containing ionic liquids show the highest aromatic distribution coefficients and aromatic/aliphatic selectivities, these ionic liquids are not suitable due to their instability in the presence of water.

Economic Evaluation An economic evaluation was made for the separation of aromatic compounds from the feed of a naphtha cracker with several ionic liquids ([C_2mim][$C_2H_5SO_4$], [1-C_4,4-C_1py][BF_4], [1-C_4,4-C_1py][CH_3SO_4], [C_4mim]Cl–1.0$AlCl_3$, [C_2mim]Cl–1.0$AlCl_3$ and later [1-C_4,3-C_1py][DCA],) and compared with that with sulfolane [75]. The separation of toluene from a mixed toluene–n-heptane stream was used to model the aromatic–aliphatic separation. Most ethylene cracker feeds contain 10–25% of aromatic components, depending on the source of the feed (naphtha or gas condensate). The aromatic compounds are not converted to olefins and even small amounts are formed during the cracking process in the cracker furnaces [76]. Therefore, they occupy a part of the capacity of the furnaces and they put an extra load on the separation section of the stream containing C_5–C_{10} aliphatic compounds. If a major part of the aromatic compounds present in the feed to the crackers could be separated upstream of the furnaces, it would offer several

6.2 Liquid Separations

Table 6.1 Overview of distribution coefficients and selectivities for aromatic–aliphatic separations.

Solvent	Separation	T (°C)	Aromatics (mol%)	D_{arom} (mol/mol)	$S_{arom/alk}$	Remarks	Ref.
Sulfolane	Benzene–n-hexane	25	7.3	0.80	48.3	Extraction	[43]
		25	7.5	0.84	46.4		[48]
		50	7.9	0.70	29.6		[48]
	Toluene–n-heptane	40	5.9	0.31	30.9	Extraction	[59]
		40	11.5	0.30	26.5		[59]
	Xylene–n-hexane	25	30	0.29	12.2	Extraction	[43]
	Xylene–n-octane	35	5.4	0.24	33.7	Extraction	[60]
$[C_2mim][BF_4]$	Toluene–n-heptane	40	7.8	0.34	46.1	Extraction	[59]
$[N_{1,1,4,2-hydroxyethyl}][BF_4]$	p-Xylene–n-hexane	25	8	0.36	28.4	Extraction	[61]
$[1-C_4,4-C_1py][BF_4]$	Benzene–n-hexane	40	10.6	0.95	55.9	Extraction	[59, 62]
	Benzene–cyclohexane	30	8.4^a	0.95^a	28.2^a	Extraction	[63]
		40	–	0.61	17.7	Act. coeff.	[64]
	Toluene–n-heptane	40	–	0.38	32.8	Act. coeff.	[64]
		40	6.45	0.45	52.9	Extraction	[59, 65]
	Ethylbenzene–n-octane	40	10	0.42	58.3	Extraction	[59, 62]
	m-Xylene–n-octane	40	8.1	0.36	42.6	Extraction	[59, 62]
$[C_1mim][Tf_2N]$	Toluene–n-heptane	40	–	0.49	29.8	Act. coeff.	[66]
$[C_2mim][Tf_2N]$	Toluene–n-heptane	25	7.0	1.37	33.6	Extraction	[67]
	Benzene–n-hexane	40	12.3	1.20	27.8	Extraction	[67]
$[C_2dmim][Tf_2N]$	Toluene–n-heptane	40	–	0.61	22.7	Act. coeff.	[68]
$[N_{1,1,2,2-hydroxyethyl}][Tf_2N]$	Benzene–n-hexane	25	41	0.9	25.6	Extraction	[69]
	p-Xylene–n-hexane	25	9	0.6	29	Extraction	[69]
	m-Xylene–n-hexane	25	20	0.2	10	Extraction	[69]
	m-Xylene–n-octane	25	10	0.3	27	Extraction	[69]
$[C_6mim][PF_6]$	Benzene–n-heptane	25	4.1	0.70	8.2^a	Extraction	[70]
		25	–	0.97	29.7	Act. coeff.	[66]

(Continued)

Table 6.1 (Continued)

Solvent	Separation	T (°C)	Aromatics (mol%)	D_{arom} (mol/mol)	$S_{arom/alk}$	Remarks	Ref.
[C$_1$mim][CH$_3$SO$_4$]	Toluene–n-heptane	40	–	0.06	16.4	Act. coeff.	[71]
		40	10.75	0.08	62.5	Extraction	[59, 65]
[C$_2$mim][C$_2$H$_5$SO$_4$]	Toluene–n-heptane	40	–	0.19	36.4	Act. coeff.	[66]
		40	10.7	0.22	50.4	Extraction	[59, 65]
[C$_4$mim][CH$_3$SO$_4$]	Toluene–n-heptane	40	8.6	0.33	31.2	Extraction	[59]
[1-C$_4$,4-C$_1$py][CH$_3$SO$_4$]	Toluene–n-heptane	40	8.1	0.61	42.3	Extraction	[59]
[C$_4$mim][DCA]	Toluene–n-heptane	30	6.5	0.63	59.0	Extraction	
[C$_4$mim][SCN]	Toluene–n-heptane	30	7.4	0.50	65.8	Extraction	
[1-C$_4$,3-C$_1$ py][DCA]	Toluene–n-heptane	30	6.4	0.86	44.7	Extraction	
[C$_4$mim]Cl–2.0AlCl$_3$	Benzene–n-heptane	20	10	1.95	80	Extraction	[72]
[Me$_3$NH]Cl–2.0AlCl$_3$	Benzene–n-heptane	20	10	1.55	35	Extraction	[72]
[Et$_3$NH]Cl–2.0AlCl$_3$	Benzene–n-heptane	20	10	1.6	35	Extraction	[72]
[C$_2$mim]Cl–1.0AlCl$_3$	Toluene–n-heptane	40	7.6	1.49	45.9	Extraction	[73]
[C$_4$mim]Cl–1.0AlCl$_3$	Toluene–n-heptane	40	6.7	1.57	35.7	Extraction	[73]
[C$_2$mim][I$_3$]	Toluene–n-heptane	45	7.5	0.84	48.6	Corrosive	[74]
[C$_4$mim][I$_3$]	Toluene–n-heptane	35	17.0	2.3	30.1	Corrosive	[74]

^{a}Calculated from the data given.

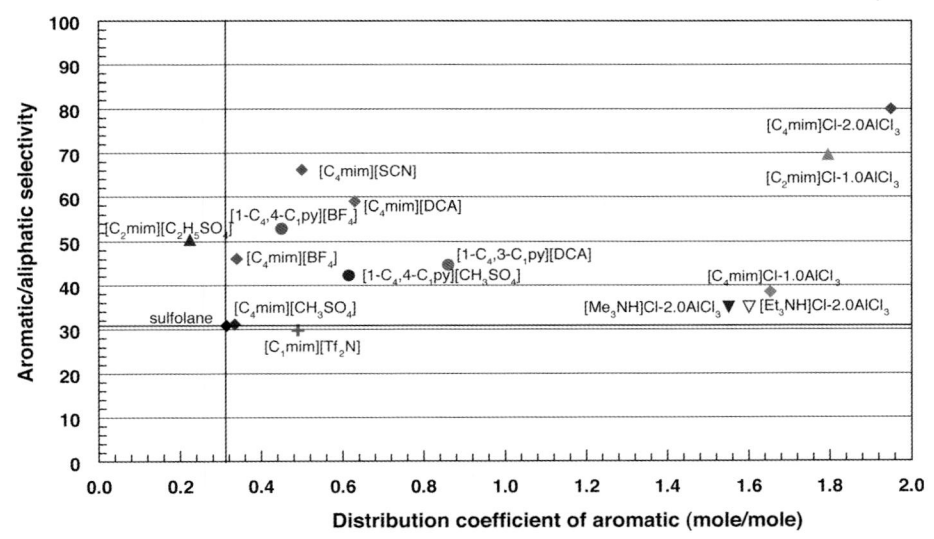

Figure 6.3 Aromatic–aliphatic separation with ionic liquids, ~10% aromatic concentration (sulfolane data for toluene–n-heptane); $T \approx 40\,°C$.

advantages: higher capacity, higher thermal efficiency and less fouling. The improved margin will be around €20 per ton of feed or €48 million per year for a naphtha cracker with a feed capacity of $300\,t\,h^{-1}$, due to lower operational costs. Figure 6.4 shows a process scheme for the extraction of aromatic hydrocarbons from naphtha feed.

For a naphtha feed of $300\,t\,h^{-1}$ containing about 10% aromatic hydrocarbons, the total investment costs in the sulfolane extraction were estimated to be about €86 million and with [1-C$_4$,4-C$_1$py][BF$_4$] about €56 million, including an ionic liquid inventory of €20 million. In the calculations, an ionic liquid price of €20 kg^{-1} was used, since BASF has indicated that it is indeed possible to reach a level of €10–25 kg^{-1} ionic liquid with production on a large scale [77–79]. After this calculation had been carried out, a more suitable ionic liquid was found by using COSMO-RS: [1-C$_4$,3-C$_1$py][DCA] with $D_{tol} = 0.86$ and $S_{tol/hept} = 44.7$.

The lower investment in the ionic liquid process results mainly from the fact that the regeneration of the ionic liquid is much simpler than that of sulfolane. Since also the energy costs are lower, the total annual costs with the [1-C$_4$,4-C$_1$py][BF$_4$] process are estimated to be €27.4 million, compared with €58.4 million for sulfolane. The investment and annual costs for the separation of 10% aromatics from a cracker feed with sulfolane and ionic liquids are shown in Figure 6.5. The loss of ionic liquid to the raffinate phase is minimal, estimated to be 0.006%. This seems to be a very small amount, but for a cracker with a capacity of $300\,t\,h^{-1}$, this amounts to $140\,t$ of ionic liquid per year. With the cost price of €20 kg^{-1}, this amounts to a loss of €2.8 million per year. Furthermore, it is unknown what will happen in the cracker with the ionic liquid and where the ionic liquid or its decomposition products will end up. However,

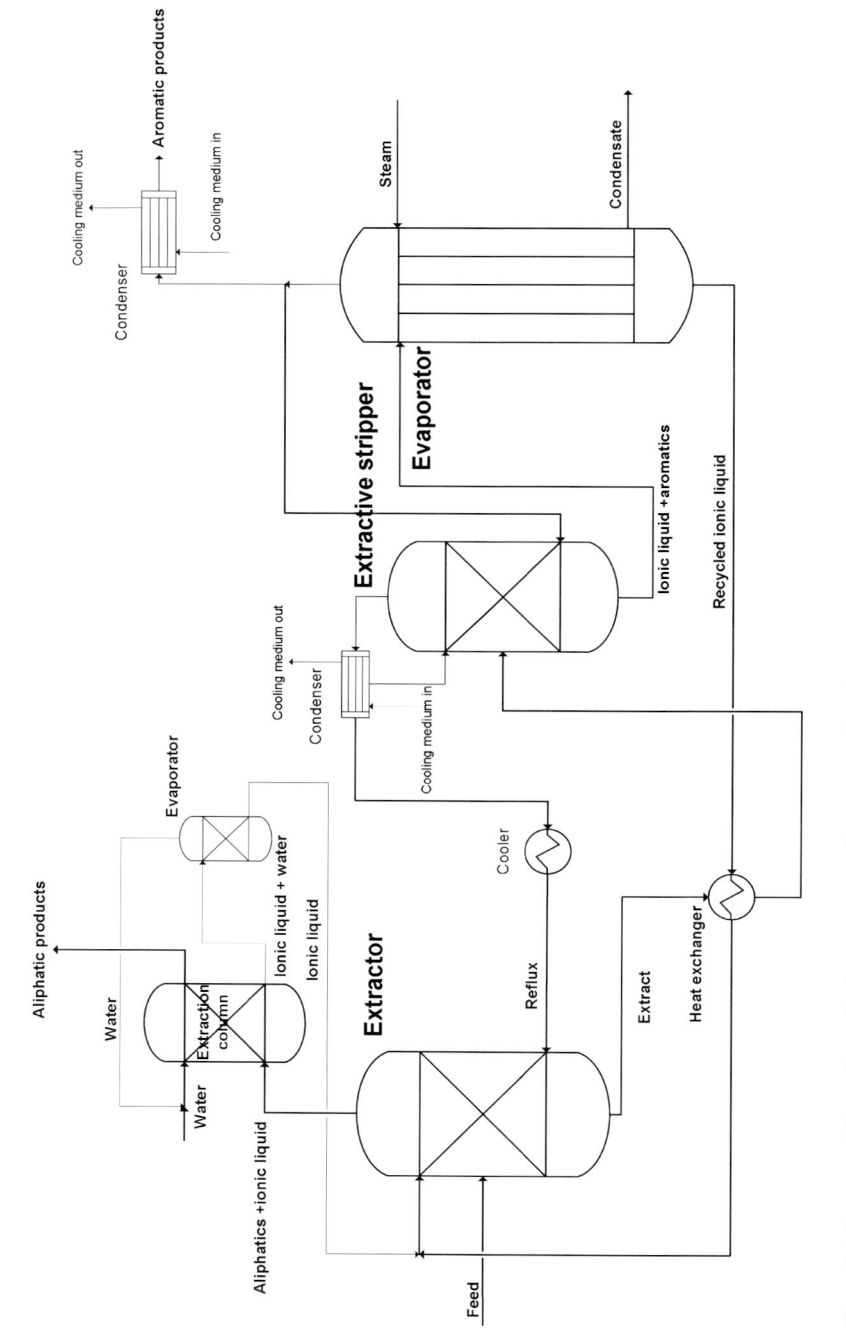

Figure 6.4 Conceptual flow scheme for the separation of aromatic and aliphatic hydrocarbons.

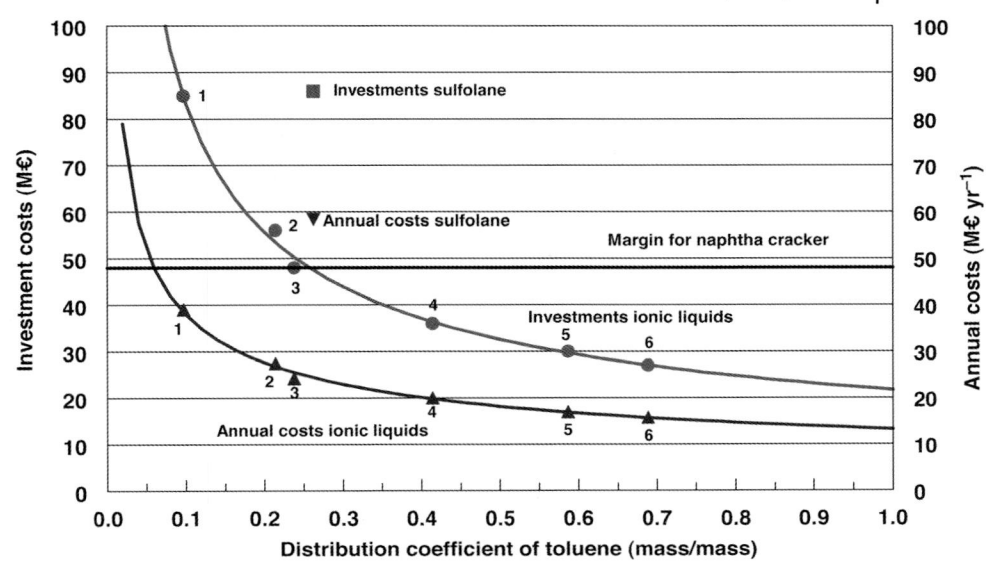

Figure 6.5 Investment and variable costs for extraction with ionic liquids.
1, [C_2mim][$C_2H_5SO_4$]; 2, [1-C_4,4-C_1py][BF_4]; 3, [1-C_4,4-C_1py][CH_3SO_4];
4, [1-C_4,3-C_1py][DCA]; 5, [C_4mim]Cl–1.0AlCl$_3$; 6, [C_2mim]Cl–1.0AlCl$_3$ [75].

the ionic liquid can be recovered from the raffinate with a simple one-stage extraction with water.

Conclusions Ionic liquids can replace conventional solvents in liquid–liquid extraction of aromatic hydrocarbons, provided that the aromatic distribution coefficient and/or the aromatic/aliphatic selectivity is higher than those with sulfolane. The main conclusion of the process evaluation is that ionic liquids showing a high aromatic distribution coefficient with a reasonable aromatic/aliphatic selectivity could reduce the investment costs of the aromatic–aliphatic separation by a factor of 2.

6.2.1.3 Proteins

Conventional Process Efficient bioseparation is very important in modern biotechnology, since the purification costs are between 20 and 60% of the total production costs and, in special cases, this can even amount to 90%. Industrially relevant separation techniques include membrane separation, adsorption and extraction [80–83]. Aqueous two-phase system (ATPS) extraction can result in a high extractability and good retention of the bioactivity [84, 85]. However, high viscosities of one or two phases complicate the scale-up of the process [86].

Extraction with Ionic Liquids Extraction of myoglobin, human serum albumin and immunoglobulin G (IgG) with the ionic liquid [$P_{14,6,6,6}$][Tf$_2$N] showed almost 100%

retention. Also, $[P_{14,6,6,6}][DCA]$ is a good solvent for myoglobin [87]. IgG is, however, not stable enough to be used in extraction with the ionic liquids investigated. The extraction of bovine serum albumin, trypsin, cytochrome c and γ-globulins in ionic liquid-based aqueous two-phase systems ($[C_4mim]Br$, $[C_6mim]Br$ or $[C_8mim]Br$ and K_2HPO_4) was effective. About 75–100% of the proteins were extracted into the ionic liquid phase in a single-step extraction [88]. Extraction efficiencies of the proteins were found to increase with increasing temperature and increasing alkyl chain length of the cation of the ionic liquids. Compared with the traditional PEG ATPS, the ionic liquid-based ATPS have the advantages of lower viscosity, little emulsion formation, rapid phase separation and others [88].

Penicillin G can be extracted from its fermentation broth with an ionic liquid ATPS consisting of the hydrophilic ionic liquid $[C_4mim][BF_4]$ and NaH_2PO_4, while leaving miscellaneous proteins in the ionic liquid-poor phase (top phase in Figure 6.6, tube A) [89]. The hydrophilic $[C_4mim][BF_4]$ can be recovered by transfer into an hydrophobic ionic liquid-rich phase of $[C_4mim][PF_6]$ (greenish bottom phase in tube C), leaving most of the penicillin in the conjugated water phase (top phase in tube D). In comparison with the butyl acetate–water system or the polymer-aqueous two-phase system, the integrated ionic liquid system shows several advantages: penicillin is efficiently extracted into the ionic liquid-rich phase at neutral pH and hydrophobic ionic liquids can separate hydrophilic ionic liquids from the penicillin-containing aqueous phase. Protein emulsification occurring in the organic solvent system is avoided and the difficulty in recovering the phase-forming material in the polymer–aqueous two-phase system is overcome [89].

An ionic liquid–aqueous two-phase system based on the hydrophilic ionic liquid $[C_4mim]Cl$ and K_2HPO_4 has been employed for direct extraction of proteins from human body fluids. Proteins present at low concentration levels were quantitatively extracted into the $[C_4mim]Cl$-rich upper phase with a distribution ratio of about 10 between the upper and lower phases and an enrichment factor of 5. Addition of an appropriate amount of K_2HPO_4 to the separated upper phase results in a further phase separation, giving rise to an improved enrichment factor

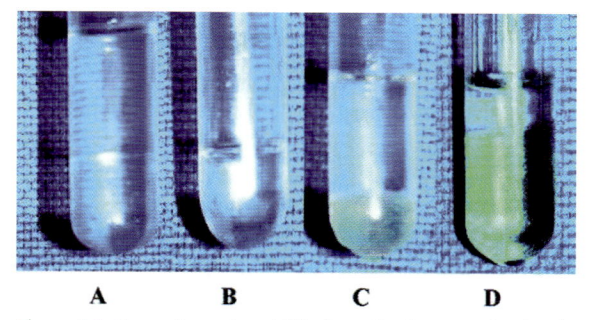

Figure 6.6 Separation of penicillin from the fermentation broth containing miscellaneous protein [89].

of 20. The use of an ionic liquid as a green solvent offers clear advantages over traditional liquid–liquid extractions, in which the use of toxic organic solvents is unavoidable [90].

The commercial ionic liquid Ammoeng 110 (Solvent Innovation) contains cations with oligoethylene glycol units in combination with an inorganic salt mixture (K_2HPO_4–KH_2PO_4) and was found to be highly effective for the formation of ATPS that can be used for the biocompatible purification of active enzymes, such as alcohol dehydrogenases (ADH), from *Lactobacillus brevis* and from a thermophilic bacterium [91]. Both enzymes were enriched in the ionic liquid phase, resulting in an increase in specific activity by a factor of 2 and 4, respectively. Furthermore, the presence of ionic liquid within the system provided the opportunity to combine the extraction process with the performance of enzyme-catalyzed reactions [91].

The transfer of lysozyme in the liquid–liquid extraction from an aqueous solution to a solution consisting of $[C_4mim]_3[CB]$ (CB = an affinity-dye, Cibacron Blue 3GA) in $[C_4mim][PF_6]$ decreased as the pH of the aqueous phase increased. An extraction level higher than 90% was observed at pH 4. At a high ionic strength, the affinity of lysozyme is higher for the aqueous phase. Lysozyme molecules were thus almost quantitatively recovered from the ionic liquid phase by contacting with aqueous solutions of 1 M KCl at pH 9–11. The resultant recoveries of lysozyme using back-extraction at pH 11 after each complete forward and backward cycle for all eight extractions ranged between 87 and 93%. Similarly, the recoveries of lysozyme at pH 8 stayed above 85% for all eight cycles. The extraction was specific for lysozyme, in contrast to cytochrome *c*, ovalbumin and bovine serum albumin [92].

Conclusions Ionic liquids can be successfully employed to extract proteins from aqueous solutions with an ATP system. The viscosity of the ionic liquid-based system is generally lower than that of traditional PEG systems. Furthermore, there is little emulsion formation, a rapid phase separation and the ionic liquids can be recycled.

6.2.2
Extractive Distillation

6.2.2.1 Conventional Process
Conventional distillation is the most commonly applied separation process in the chemical industry. However, mixtures with close-boiling compounds and mixtures with a low relative volatility are difficult, impossible or economically unattractive to separate by ordinary distillation. One of the most useful ways to separate chemicals that cannot be distilled is to employ selective solvents. These solvents take advantage of the non-ideality of a mixture of components having different chemical structures. Liquid–liquid extraction, azeotropic distillation and extractive distillation are examples of the use of solvents in a separation process. Volatile entrainers are used in azeotropic distillation and non-volatile solvents in extractive

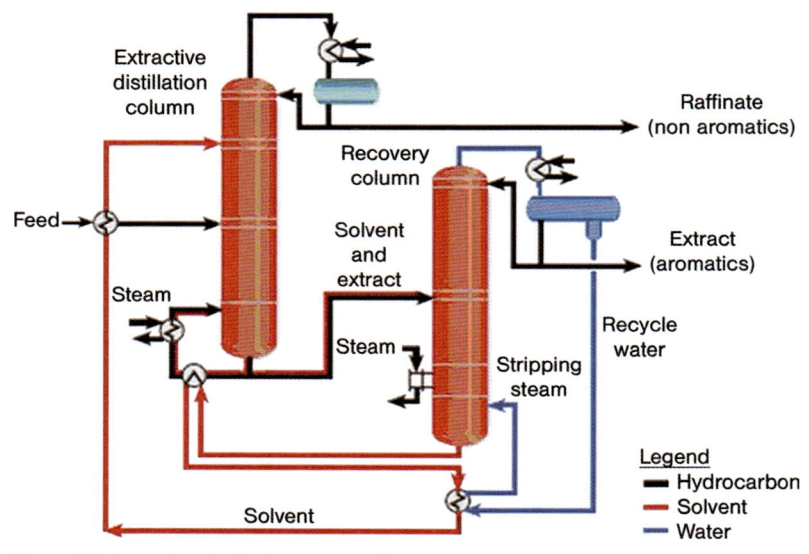

Figure 6.7 Extractive distillation process [94].

distillation. Of course, the added selective solvent has to be isolated from process mixtures in order to allow its re-use. This implicates additional investments and energy costs. In extractive distillation, selective solvents, such as sulfolane, N-methylpyrrolidone (NMP), dimethylformamide (DMF) or acetonitrile, are added at the top of the distillation column to increase the relative volatility [93]. Therefore, extractive distillation is, in principle, more attractive than azeotropic distillation, because the solvent is not evaporated and the energy requirements are much lower.

Compounds with no or little interaction with the solvent leave the extraction column via the top and the compounds with a strong interaction with the solvent leave the column via the bottom. In the next step, the solvent is separated from the compounds and is recycled to the top of the extraction column. Figure 6.7 shows an extractive distillation process for the separation and purification of benzene and toluene.

The solvent-to-feed ratio (S/F) in extractive distillation is usually 5–8, which implicates high additional costs due to larger equipment and a high energy requirement. Sometimes, a salt is added to increase the relative volatility of the compounds through the salting-out effect [93, 95–101]. The advantages of adding a salt are that a lower solvent-to-feed ratio is required. Additionally, the salt possesses no vapor pressure and is, therefore, not present in the vapor phase, unless it is entrained. Furthermore, the salt can easily be separated from the bottom product. Disadvantages are, however, that salts are often corrosive, that precipitation of the salt can occur, that the salt can cause decomposition of components at higher temperatures and that salt can only be applied in aqueous systems, such as in the separation of alcohols and water.

6.2.2.2 Ionic Liquids in Extractive Distillation

Ionic liquids as solvents combine the advantages of both organic solvents and salts: increasing the relative volatility of one of the components and reducing the solvent-to-feed ratio by the salting-out effect without the disadvantages of a solid salt [93, 102–113]. Ionic liquids are suitable as entrainers for a whole range of azeotropic systems [103, 113, 114].

The selected solvent for an extractive distillation process should have the following characteristics:

- High boiling point to avoid solvent losses.
- High selectivity and solvency to obtain high-purity products and to reduce the quantity of solvent used. The high solvency is reflected in the formation of a second liquid phase, which improves the operational behavior of the extractive distillation column.
- Easy recovery to increase the efficiency in the solvent recovery process and to reduce the recovery costs.

Figure 6.8 shows a process scheme with ionic liquids as solvents.

The best option for the regeneration column will probably be a multi-effect evaporator, which requires a low amount of energy, possibly followed by a strip column for the removal of the last traces of products.

Olefin–Paraffin Separation Since the boiling points of olefins and paraffins lie within a close range, separation by conventional distillation is difficult and expensive. The separation of 1-hexene and n-hexane is taken as an example for the separation of olefins and paraffins. The highest selectivity of n-hexane to 1-hexene with a solvent concentration of 0.3 in the liquid phase is with the conventional solvent NMP ($S = 1.1613$ at 313.15 K and $S = 1.632$ at 333.15 K), measured by headspace gas chromatography (GC). From a series of ionic liquids tested, the best ionic liquid at 313.15 K was [C$_2$mim][Tf$_2$N] ($S = 1.0576$) and at 333.15 K [C$_8$quin][Tf$_2$N] ($S = 1.0390$) [115]. A suitable solvent should possess both a high selectivity and a

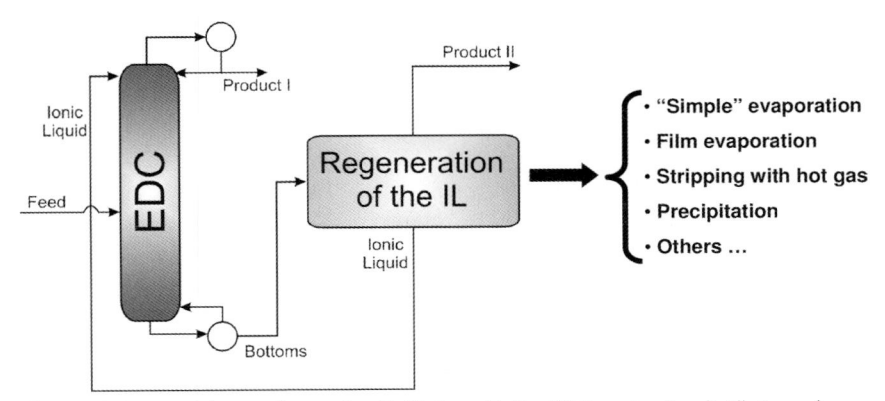

Figure 6.8 Process scheme of extractive distillation with ILs. EDC = extractive distillation column.

high capacity for the components to be separated. The selectivity and capacity at infinite dilution were calculated with a COSMO-RS model. NMP showed a selectivity of $S^\infty = 1.50$ and a capacity of $1/\gamma^\infty = 0.41$. The best ionic liquid was [C$_8$quin][Tf$_2$N] with $S^\infty = 1.93$ and $1/\gamma^\infty = 0.16$, although it exhibited the lowest selectivity of the ionic liquids tested, but it had the highest capacity [116]. It is clear that there is a large difference in the selectivity at finite and at infinite dilution, especially for the use of ionic liquids. At a finite dilution, $x_{solvent} = 0.3$, the best solvent is NMP, but at infinite dilution, an ionic liquid exhibits the highest selectivity.

Since both the selectivity and the capacity at finite dilution of the conventional solvent NMP are considerably higher than those of the ionic liquids tested, replacement of NMP for this separation would only be advantageous if the thermal and chemical stability of NMP would cause a problem under the process conditions of the extractive distillation.

Aromatic–aliphatic Hydrocarbon Separation The separation of aromatic from non-aromatic hydrocarbons is challenging because of close boiling points and azeotrope formation. The mixture of toluene and methylcyclohexane (MCH) has been used as representative of the compounds found in an industrial stream. Promising ionic liquids for the separation of this mixture are [C$_6$mim][Tf$_2$N] ($S^\infty_{MCH/tol} = 6.97$ and $1/\gamma^\infty_{tol} = 1.06$ at 313.15 K) [117] and [C$_8$quin][Tf$_2$N] ($S^\infty_{MCH/tol} = 3.24$ and $1/\gamma^\infty_{tol}$ 1.52 at 373.15 K) [118]. The selectivities and capacities with conventional solvents at approximately 333 K for this separation are as follows: sulfolane, $S^\infty_{MCH/tol} = 7.43$ and $1/\gamma^\infty_{tol} = 0.28$ [119]; NFM, $S^\infty_{MCH/tol} = 6.98$ and $1/\gamma^\infty_{tol} = 0.38$; NMP, $S^\infty_{MCH/tol} = 5.47$ and $1/\gamma^\infty_{tol} = 0.60$ [120]; DMF, $S^\infty_{MCH/tol} = 3.12$ and $1/\gamma^\infty_{tol} = 0.37$; and furfural, $S^\infty_{MCH/tol} = 3.37$ and $1/\gamma^\infty_{tol} = 0.23$ [121]. Ionic liquids exhibit selectivities comparable to those of the conventional solvents, but their capacities are higher than those of the conventional solvents.

Organic Compound–Water Separation Especially when water is part of the azeotropic mixture, very high separation factors can be achieved, because many ionic liquids are hygroscopic materials with a strong affinity to water. Figure 6.9 displays the classic vapor–liquid diagram for THF–water [122]. The data points indicate the change after the addition of the ionic liquid. Suitable ionic liquids for this separation are [C$_4$mim]Cl, [C$_4$mim][BF$_4$], [C$_2$mim][BF$_4$] and [C$_8$mim][BF$_4$]. To afford a sufficient separation, the amount of ionic liquid added has to be in the range of 30–50 wt%.

As the ionic liquid has no relevant vapor pressure, a second separation column for entrainer distillation is not required and energy can be saved. For the separation of ethanol from water, a process with [C$_2$mim][BF$_4$] as solvent was compared with a conventional process with 1,2-ethanediol as solvent. The overall heat duty for the conventional process is 2213 kJ kg^{-1} ethanol and for the ionic liquid process 1656 kJ kg^{-1} ethanol, a reduction of 25% [106].

Separation of Trimethyl Borate (TMB) and Methanol BASF has carried out an extractive distillation in a miniplant with a column diameter of 30 mm with either

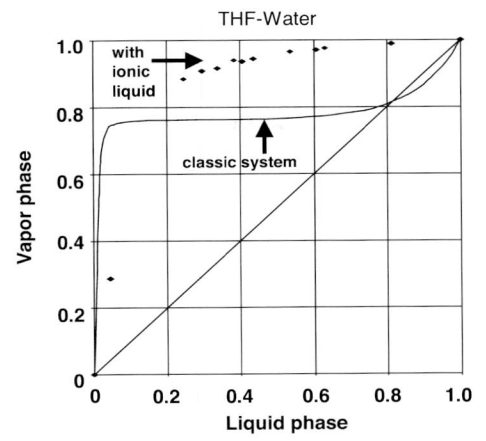

Figure 6.9 Equilibrium phase diagram for the system THF–water. The solid line shows the classic azeotropic mixture. The data points indicate how the azeotrope has been broken after addition of the ionic liquid. The amount of THF in the vapor phase is always higher than that in the liquid phase [122].

dimethylformamide (DMF), [C_2mim][CH_3SO_3] or [C_2mim][OTs] (OTs = tosylate = p-toluenesulfonate, $CH_3C_6H_4SO_3^-$) as solvent. Either a stripper or an evaporator replaced the distillation column for the regeneration of the DMF in the ionic liquid process. The S/F ratio for [C_2mim][CH_3SO_3]/[C_2mim][OTs]/DMF was 1.0/1.7/4.7 kg S/kg F: almost a factor of five lower for the ionic liquid [C_2mim][CH_3SO_3] than for DMF. The ED column with DMF as solvent required more than twice the number of theoretical stages compared with the ionic liquid process. A benchmark calculation for a capacity of 5000 t per year has revealed that a saving potential of 37–59% on energy cost and 22–35% on the investment can be achieved, depending on the IL used and the regeneration process of the IL. The total savings are in the range 25–35%.

BASF has been conducting this extractive distillation process in the miniplant continuously for 3 months. Although the ionic liquid faced a severe thermal treatment of about 250 °C in the recycling step, its performance was retained completely without a purge [109].

6.2.2.3 Conclusions

Extractive distillation with ionic liquids can be attractive, because of the potential savings in investments, fewer stages in the extraction column, lower energy costs, lower solvent-to-feed ratio and an easier regeneration, but in some cases conventional solvents perform better. The use of activity coefficients at infinite dilution sometimes provides results that differ from the distribution coefficients and selectivities at finite solutions. Therefore, for the design of an extractive distillation, experimental values of the distribution coefficients and selectivities will give more reliable results than activity coefficients at infinite dilution.

6.3
Environmental Separations

6.3.1
Desulfurization and Denitrogenation of Fuels

6.3.1.1 Conventional Desulfurization

Conventional desulfurization of fuels is achieved by catalytic hydroprocessing. However, further or deep hydrodesulfurization (HDS) leads to a high consumption of energy and hydrogen. The sulfur compounds are converted into hydrogen sulfide, which is easy to remove. Ultra-low sulfur gasoline and diesel oil (<15 ppm S in the USA and <10 ppm in the EU) are needed for new engines and catalysts and for further reduction of CO and NO_x emissions. The HDS process is normally effective only for the removal of aliphatic and alicyclic organosulfur compounds. The aromatic sulfur compounds such as thiophenes, benzothiophenes (BTs), dibenzothiophenes (DBTs) and their alkylated derivatives (e.g. 3-methylthiophene, 3-MT) are very difficult to convert to H_2S through HDS catalysts and require high operating and investment costs. Therefore, alternative processes for deep desulfurization are desirable. However, in a review of novel processes for removing sulfur compounds from refinery streams by Ito and van Veen of Shell Research, it was concluded that the classical hydrotreating options and their offshoots are competitive in transport fuel desulfurization, considering the ongoing development of these processes [123]. Only if the sulfur levels have to be below 1 ppmv will polishing processes become interesting.

6.3.1.2 Desulfurization with Ionic Liquids

Liquid–liquid extraction with chloroaluminate-containing ionic liquids has shown promising results, but their industrial use is not desirable with respect to corrosion, environmental concerns, hydrolytic stability and regeneration aspects. With [C_4mim]Cl–$AlCl_3$ (0.65 : 0.35) a reduction in sulfur content in real diesel of 80% was obtained in a five-stage extraction at 60 °C [124]. The ionic liquid [C_4mim]Cl–$1.0AlCl_3$ is partly effective in removing sulfur compounds from real fuels, both diesel and gasoline [125]. Nitrogen levels were significantly reduced, mainly due to a pretreatment with a molecular sieve. The fuels had to be dried before the extraction with activated 13X molecular sieve in order to prevent decomposition of the ionic liquid to form HCl. The ionic liquid:fuel ratio was 1 : 6 and the extraction was carried out in four stages. After the extraction of partially desulfurized gasoline, almost no thiophenes (C_6-thiophenes: 0.71 ppm) and no sulfides were detected and the amount of other sulfur species was 3.84 ppm. However, C_1–C_6 benzothiophenes, which were not detectable in the feed, were present in the final product, most likely due to a Lewis acid-catalyzed Diels–Alder reaction. Also with untreated gasoline, C_4-benzothiophenes and C_{3+}-dibenzothiophenes were formed during the extraction. Diesels did not show any formation of new sulfur compounds. The level of sulfur compounds after the four-stage extraction is still high: 129.15 ppm for regular diesel and 239.08 ppm for regular gasoline. The levels are lower for the fuels pretreated with

Desulfurization of regular diesel & gasoline

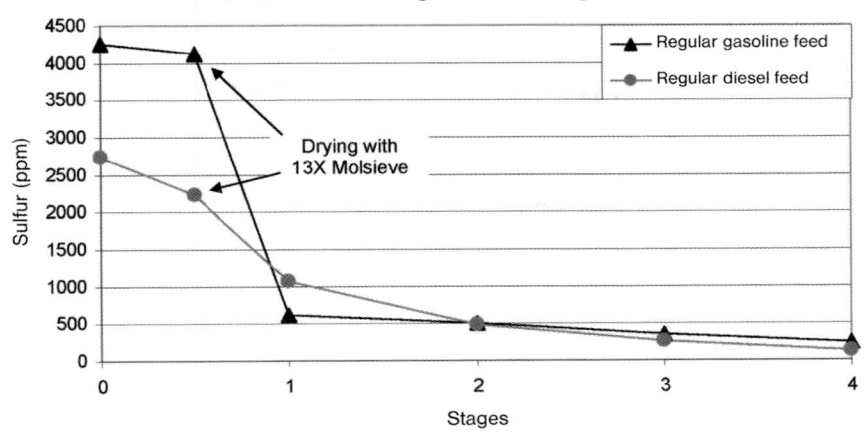

[Graph showing Sulfur (ppm) vs Stages (0–4) for Regular gasoline feed and Regular diesel feed, with annotation "Drying with 13X Molsieve" between stages 0 and 1.]

Desulfurization of pretreated diesel & gasoline

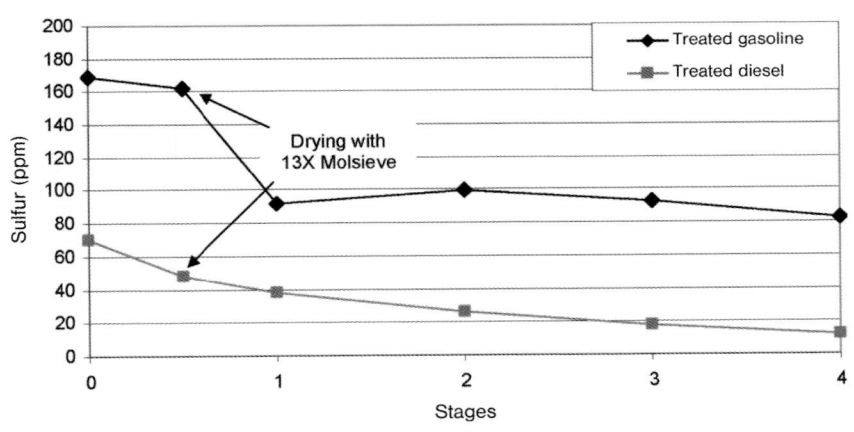

[Graph showing Sulfur (ppm) vs Stages (0–4) for Treated gasoline and Treated diesel, with annotation "Drying with 13X Molsieve" between stages 0 and 1.]

Figure 6.10 Four-stage extraction of diesel and gasoline. Ionic liquid:fuel ratio = 1:6, room temperature extraction [125].

a molecular sieve: 11.56 ppm for treated diesel and 82.33 ppm for treated gasoline (Figure 6.10). Regeneration of the ionic liquid was not carried out.

The ionic liquid [C_4mim][Cu_2Cl_3] showed a higher sulfur removal rate from a model oil [dibenzothiophene (DBT) and piperidine in n-dodecane, 500–764 ppm S] than [C_4mim]Cl–1.0AlCl$_3$ or [C_4mim][BF$_4$]: 23.4, 16 and 11%, respectively. The desulfurization rate with gasoline with different sulfur contents with [C_4mim][Cu_2Cl_3] varied from 16.2% (950 ppm S in the feed) to 37.4% (196 ppm S in the feed). The sulfur removal rate is lower with gasoline (21.6%) than with the model oil (23.4%) with the same S content of 680 ppm in the feed [126].

The ionic liquid [C_4mim]Cl can remove 50% of N-compounds from straight-run diesel (13240 ppm S and 105 ppm N), but only 5% of the S compounds [127].

N-compounds are inhibitors for the HDS reaction and removal of these compounds can improve the reaction. Regeneration of the ionic liquid was carried out by back-extraction with water (50 wt%) or methanol (10 wt%).

Effective removal of DBT can be achieved by addition of $FeCl_3$ to [C_4mim]Cl. The best results were obtained with [C_4mim]Cl–2.0 $FeCl_3$: DBT was completely removed from a model oil containing 5000 ppm DBT. This combination was also effective in removing 1180 ppm sulfur compounds from diesel oil [128]. However, removal of other sulfur compounds and the regeneration of the ionic liquid were not mentioned.

Zhang's group used the ionic liquids [C_2mim][BF_4], [C_4mim][BF_4], [C_4mim][PF_6] [129], [C_6mim][PF_6], [C_8mim][BF_4], [Me_3NH]Cl–1.5$AlCl_3$ and [Me_3NH]Cl–2.0$AlCl_3$ [130] to lower the sulfur content. With a low (240 ppm S) or a high sulfur (820 ppm S) gasoline, only 10–30% of the sulfur compounds were removed [129]. After 10 extraction cycles with [C_6mim][PF_6] and [C_8mim][BF_4], the sulfur content in gasoline was lowered from 820 to 400 and 320 ppm, respectively. In addition to the sulfur compounds, aromatics were also removed, to about 8% after 10 cycles [130]. In a three-stage extraction of a model oil with [C_8mim][BF_4], the thiophene and dibenzothiophene contents were reduced by 79 and 87%, respectively [131]. With the ILs [Me_3NH]Cl–2.0$AlCl_3$ and [Me_3NH]Cl–1.5$AlCl_3$, the removal rate of sulfur compounds was only 15–20%. The regeneration of the 1,3-dialkyimidazolium-based ionic liquids was carried out by evaporating the absorbed compounds and the removal rate was retained. However, the [Me_3NH]Cl–$AlCl_3$-based ionic liquids could not be regenerated. After extraction with 1-(4-sulfonic acid)butyl-3-methylimidazolium p-toluenesulfonate, 250 ppm DBT was removed from a model oil (500 ppmw DBT in n-tetradecane) at 60 °C and an oil:ionic liquid mass ratio of 4 : 1. After a five-stage extraction with this ionic liquid of a pre-desulfurized oil with a sulfur content of 438 ppmw, the sulfur content was 45 ppmw. The process conditions were $T = 80$ °C, diesel:ionic liquid mass ratio $= 4:1$ and extraction time 25 min [132].

1,3-Dialkylimidazolium-based dialkylphosphate ionic liquids are useful for extractive desulfurization of fuels [133–136]. The best results were obtained with [C_4mim][DMP], although the solubility of gasoline in this ionic liquid is 35.3 mg g^{-1} ionic liquid at 298.15 K, which may lead to increased separation costs. The ionic liquid is not soluble in the gasoline. The mass-based distribution coefficients of sulfur compounds in straight-run gasoline with this ionic liquid at 298.15 K are $D_{3\text{-MT}} = 0.59$, $D_{BT} = 1.37$ and $D_{DBT} = 1.59$ [136]. The removal of S-compounds from gasoline is more difficult than for straight-run gasoline, which is likely due to co-extraction of aromatic compounds from the gasoline. The regeneration of these ionic liquids still needs more investigation.

With pyridinium-containing ionic liquids, the sulfur removal rate of a model oil (thiophene in heptane) is 45.5% with [1-C_4py][BF_4] and lower for other pyridinium-containing ionic liquids [137]. The mass ratio of ionic liquid to model oil was 1 : 1 and the extraction time was 30–40 min. The sulfur removal rate increased somewhat with temperature: 48.3% at 60 °C. After six extraction cycles, the sulfur content was decreased from 498 to 18 ppm. Regeneration of the ionic liquids was carried out by evaporation of thiophene at 100 °C or by re-extraction with CCl_4.

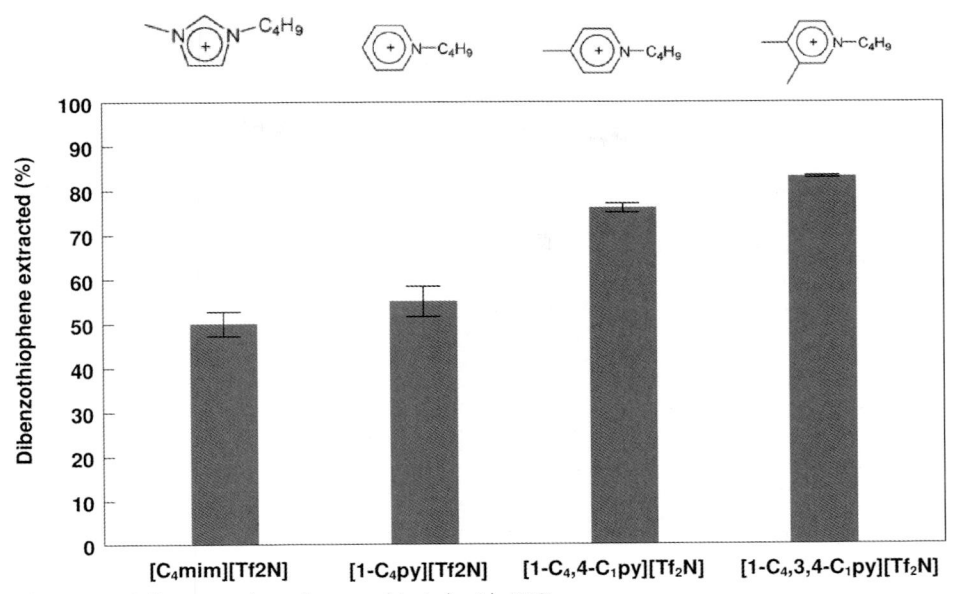

Figure 6.11 Sulfur removal rate for several ionic liquids [138].

The distribution coefficient of DBT in its extraction from dodecane showed a clear variation in cation types: 1,4-dimethylpyridinium > methylpyridinium > pyridinium ≈ imidazolium ≈ pyrrolidinium and much less significant variation with the anion type (Figure 6.11) [138]. Polyaromatic quinolinium-based ionic liquids showed even greater S removal rates (90% at 60 °C), but those ionic liquids have higher melting points.

Other promising ionic liquids are [C_4mim][$C_8H_{17}SO_4$] and [C_2mim][$C_2H_5SO_4$], with DBT distribution coefficients of 1.9 and 0.8, respectively (500 ppm DBT in dodecane, ionic liquid:dodecane = 1 : 1) [139]. With diesel and FCC (fluid catalytic cracked) gasoline, the distribution coefficient of S-compounds was lower for [C_4mim][$C_8H_{17}SO_4$]: 0.3–0.8 (200–400 ppm S in the feed) and 0.5 (300 ppm in the feed). The process conditions were mass ratio 1 : 1, mixing time 15 min at room temperature. Regeneration of [C_2mim][$C_2H_5SO_4$] containing 20 mg kg^{-1} tetrahydothiophene and 2.5 wt% cyclohexane was carried out by stripping with air at 100 °C for 30 min. However, regeneration of S-loaded ionic liquids from diesel oil was not possible by stripping alone and additional re-extraction and distillation steps will be needed.

A major drawback of these ionic liquids is the cross-solubility of hydrocarbons in these ionic liquids, as the co-extracted hydrocarbons must be separated from the ionic liquid together with the S-compounds. The cross-solubility is lower for [C_2mim][$C_2H_5SO_4$] than for [C_4mim][$C_8H_{17}SO_4$]. In Figure 6.12, a possible process scheme of an ionic liquid extraction implemented into an existing refinery for the desulfurization of diesel is depicted [139].

The Institut Français du Pétrole has obtained a patent in which S- and N-compounds can be removed from fuels with ionic liquids containing an alkylation

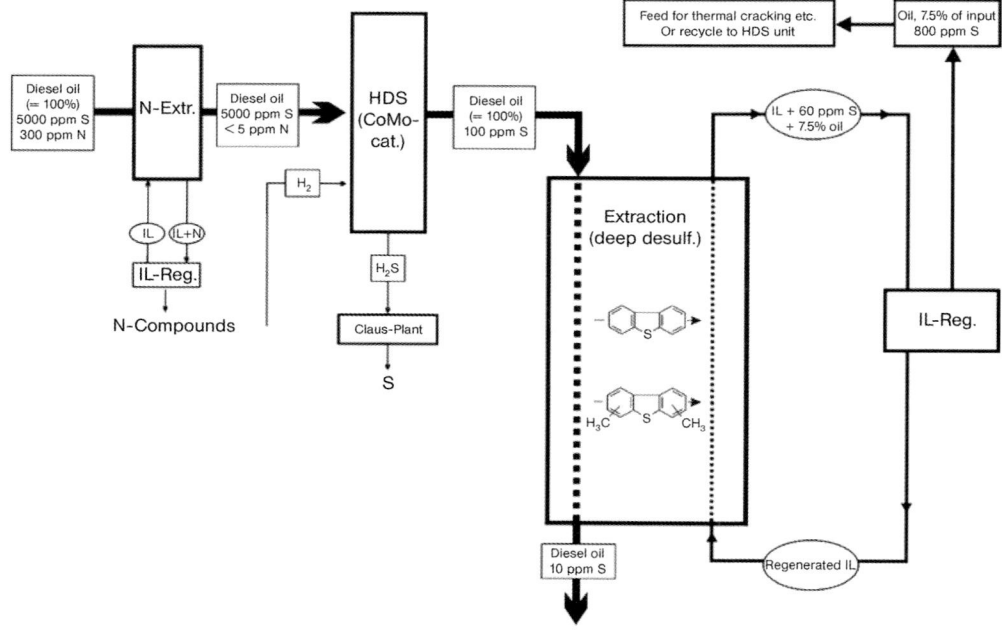

Figure 6.12 Extractive desulfurization of diesel with [C$_4$mim][C$_8$H$_{17}$SO$_4$], cross-solubility of oil 5%, IL:oil mass ratio = 1.5, D_S = 0.8, six stages [139].

agent allowing the formation of ionic S- and N-derivatives that are soluble in the ionic liquid [140]. The ionic liquid is separated from the oil by decantation.

6.3.1.3 Oxidative Desulfurization

A low removal rate of sulfur compounds of 7–8% was found by Lo et al., but the rate was increased after oxidation of the sulfur compounds by H$_2$O$_2$–acetic acid to form sulfones, to 55% with [C$_4$mim][BF$_4$] and to 85% with [C$_4$mim][PF$_6$] [141]. Zhao et al. found that DBT was completely removed after 50 min at 60 °C with the ionic liquid [1-C$_4$,4-C$_1$py][BF$_4$] (Figure 6.13) in combination with oxidation by H$_2$O$_2$ (H$_2$O$_2$:S = 3) [142]. The ionic liquid can be used seven times without a

Figure 6.13 Extraction and oxidation reaction of DBT [142]. [Hnmp]BF$_4$ = [1-C$_4$,4-C$_1$py][BF$_4$].

significant decrease in activity (99.6% sulfur removal). The sulfur content of an actual diesel fuel with 3240 ppm S was reduced to 20 ppm after extraction with the ionic liquid, oxidation and extraction with DMF (99.4% removal). The process conditions were $T = 60\,°C$, $V_{oil}/V_{IL} = 1$, H_2O_2:S $= 6$ and reaction time 2 h. Zhao's group also investigated another ionic liquid, $[N_{4,4,4,4}]Br \cdot 2C_6H_{11}NO$, as an effective catalyst for the oxidative desulfurization of thiophene with H_2O_2–acetic acid [143]. Most of the oxidation products were transferred to the aqueous phase owing to their higher polarity. The desulfurization rates of thiophene-containing model oil and actual FCC gasoline were 98.8 and 95.3%, respectively.

A simple liquid–liquid extraction and catalytic desulfurization system, composed of a molybdic compound, 30% H_2O_2 and $[C_4mim][BF_4]$ is effective in removing 99% of sulfur compounds (BT, DBT and 4,6-dimethyldibenzothiophene) from a model oil. The desulfurization system was recyclable for five times with very little decrease in activity [144]. Ultraclean Fuel in Australia has claimed an oxidative desulfurization process, which is capable of reducing the S content from 500 to 30 ppm [145].

6.3.1.4 Conclusions

Desulfurization of fuels by extraction with ionic liquids as the only process step does not lead to sulfur levels of 10 ppm and lower at the present state of research. Only if pre-desulfurized fuels are used or if extraction is combined with other processes, such as oxidation, can the required S levels can be reached. Regeneration of the ionic liquid needs attention, as in most reports regeneration processes are only suggested and not tested. Some authors report a decrease in the S removal rate after several extraction cycles. Following experiments with a model oil, measurements with real fuels must be carried out, because the sulfur removal rates will differ, due to the presence of other compounds in the fuel.

6.4
Combination of Separations in the Liquid Phase with Membranes

The separation of several compounds with ionic liquids can also be combined with membranes, either as a bulk liquid membrane (BILM) [146, 147], as a support for the ionic liquid [supported ionic liquid membrane (SILM)] [148–152] or as a separating barrier between the ionic liquid and product phase [153–157].

The toluene/n-heptane selectivity in a BILM with $[C_8mim]Cl$ at 25 °C was around 4–9, depending on the initial toluene concentration [146]. The toluene/n-heptane selectivity with extraction, calculated from activity coefficients at infinite dilution, is around 8 at 35 °C [158]. The separation factor at 25 °C with a BILM with $[C_2mim][C_2H_5SO_4]$, is 2.75 after 40 h, but decreases to 1.5 at 80 h. With extraction using the same ionic liquid, the selectivity is around 50 at 40 °C [59].

SILMs combine extraction and stripping and the amount of solvent in the SILM process is much less than in a solvent extraction process. $[C_4mim][PF_6]$ in porous poly(vinylidene fluoride) film showed a benzene/n-heptane selectivity of 67 and a toluene/n-heptane selectivity of 11 and the mass transfer coefficients of benzene and toluene were 6.2×10^{-4} and 9.9×10^{-4} m h^{-1}, respectively [151]. The receiving phase was hexadecane. The toluene/n-heptane selectivity is lower than for several other ionic liquids (see Table 6.1) and certainly lower than for sulfolane. Moreover, the product has to be separated from the receiving phase. Ionic liquids with silver salts impregnated on porous membranes facilitate the transport of olefinic hydrocarbons through the membrane [152, 159, 160]. In these processes, the combination of the silver salt and the ionic liquid determines the selectivity for the olefin.

Nanofiltration is a suitable process for the separation of non-volatile products from the ionic liquid [154, 155]. However, the feed phase must be diluted in order to decrease the viscosity and this involves an extra energy requirement to concentrate the product afterwards.

6.4.1
Conclusions

The selectivity in both BILM and SILM processes is determined by the nature of the ionic liquids used. In general, the capacity or mass transfer rates in the BILM and SILM processes are relatively low. The advantages claimed by several authors are that the solvent inventory in BILM and SILM processes is lower than that for extraction processes and that extraction and stripping, including continuous regeneration of the ionic liquid, are combined in one process. However, it remains to be seen if these applications are useful, as a membrane poses an extra barrier for transport of the desired compounds from the feed phase to the ionic liquid, resulting in a lower capacity or lower transfer rates. Moreover, the product has to be separated from the receiving phase by another separation process and the feed phase has to be diluted to lower the viscosity in case of separating the products from the ionic liquids. Only in the case of separating a heat-sensitive or non-volatile product from the ionic liquid phase could a separation process with a membrane be useful.

6.5
Gas Separations

6.5.1
Conventional Processes

Chemical and physical absorption processes are extensively used in the natural gas, petroleum and chemical industries for the separation of CO_2 [161]. Physical absorption is preferred when acidic gases (H_2S, CO_2) are present at elevated concentrations in the gas stream. Physical solvents are non-reactive polar organic compounds with an acidic gas affinity. Chemical absorption is typically used for the removal of remaining acidic impurities and when gas purity is a downstream constraint. For chemical CO_2 removal, aqueous solutions of primary, secondary, tertiary, sterically hindered amines and formulated amine mixtures are the most widely used solvents. About 75–90% of the CO_2 is captured using a monoethanolamine (MEA)-based technology producing a gas stream of high CO_2 content (>99%) after desorption [162].

The major drawbacks of the traditional gas absorption separation processes are mainly caused by the nature of the solvent and by the type of interactions given between the solute and the solvent. In an industrial gas absorption process, it is desirable to achieve high absorption rates and high solute capacities of a solvent that is easily regenerated and for which volume make-up is minimized.

6.5.2
CO_2 Separation with Standard Ionic Liquids

Ionic liquids can be used for gas separations and the removal of CO_2 from several gases [163–176] has mainly been investigated. The CO_2 solubility is higher in ionic liquids with anions containing fluoroalkyl groups, such as $[Tf_2N]^-$ and $[Tf_3C]^-$, regardless of whether the cation is 1,3-dialkylimidazolium, N,N-dialkylpyrrolidinium or tetraalkylammonium. These results suggest that the nature of the anion has the most significant influence on the gas solubilities [166, 167, 169, 171, 176].

An increase in the alkyl chain length on the cation increases the CO_2 solubility marginally. Oxygen is hardly soluble in several ionic liquids and H_2, N_2 and CO have a solubility below the detection limit in several studies [163, 164, 171]. Figure 6.14 shows the solubility of a number of gases in $[C_4mim][PF_6]$ at 298 K [164]. The CO_2/C_2H_4 selectivity is around 3 and the CO_2/CH_4 selectivity is about 30.

6.5.3
CO_2 Separation with Functionalized Ionic Liquids

In order to increase the CO_2 solubility in ionic liquids, functionalized task-specific ionic liquids can be used. Since the conventional solvents for CO_2 absorption are amine-based, the functionalization with amine groups seemed obvious [175, 177–181]. Figure 6.15 shows the CO_2 absorption in MEA, Sulfinol, methyldiethanolamine

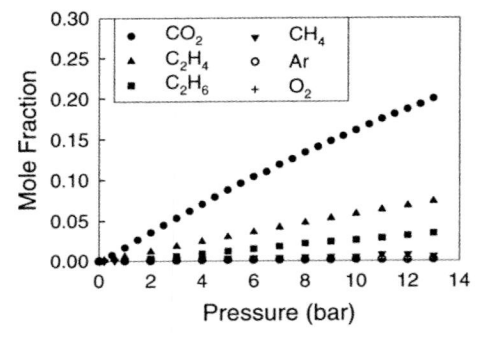

Figure 6.14 Solubilities of gases in $[C_4mim][PF_6]$ at 298 K [164].

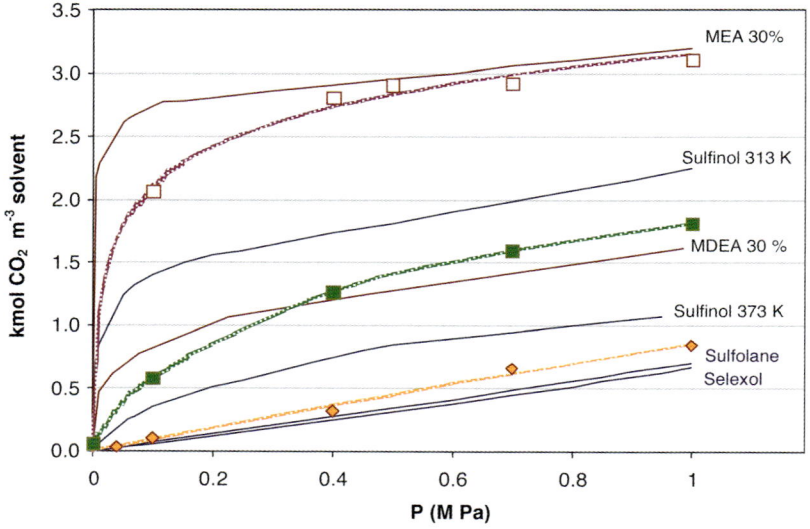

Figure 6.15 Volumetric $_2$ loads in several liquids at 333 K [180, 181].

(MDEA), Selexol and several NH_2-functionalized ILs: $[NH_2C_2H_4pyrr][BF_4]$, $[NH_2C_3H_6mim][BF_4]$ and $[NH_2C_3H_6mim][Tf_2N]$.

The ionic liquid $[NH_2C_3H_6mim][BF_4]$ has about the same performance as a 30% MEA solution [180, 181]. The functionalized ionic liquids shows both a physical and a chemical absorption behavior.

6.5.4
CO_2 Separation with Ionic Liquid (Supported) Membranes

Copolymers of polymerizable RTILs [poly(RTILs)] and different lengths of polyethylene glycol (PEG) polymers were found to a have CO_2/N_2 separation performance exceeding the upper bound of the "Robeson plot" [182, 183]. The ideal CO_2/N_2 selectivities with $[C_nmim][Tf_2N]$ ionic liquids are of the order of 20, while the selectivity with the poly(RTIL) $[P_nmim][Tf_2N]$ is around 30.

The CO_2/CH_4 selectivity is only slightly higher with the poly(RTILs) than with the corresponding 1,3-dialkylimidazolium-based ionic liquids, i.e. 12 versus 9. Alkyl-terminated nitrile groups are also of interest for gas separations in poly(RTILs) [183]. Both the CO_2/N_2 and the CO_2/CH_4 selectivities are higher with these poly(RTILs), 40 and 32, respectively (Figure 6.16).

Supported liquid membranes (SLMs) based on functionalized ionic liquids, such as $[H_2N-C_3mim][Tf_2N]$ in a PTFE membrane, showed a high CO_2/CH_4 selectivity of the order of 65 and a high stability for more than 260 days. The CO_2 permeability decreased gradually from 690 to 560 Barrer during that period [184]. $[C_2mim][DCA]$ in a PES membrane showed a CO_2/CH_4 selectivity of 20 and a CO_2/N_2 selectivity of 60. The CO_2 permeability was 610 Barrer [185].

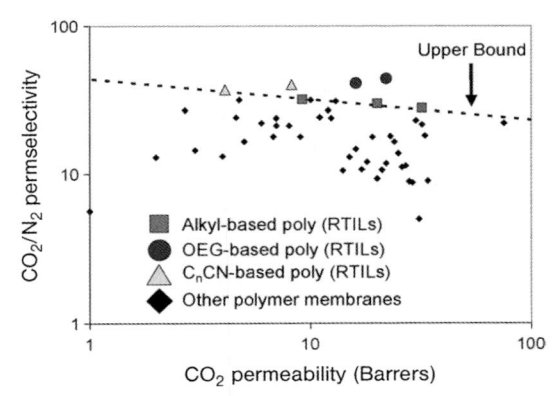

Figure 6.16 Robeson plot for CO_2–N_2 separation [183]. OEG = oligoethylene glycol.

Porous alumina membranes saturated with [C_2mim][Tf_2N] showed a CO_2/N_2 selectivity of 127. A cost comparison with conventional amine scrubbing was carried out [186]. The measured CO_2 permeance in [C_4mim][Tf_2N] was 4×10^{-5} mol bar^{-1} m^{-2} s^{-1}. From Figure 6.17, it is clear that a cost for the ionic liquid of the order of \$1000 kg^{-1} is required at a permeance of at least a factor 15 higher than the permeance in [C_4mim][Tf_2N].

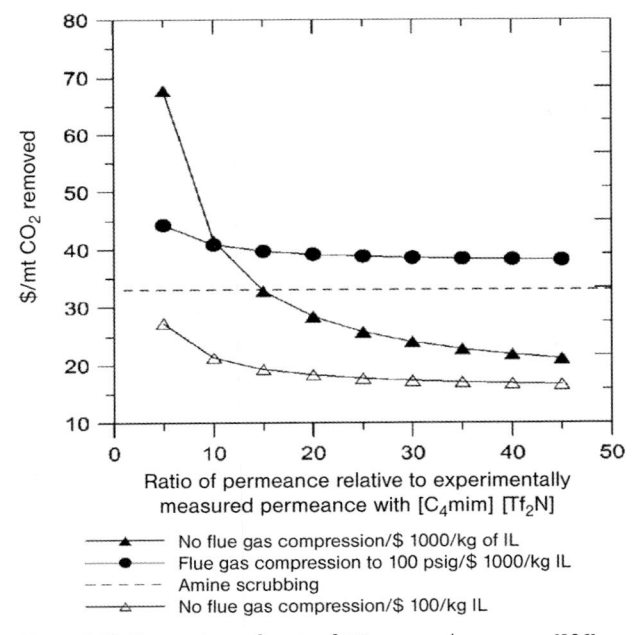

Figure 6.17 Comparison of costs of CO_2 removal processes [186].

6.5.5
Olefin–Paraffin Separations with Ionic Liquids

Except for the removal of CO_2 from other gases, ionic liquids can also be used for ethylene–ethane and propylene–propane separations [174, 187–198]. For the separation of hydrocarbon gases, it was found that the solubility increased as the number of carbon atoms increased and also as the number of carbon–carbon double bonds increased [190]. The separation factors with standard ionic liquids for ethylene–ethane and propylene–propane separation are relatively small, as the solubilities of these gases in these ionic liquids are very low (Figure 6.14). Therefore, ionic liquids with functional groups or dissolved carrier molecules are more effective, such as the use of Ag salts, e.g. $Ag[BF_4]$ in an ionic liquid with the $[BF_4]^-$ anion [187–189, 195, 196, 198]. The propylene/propane selectivity with 0.25 M $Ag[BF_4]$ containing $[C_4mim][BF_4]$ is 20 at a low partial C_3H_x pressure of 1 bar and decreases at higher pressures [195]. The flux rates with liquid functionalized ionic liquid membranes are of the order of $2 \times 10^{-2}\,l\,m^{-2}\,s^{-1}$ and the selectivity can vary between 100 and 540 for 1-hexene/hexane and 1-pentene/pentane [196].

6.5.6
Conclusions

Standard ionic liquids often do not perform better than conventional solvents in gas separations. However, functionalized ionic liquids show a higher solubility and a higher selectivity than conventional solvents. Functionalized ionic liquids combine the physical and chemical absorption of gases.

6.6
Engineering Aspects

6.6.1
Equipment

Successful introduction of ionic liquids into extraction operations also requires knowledge of their hydrodynamic and mass transfer characteristics, because the viscosity and the density of ionic liquids are usually higher than those of conventional solvents. Common extraction contactors may not be suitable for separation processes with ionic liquids as extractants. Therefore, mechanical energy has to be used to enhance mass transfer into ionic liquids.

A centrifugal extractor was used for the separation of ethylbenzene from *n*-octane with $[C_4mim][PF_6]$ [199]. The centrifugal extraction system contained four 50 mm diameter annular space extractors. The density of the ionic liquid was $1030\,kg\,m^{-3}$ and its viscosity was 450 mPa s at 298.15 K and 80 mPa s at 313.15 K. Under the optimum process conditions (a rotation speed of 3500 rpm for this extractor), a single-phase efficiency of 90% was obtained.

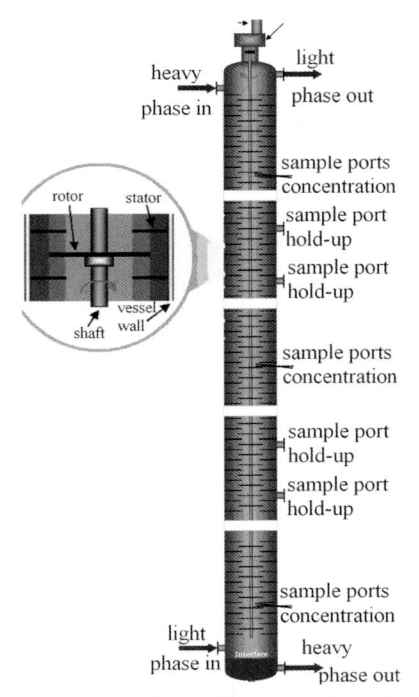

Figure 6.18 Scheme of a rotating disc column.

Extraction of aromatics from aliphatics in a pilot plant rotating disc contactor (RDC) with [1-C_4,4-C_1py][BF_4] as the solvent is currently under investigation [200]. In the extraction process, hydrodynamics (drop size, hold-up and operational window) and mass transfer efficiency determine the column performance [201]. The investigated parameters concern the total flux, the rotation speed of the RDC [202, 203] and the concentration of toluene in the organic (*n*-heptane) phase.

The pilot RDC extraction column is shown schematically in Figure 6.18. The column consists of five jacketed glass segments each 360 mm in length and with an inside diameter of 60 mm, and with eight stirred compartments each. Settlers of 240 mm (bottom) and 210 mm (top) with an internal diameter of 90 mm enclose the stirred segments. The solvent (ionic liquid or sulfolane) is the dispersed phase, which is fed at the top of the column and the extract phase is collected from the bottom settler. The heptane–toluene phase is fed from the bottom and the raffinate phase is collected from the top settler. Regenerated ionic liquid was used in the extraction experiments.

6.6.2
Hydrodynamics

Figure 6.19 shows that the drop size for the extraction of 10 and 75 wt% toluene from *n*-heptane decreases with increasing rotor speed. The differences in drop size can be explained by the higher concentration of toluene (75 versus 10%) in [1-C_4,4-C_1py]

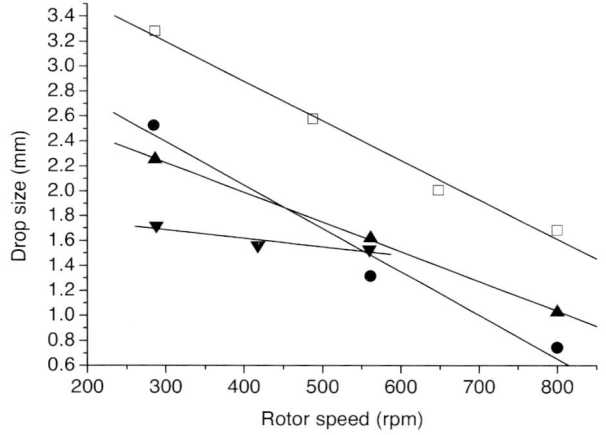

Figure 6.19 Drop sizes for extraction of 10 wt% toluene from n-heptane with [1-C$_4$,4-C$_1$py][BF$_4$] (flux = 2.8 m h^{-1}, □) and of 75 wt% toluene with varying flux of 3.7 (●), 8.2 (▲) and 10.5 m^3 m^{-2} h^{-1} (▼) [200].

[BF$_4$]. The higher concentration reduces the viscosity of the ionic liquid, which results in easier break-up and hence smaller droplets.

At low fluxes with constant flow rates of both phases, the hold-up decreases with increasing rotor speed when [1-C$_4$,4-C$_1$py][BF$_4$] is used as solvent (Figure 6.20). This is not expected, since an increase in rotor speed should result in a greater toroidal motion of the dispersed phase, which results in a longer residence time of the dispersed phase. This was also found by others [204]. The explanation for this is that cohesion of droplets at the stator rings causes an increase in the hold-up at lower fluxes. At higher fluxes, there is less cohesion, which is confirmed by the decrease in drop size with increasing flux (Figure 6.19). Because sulfolane has a much lower viscosity (8 mPa s at 313 K) than [1-C$_4$,4-C$_1$py][BF$_4$] (80 mPa s at 313 K), there is less

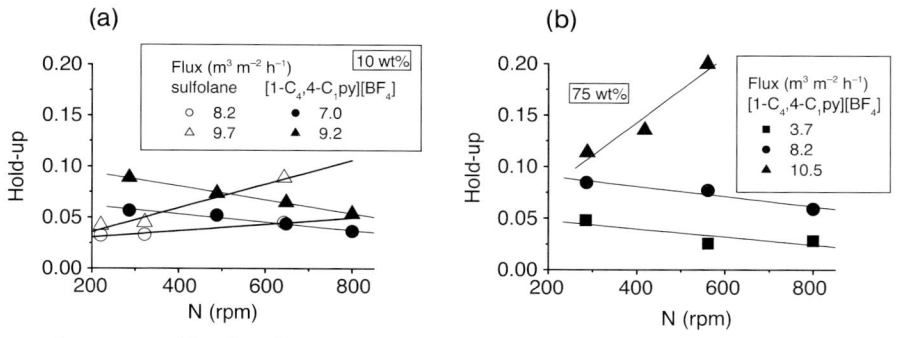

Figure 6.20 Hold-up for sulfolane (open symbols, S/F = 4) and [1-C$_4$,4-C$_1$py][BF$_4$] (closed symbols, S/F = 6) extraction of 10 wt% toluene (a) and for [1-C$_4$,4-C$_1$py][BF$_4$] extraction of 75 wt% toluene [(b), S/F = 7.5] [200].

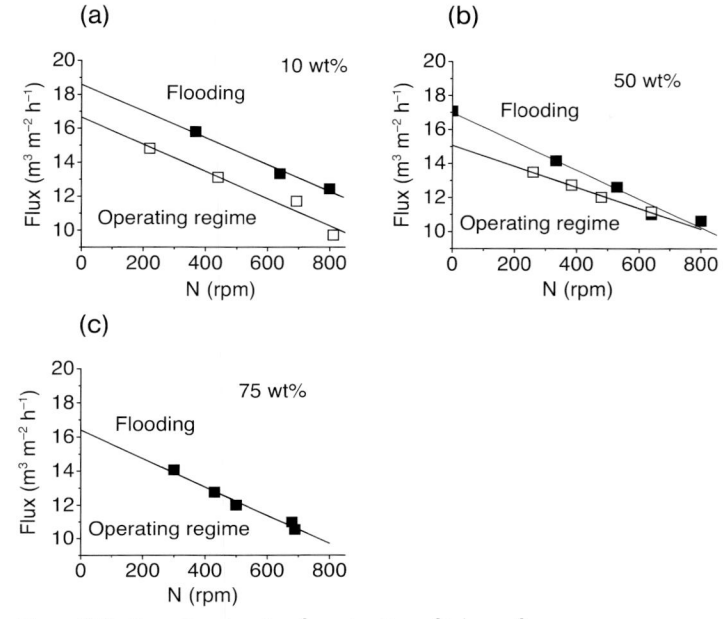

Figure 6.21 Operational region for extraction of toluene from n-heptane in RDC at 313 K with sulfolane (open symbols) and [1-C$_4$,4-C$_1$py][BF$_4$] (closed symbols) [200].

cohesion when sulfolane is used and, therefore, the sulfolane hold-up increases with increasing rotor speed, as expected.

For sulfolane, solvent-to-feed (S/F) ratios of 4 (10 wt%) and 6 (50 wt%) and for [1-C$_4$,4-C$_1$py][BF$_4$], S/F ratios of 6 (10 wt%) and 7.5 (50 and 75 wt%) were used for the extractive removal of toluene from a mixture of 10 to 75 wt% toluene in n-heptane at 40 °C. From Figure 6.21, it can be concluded that for extraction of 10 wt% toluene and also for the extraction of 50 wt% toluene in the RDC, the maximum achievable flux is higher for [1-C$_4$,4-C$_1$py][BF$_4$] than for sulfolane.

The difference in maximum flux can be explained by the difference in drop size. Due to the higher viscosity of [1-C$_4$,4-C$_1$py][BF$_4$] than sulfolane, the ionic liquid has larger drop sizes than sulfolane and consequently a higher gravitational force and less resistance. The [1-C$_4$,4-C$_1$py][BF$_4$] droplets fall faster through the heptane and higher counteracting forces at higher fluxes can be overcome. Extraction of more than 65 wt% toluene from n-heptane is not possible with sulfolane, since this composition approaches the plait point in the ternary diagram [59, 65].

6.6.3
Mass Transfer

Figure 6.22 shows the concentration profiles of toluene over the column length. From Figure 6.22a, it can be concluded that with sulfolane as solvent more toluene is extracted

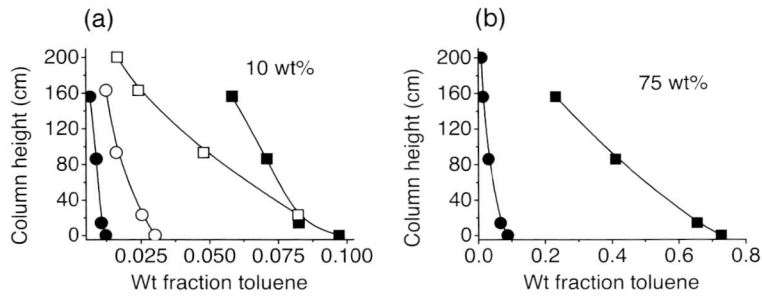

Figure 6.22 Concentration profile of toluene over the column. $T = 313$ K; $N = 640$ rpm; closed symbols for [1-C_4,4-C_1py][BF_4] as solvent and open symbols for sulfolane; (■, □) raffinate, (●, ○) extract [200].

than with [1-C_4,4-C_1py][BF_4], although the distribution ratios of toluene in sulfolane and [1-C_4,4-C_1py][BF_4] are comparable on a weight basis (0.22 w/w for [1-C_4,4-C_1py][BF_4] and 0.26 w/w for sulfolane). Extraction of more than 65 wt% toluene is not possible with sulfolane [59, 65], but from Figure 6.22b it can be concluded that with [1-C_4,4-C_1py][BF_4] a rapid decrease in toluene concentration in the raffinate phase occurs.

From a hydrodynamic point of view, [1-C_4,4-C_1py][BF_4] is a better solvent than sulfolane, because the operational region is larger. However, based on mass transfer characteristics, sulfolane outperforms [1-C_4,4-C_1py][BF_4]. This is due to the large difference in the viscosities of the two solvents. Ionic liquids such as [1-C_4,4-C_1py][BF_4] offer the advantage of extracting feeds containing high concentrations (>65%) of aromatics, which is not possible with the currently used solvents.

6.6.4
Conclusions

Ionic liquids can be suitable extractants for several compounds, as proven in batch equilibrium experiments. It has also been shown that a continuous extraction using ionic liquids in a pilot plant produces good results. The extraction with ionic liquids was compared with an extraction with a well-known solvent in industry, sulfolane, and the results were that the ionic liquid has a larger operational window, but that sulfolane has better mass transfer properties than the more viscous ionic liquid [1-C_4,4-C_1py][BF_4]. Less viscous ionic liquids should produce better results.

6.7
Design of a Separation Process

6.7.1
Introduction

Since the properties of an ionic liquid are defined by the combination of the cation and anion, so-called tailoring offers the possibility to create a special solvent for a specific task. Because of the large number of combinations ($>10^{14}$), it is impossible to synthesize all ionic liquids and measure their properties [205]. Therefore, to determine suitable ionic liquids for a certain problem, simulation tools will be very useful. The use of group contribution methods such as UNIFAC is difficult because the specific interaction energy parameters of ionic liquids are not yet always available. Therefore, a dielectric continuum model (COSMO-RS: COnductor-like Screening MOdel for Real Solvents) has been chosen by a large number of authors. COSMO-RS is a quantum chemical approach, recently proposed by Klamt and Eckert [206–208] for the *a priori* prediction of activity coefficients and other thermophysical data using only structural information of the molecules. This method enables one to screen ionic liquids based on the surface charge, the polarity. In addition, it is possible to calculate activity coefficients at infinite dilution.

6.7.2
Application of COSMO-RS

Several authors have used COSMO-RS for the prediction of activity coefficients at infinite dilution with various ionic liquids [209–211]. In aqueous systems, COSMO-RS provides good results for systems with alkyl halides or aromatics as solutes in water, but it was less successful for non-aqueous systems. The predictions were evaluated with experimental data and the deviations varied by up to 16% [211].

Liquid–liquid and vapor–liquid equilibria were also evaluated with COSMO-RS [212–219] and ionic liquids can be used as solvents in extractive distillation [111, 116, 118]. COSMO-RS calculations for a non-polar mixture of methylcyclohexane and toluene showed that the selection of the solvent is complicated.

Figure 6.23 shows the COSMO-RS predictions and the experimental values for the separation factors of methylcyclohexane and toluene with a molar fraction of 0.30 of

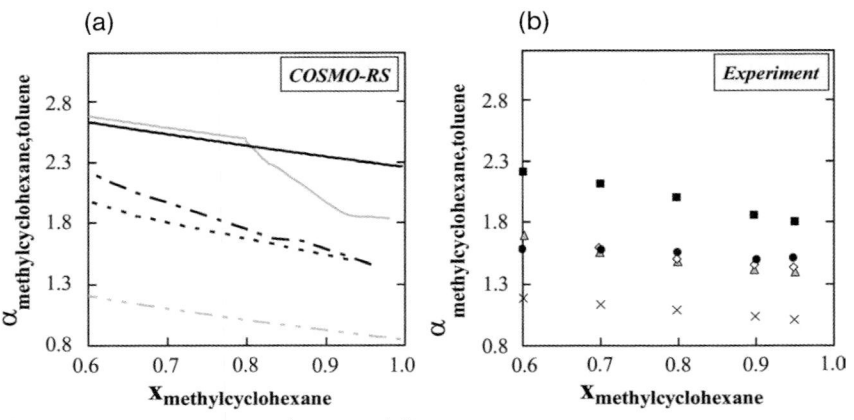

Figure 6.23 (a) COSMO-RS predictions and (b) experiments. $T = 373.15$ K. (■, black line —) [C$_8$quin][Tf$_2$N], (△, gray line —) [C$_8$quin][BBB], (◇, –··–) [C$_4$mim][Tf$_2$N], (●, ···) ECOENG™ 500, (×, –···–) binary [118].

the selected ionic liquids in the liquid phase at 373.15 K [118]. COSMO-RS overpredicts the separation factor values for this system by about 20%.

Not only activity coefficients but also other thermophysical data can be predicted by COSMO-RS, such as vapor pressure and vaporization enthalpy [220], densities and molar volumes [221] and water solubilities [222, 223]. The calculated enthalpies are in good agreement with the experimental data, but the vapor pressure is underestimated.

Comparison of COSMO-RS with other methods has been carried out by several authors, e.g. COSMOSPACE [224], UNIFAC [225] and UNIQUAC [226]. COSMOSPACE yielded better results than UNIQUAC and COSMO-RS. On the other hand, calculations with COSMO-RS of the miscibility gap of $[C_4mim][PF_6]$ with an alkanol differed widely from the experimental results and a far better agreement with experimental data was found with a UNIQUAC-based correlation [226].

Provided that the appropriate interaction parameters are available, the modified UNIFAC (Dortmund) can be applied successfully to systems with ionic liquids [225, 227]. COSMO-RS can also be used as a tool to screen suitable ionic liquids for the absorption of CO_2 [228]. Figure 6.24 shows a comparison of the Henry constants of CO_2 in $[C_4mim][PF_6]$.

6.7.3
Conclusions

The general conclusion is that γ^∞ values determined with COSMO-RS can be used for a first screening. However, quantitative predictions are still inaccurate and COSMO-RS does not give sound predictions of γ values for finite concentrations. Therefore, it cannot be used yet for the calculation of distribution coefficients and selectivities in real solutions. COSMO-RS appears to be a very promising tool to support the design of suitable ionic liquids for specific (separation) problems.

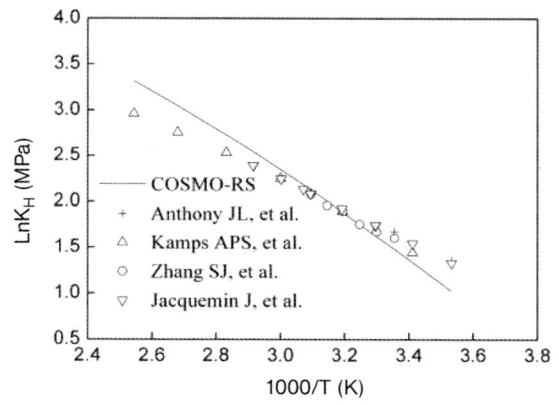

Figure 6.24 Henry constants of CO_2 in $[C_4mim][PF_6]$ [228].

6.8
Conclusions

There are no industrial applications for separations using ionic liquids yet. In order to introduce separations with ionic liquids in industry, more applied research is required, especially on a pilot-plant scale to determine optimal process conditions.

Development of a complete separation process with ionic liquids, i.e. including the primary separation, recovery of the products and the regeneration of the ionic liquid for re-use, must be carried out.

Recovery of the separated products is rarely investigated, but it is a required aspect for a complete separation process. Regeneration of ionic liquids is hardly reported, but is also a very important issue for green industrial use as only ionic liquids that can be regenerated will be used.

Activity coefficients at infinite dilution of ionic liquids are useful for screening purposes, but for separations with ionic liquids, real distribution coefficient and selectivity values at finite dilutions will have to be obtained, as these are concentration dependent.

Ionic liquids are not always better than conventional solvents but they can show advantages in some specific cases. Unfortunately, no benchmarks with conventional solvents are mentioned in a large number of publications.

Too much emphasis still exists on ionic liquids with $[PF_6]^-$ as anion, because these are versatile hydrophobic ionic liquids, but in industry these ionic liquids will never be used owing to their instability towards water combined with HF formation.

Task-specific ionic liquids are generally more selective than standard ionic liquids and, therefore, more focus must be directed to the development of these ionic liquids.

Abbreviations and Symbols

[allyl-mim]$^+$	1-allyl-3-methylimidazolium
[$N_{1,1,4,2\text{-hydroxyethyl}}$]	butyl(2-hydroxyethyl)dimethylammonium
[$N_{1,1,2,2\text{-hydroxyethyl}}$]	ethyl(2-hydroxyethyl)dimethylammonium
poly(RTIL)	polymerizable room-temperature ionic liquid
[$P_n$mim]$^+$	polymerized dialkylimidazolium cation
[BBB]	bis[1,2-benzenediolato(2–)-O,O']borate
[DMP]	dimethylphosphate
ADH	alcohol dehydrogenases
ATPS	aqueous two-phase system
BILM	bulk liquid membrane
BT	benzothiophene
CE	crown ether
CMPO	octyl(phenyl)-N,N-diisobutylcarbamoylmethylphosphine oxide
COSMO-RS	COnductor-like Screening MOdel for Real Solvents
D_i	distribution coefficient, $D_i = C^E{}_i / C^R{}_i$, Equation 6.1

DC18C6	dicyclohexyl-18-crown-6 ether
DBT	dibenzothiophene
DMF	N,N-dimethylformamide
EDC	extractive distillation column
EDTA	ethylenediaminetetraacetic acid
FCC	fluid catalytic cracked
GC	gas chromatography
HDS	hydrodesulfurization
IgG	immunoglobulin G
MDEA	methyldiethanolamine
MEA	monoethanolamine
MCH	methylcyclohexane
3-MT	3-methylthiophene
NFM	N-formylmorpholine
NMP	N-methylpyrrolidone
OEG	oligoethylene glycol
PEG	polyethylene glycol
PES	polyethersulfone
PTFE	polytetrafluoroethylene
RDC	rotating disc contactor
$S_{1/2}$	selectivity, $S_{1/2} = D_1/D_2 = (C^E_1/C^R_1)/(C^E_2/C^R_2)$, Equation 6.2
SLM	supported liquid membranes
SILM	supported ionic liquid membrane
THF	tetrahydrofuran
TMB	trimethyl borate
TSIL	task-specific ionic liquid

References

1 Huddleston, J.G., Visser, A.E., Reichert, W.M., Willauer, H.D., Broker, G.A. and Rogers, R.D. (2001) Characterization and comparison of hydrophilic and hydrophobic room temperature ionic liquids incorporating the imidazolium cation. Green Chemistry, 3 (4), 156–164.

2 Seddon, K.R., Stark, A. and Torres, M.-J. (2000) Influence of chloride, water and organic solvents on the physical properties of ionic liquids. Pure and Applied Chemistry, 72 (12), 2275–2287.

3 Swatloski, R.P., Holbrey, J.D. and Rogers, R.D. (2003) Ionic liquids are not always green: hydrolysis of 1-butyl-3-methylimidazolium hexafluorophosphate. Green Chemistry, 5 (4), 361–363.

4 Filiz, M., Sayar, N.A. and Sayar, A.A. (2006) Extraction of cobalt(II) from aqueous hydrochloric acid solutions into Alamine 336–m-xylene mixtures. Hydrometallurgy, 81 (3–4), 167–173.

5 Bradshaw, J.S. and Izatt, R.M. (1997) Crown ethers: the search for selective ion ligating agents. Accounts of Chemical Research, 30 (8), 338–345.

6 Dai, S., Ju, Y.H. and Barnes, C.E. (1999) Solvent extraction of strontium nitrate by a crown ether using room-temperature ionic liquids. Journal of The Chemical Society-Dalton Transactions (8) 1201–1202.

7 Visser, A.E., Swatloski, R.P., Reichert, W.M., Griffin, S.T. and Rogers, R.D. (2000) Traditional extractants in nontraditional solvents: groups 1 and 2 extraction by crown ethers in room-temperature ionic liquids. *Industrial & Engineering Chemistry Research*, **39** (10), 3596–3604.

8 Visser, A.E., Swatloski, R.P., Griffin, S.T., Hartman, D.H. and Rogers, R.D. (2001) Liquid/liquid extraction of metal ions in room temperature ionic liquids. *Separation Science Technololgy*, **36** (5–6), 785–804.

9 Bartsch, R.A., Chun, S. and Dzyuba, S.V. (2002) Ionic liquids as novel diluents for solvent extraction of metal salts by crown ethers, in *Ionic Liquids: Industrial Applications to Green Chemistry*, ACS Symposium Series, Vol. 818, (eds R.D. Rogers and K.R. Seddon), American Chemical Society, Washington, DC, pp. 58–68.

10 Hirayama, N., Okamura, H., Kidani, K. and Imura, H. (2008) Ionic liquid synergistic cation-exchange system for the selective extraction of lanthanum(III) using 2-thenoyltrifluoroacetone and 18-crown-6. *Analytical Sciences*, **24** (6), 697–699.

11 Luo, H., Yu, M. and Dai, S. (2007) Solvent extraction of Sr^{2+} and Cs^+ based on hydrophobic protic ionic liquids. *Zeitschrift fur Naturforschung. C: Physical Science*, **62** (5/6), 281–291.

12 Shimojo, K. and Goto, M. (2004) First application of calixarenes as extractants in room-temperature ionic liquids. *Chemistry Letters*, **33** (3), 320–321.

13 Visser, A.E., Swatloski, R.P., Hartman, D.H., Huddleston, J.G. and Rogers, R.D. (2000) Calixarenes as ligands in environmentally-benign liquid–liquid extraction media, aqueous biphasic systems and room temperature Ionic liquid, in *Calixarenes for Separations*, (eds G.J. Lunetta, R.D. Rogers and A.S. Gopalan), ACS Symposium Series, Vol. 757, American Chemical Society, Washington, DC, pp. 223–236.

14 Shimojo, K. and Goto, M. (2004) Solvent extraction and stripping of silver ions in room-temperature ionic liquids containing calixarenes. *Analytical Chemistry*, **76** (17), 5039–5044.

15 Sieffert, N. and Wipff, G. (2006) Alkali cation extraction by calix[4]crown-6 to room-temperature ionic liquids. The effect of solvent anion and humidity investigated by molecular dynamics simulations. *Journal of Physical Chemistry A*, **110** (3), 1106–1117.

16 Sieffert, N. and Wipff, G. (2006) Comparing an ionic liquid to a molecular solvent in the cesium cation extraction by a calixarene: a molecular dynamics study of the aqueous interfaces. *The Journal of Physical Chemistry. B*, **110** (39), 19497–19506.

17 Stepinski, D.C., Jensen, M.P., Dzielawa, J.A. and Dietz, M.L. (2005) Synergistic effects in the facilitated transfer of metal ions into room-temperature ionic liquids. *Green Chemistry*, **7** (3), 151–158.

18 Davis, J.H. Jr. (2004) Task-specific ionic liquids. *Chemistry Letters*, **33** (9), 1072–1077.

19 Zhao, H., Xia, S. and Ma, P. (2005) Use of ionic liquids as "green" solvents for extractions. *Journal of Chemical Technology and Biotechnology*, **80** (10), 1089–1096.

20 Chun, S., Dzyuba, S.V. and Bartsch, R.A. (2001) Influence of structural variation in room-temperature ionic liquids on the selectivity and efficiency of competitive alkali metal salt extraction by a crown ether. *Analytical Chemistry*, **73** (15), 3737–3741.

21 Keskin, S., Kayrak-Talay, D., Akman, U. and Hortacsu, O. (2007) A review of ionic liquids towards supercritical fluid applications. *Journal of Supercritical Fluids*, **43** (1), 150–180.

22 Dietz, M.L., Dzielawa, J.A., Laszak, I., Young, B.A. and Jensen, M.P. (2003) Influence of solvent structural variations on the mechanism of facilitated ion transfer into room-temperature ionic liquids. *Green Chemistry*, **5** (6), 682–685.

23 Dietz, M.L., Jakab, S., Yamato, K. and Bartsch, R.A. (2008) Stereochemical effects on the mode of facilitated ion transfer into room-temperature ionic liquids. *Green Chemistry*, **10** (2), 174–176.

24 Wei, G.-T., Yang, Z. and Chen, C.-J. (2003) Room temperature ionic liquid as a novel medium for liquid/liquid extraction of metal ions. *Analytica Chimica Acta*, **488** (2), 183–192.

25 Cocalia, V.A., Jensen, M.P., Holbrey, J.D., Spear, S.K., Stepinski, D.C. and Rogers, R.D. (2005) Identical extraction behavior and coordination of trivalent or hexavalent f-element cations using ionic liquid and molecular solvents. *Dalton Transactions*, **(11)**, 1966–1971.

26 Cocalia, V.A., Holbrey, J.D., Gutowski, K.E., Bridges, N.J. and Rogers, R.D. (2006) Separations of metal ions using ionic liquids: the challenges of multiple mechanisms. *Tsinghua Science and Technology*, **11** (2), 188–193.

27 Dietz, M.L. and Stepinski, D.C. (2005) A ternary mechanism for the facilitated transfer of metal ions into room-temperature ionic liquids (RTILs): implications for the greenness of RTILs as extraction solvents. *Green Chemistry*, **7** (10), 747–750.

28 Nakashima, K., Kubota, F., Maruyama, T. and Goto, M. (2005) Feasibility of ionic liquids as alternative separation media for industrial solvent extraction processes. *Industrial & Engineering Chemistry Research*, **44** (12), 4368–4372.

29 Sun, X.Q., Peng, B., Chen, J., Li, D.Q. and Luo, F. (2008) An effective method for enhancing metal-ions' selectivity of ionic liquid-based extraction system: adding water-soluble complexing agent. *Talanta*, **74** (4), 1071–1074.

30 Dietz, M.L. (2006) Ionic liquids as extraction solvents: where do we stand? *Separation Science Technology*, **41** (10), 2047–2063.

31 Dietz, M.L. and Stepinski, D.C. (2008) Anion concentration-dependent partitioning mechanism in the extraction of uranium into room-temperature ionic liquids. *Talanta*, **75** (2), 598–603.

32 Zuo, Y., Liu, Y., Chen, J. and Li, D.Q. (2008) The separation of cerium(IV) from nitric acid solutions containing thorium (IV) and lanthanides(III) using pure [C8mim]PF6 as extracting phase. *Industrial & Engineering Chemistry Research*, **47** (7), 2349–2355.

33 Kozonoi, N. and Ikeda, Y. (2007) Extraction mechanism of metal ions from aqueous solution to the hydrophobic ionic liquid, 1-butyl-3-methylimidazolium nonafluorobutanesulfonate. *Monatshefte fur Chemie*, **138** (11), 1145–1151.

34 Jensen, M.P., Neuefeind, J., Beitz, J.V., Skanthakumar, S. and Soderholm, L. (2003) Mechanisms of metal ion transfer into room-temperature ionic liquids: the role of anion exchange. *Journal of the American Chemical Society*, **125** (50), 15466–15473.

35 Dietz, M.L. and Dzielawa, J.A. (2001) Ion-exchange as a mode of cation transfer into room-temperature ionic liquids containing crown ethers: implications for the 'greenness' of ionic liquids as diluents in liquid-liquid extraction. *Chemical Communications*, (20), 2124–2125.

36 Visser, A.E., Swatloski, R.P., Reichert, W.M., Mayton, R., Sheff, S., Wierzbicki, A., Davis, J.H., Jr. and Rogers, R.D. (2002) Task-specific ionic liquids incorporating novel cations for the coordination and extraction of Hg^{2+} and Cd^{2+}: synthesis, characterization and extraction studies. *Environmental Science & Technology*, **36** (11), 2523–2529.

37 Nockemann, P., Thijs, B., Pittois, S., Thoen, J., Glorieux, C., VanHecke, K., VanMeervelt, L., Kirchner, B. and Binnemans, K. (2006) Task-specific ionic liquid for solubilizing metal oxides. *The Journal of Physical Chemistry. B*, **110** (42), 20978–20992.

38 Ajioka, T., Oshima, S. and Hirayama, N. (2008) Use of 8-sulfonamidoquinoline derivatives as chelate extraction reagents

in ionic liquid extraction system. *Talanta*, **74** (4), 903–908.
39 Harjani, J.R., Friscic, T., MacGillivray, L.R. and Singer, R.D. (2006) Metal chelate formation using a task-specific ionic liquid. *Inorganic Chemistry*, **45** (25), 10025–10027.
40 Harjani, J.R., Friscic, T., MacGillivray, L.R. and Singer, R.D. (2008) Removal of metal ions from aqueous solutions using chelating task-specific ionic liquids. *Dalton Transactions*, (34): 4595–4601.
41 Li, M., Wang, T., Pham, P.J., Pittman, C.U., Jr. and Li, T. (2008) Liquid phase extraction and separation of noble organometallic catalysts by functionalized ionic liquids. *Separation Science Technology*, **43** (4), 828–841.
42 Weissermel, K. and Arpe, H.-J., (2003) Aromatics – production and conversion, in *Industrial Organic Chemistry*, 4th edn, Wiley-VCH Verlag GmbH, Weinheim, pp. 313–336.
43 Chen, J., Duan, L.-P., Mi, J.-G., Fei, W.-Y. and Li, Z.-C. (2000) Liquid–liquid equilibria of multi-component systems including n-hexane, n-octane, benzene, toluene, xylene and sulfolane at 298.15 K and atmospheric pressure. *Fluid Phase Equilibrium*, **173** (1), 109–119.
44 Chen, J., Li, Z. and Duan, L. (2000) Liquid–liquid equilibria of ternary and quaternary systems including cyclohexane, 1-heptene, benzene, toluene and sulfolane at 298.15 K. *Journal of Chemical and Engineering Data*, **45** (4), 689–692.
45 Choi, Y.J., Cho, K.W., Cho, B.W. and Yeo, Y.K. (2002) Optimization of the sulfolane extraction plant based on modeling and simulation. *Industrial & Engineering Chemistry Research*, **41** (22), 5504–5509.
46 Krishna, R., Goswami, A.N., Nanoti, S.M., Rawat, B.S., Khanna, M.K. and Dobhal, J. (1987) Extraction of aromatics from 63–69 °C naphtha fraction for food-grade hexane production using sulfolane and NMP as solvents. *Indian Journal of Technology*, **25** (12), 602–606.
47 Yorulmaz, Y. and Karpuzcu, F. (1985) Sulfolane versus diethylene glycol in recovery of aromatics. *Chemical Engineering Research & Design*, **63** (3), 184–190.
48 De Fre, R.M. and Verhoeye, L.A. (1976) Phase equilibria in systems composed of an aliphatic and an aromatic hydrocarbon and sulfolane. *Journal of Applied Chemistry and Biotechnology*, **26** (9), 469–487.
49 Wang, W., Gou, Z. and Zhu, S. (1998) Liquid–liquid equilibria for aromatics extraction systems with tetraethylene glycol. *Journal of Chemical and Engineering Data*, **43** (1), 81–83.
50 Al-Sahhaf, T.A. and Kapetanovic, E. (1996) Measurement and prediction of phase equilibria in the extraction of aromatics from naphtha reformate by tetraethylene glycol. *Fluid Phase Equilibrium*, **118** (2), 271–285.
51 Ali, S.H., Lababidi, H.M.S., Merchant, S.Q. and Fahim, M.A. (2003) Extraction of aromatics from naphtha reformate using propylene carbonate. *Fluid Phase Equilibrium*, **214** (1), 25–38.
52 Schneider, D.F. (2004) Avoid sulfolane regeneration problems. *Chemical Engineering Progress*, **100** (7), 34–39.
53 Firnhaber, B., Emmrich, G., Ennenbach, F. and Ranke, U. (2000) Separation process for the recovery of pure aromatics. *Erdöl Erdgas Kohle*, **116** (5), 254–260.
54 Hombourger, T., Gouzien, L., Mikitenko, P. and Bonfils, P. (2000) Solvent extraction in the oil industry, In: *Petroleum Refining, 2. Separation Processes*, (ed. J.P. Wauqier), Editions Technip, Paris, pp. 359–456.
55 Hamid, S.H. and Ali, M.A. (1996) Comparative study of solvents for the extraction of aromatics from naphtha. *Energy Sources, Part A: Recov. Utiliz. Environ. Effects*, **18** (1), 65–84.
56 Rawat, B.S. and Gulati, I.B. (1976) Liquid–liquid equilibrium studies for separation of aromatics. *Journal of Applied Chemistry and Biotechnology*, **26** (8), 425–435.

57 Huddleston, J.G., Willauer, H.D., Swatloski, R.P., Visser, A.E. and Rogers, R.D. (1998) Room temperature ionic liquids as novel media for 'clean' liquid–liquid extraction. *Chemical Communications*, (16), 1765–1766.

58 Meindersma, G.W., Podt, A., Meseguer, M.G. and de Haan, A.B. (2005) Ionic liquids as alternatives to organic solvents in liquid–liquid extraction of aromatics, In: *Ionic Liquids IIIB, Fundamentals, Progress, Challenges and Opportunities*, (ed. R.D. Rogers and K.R. Seddon), ACS Symposium Series, Vol. 902, American Chemical Society, Washington. DC, pp. 57–71.

59 Meindersma, G.W., Podt, A. and de Haan, A.B. (2005) Selection of ionic liquids for the extraction of aromatic hydrocarbons from aromatic/aliphatic mixtures. *Fuel Processing Technology*, **87** (1), 59–70.

60 Lee, S. and Kim, H. (1995) Liquid–liquid equilibria for the ternary systems sulfolane + octane + benzene, sulfolane + octane + toluene and sulfolane + octane + p-xylene. *Journal of Chemical and Engineering Data*, **40** (2), 499–503.

61 Domanska, U., Pobudkowska, A. and Zolek-Trynowska, Z. (2007) Effect of an ionic liquid (IL) cation on the ternary system (IL + p-xylene + hexane) at $T = 298.15$ K. *Journal of Chemical and Engineering Data*, **52** (6), 2345–2349.

62 Meindersma, G.W., Podt, A. and de Haan, A.B. (2006) Ternary liquid–liquid equilibria for mixtures of an aromatic + an aliphatic hydrocarbon + 4-methyl-N-butylpyridinium tetrafluoroborate. *Journal of Chemical and Engineering Data*, **51** (5), 1814–1819.

63 Abu-Eishah, S.I. and Dowaidar, A.M. (2008) Liquid–liquid equilibrium of ternary systems of cyclohexane + (benzene, + toluene, + ethylbenzene, or + o-xylene) + 4-methyl-N-butylpyridinium tetrafluoroborate ionic liquid at 303.15 K. *Journal of Chemical and Engineering Data*, **53** (8), 1708–1712.

64 Heintz, A., Kulikov, D.V. and Verevkin, S.P. (2001) Thermodynamic properties of mixtures containing ionic liquids. 1. Activity coefficients at infinite dilution of alkanes, alkenes and alkylbenzenes in 4-methyl-N-butylpyridinium tetrafluoroborate using gas–liquid chromatography. *Journal of Chemical and Engineering Data*, **46** (6), 1526–1529.

65 Meindersma, G.W., Podt, A.J.G. and de Haan, A.B. (2006) Ternary liquid–liquid equilibria for mixtures of toluene + n-heptane + an ionic liquid. *Fluid Phase Equilibrium*, **247** (1–2), 158–168.

66 Krummen, M., Wasserscheid, P. and Gmehling, J. (2002) Measurement of activity coefficients at infinite dilution in ionic liquids using the dilutor technique. *Journal of Chemical and Engineering Data*, **47** (6), 1411–1417.

67 Arce, A., Earle, M.J., Rodriguez, H. and Seddon, K.R. (2007) Separation of aromatic hydrocarbons from alkanes using the ionic liquid 1-ethyl-3-methylimidazolium bis{(trifluoromethyl)sulfonyl}amide. *Green Chemistry*, **9** (1), 70–74.

68 Heintz, A., Kulikov, D.V. and Verevkin, S.P. (2002) Thermodynamic properties of mixtures containing ionic liquids. 2. Activity coefficients at infinite dilution of hydrocarbons and polar solutes in 1-methyl-3-ethylimidazolium bis(trifluoromethylsulfonyl)amide and in 1,2-dimethyl-3-ethylimidazolium bis(trifluoromethylsulfonyl)amide using gas–liquid chromatography. *Journal of Chemical and Engineering Data*, **47** (4), 894–899.

69 Domanska, U., Pobudkowska, A. and Krolikowski, M. (2007) Separation of aromatic hydrocarbons from alkanes using ammonium ionic liquid C_2NTf_2 at $T = 298.15$ K. *Fluid Phase Equilibrium*, **259** (2), 173–179.

70 Letcher, T.M. and Reddy, P. (2005) Ternary (liquid + liquid) equilibria for mixtures of 1-hexyl-3-methylimidazolium (tetrafluoroborate or hexafluorophosphate) +

benzene + an alkane at $T = 298.2$ K and $p = 0.1$ MPa. *Journal of Chemical Thermodynamics*, **37** (5), 415–421.

71 Kato, R. and Gmehling, J. (2004) Activity coefficients at infinite dilution of various solutes in the ionic liquids $[MMIM]^+[CH_3SO_4]^-$, $[MMIM]^+[CH_3OC_2H_4SO_4]^-$, $[MMIM]^+[(CH_3)_2PO_4]^-$, $[C_5H_5NC_2H_5]^+[(CF_3SO_2)_2N]^-$ and $[C_5H_5NH]^+[C_2H_5OC_2H_4OSO_3]^-$. *Fluid Phase Equilibrium*, **226** 37–44.

72 Zhang, J., Huang, C., Chen, B., Ren, P. and Lei, Z. (2007) Extraction of aromatic hydrocarbons from aromatic/aliphatic mixtures using chloroaluminate room-temperature ionic liquids as extractants. *Energy Fuels*, **21** (3), 1724–1730.

73 Meindersma, G.W., Galán Sánchez, L.M., Hansmeier, A.R. and de Haan, A.B. (2007) Invited review. Application of task-specific ionic liquids for intensified separations. *Monatshefte fur Chemie*, **138** (11), 1125–1136.

74 Selvan, M.S., McKinley, M.D., Dubois, R.H. and Atwood, J.L. (2000) Liquid–liquid equilibria for toluene + heptane + 1-ethyl-3-methylimidazolium triiodide and toluene + heptane + 1-butyl-3-methylimidazolium triiodide. *Journal of Chemical and Engineering Data*, **45** (5), 841–845.

75 Meindersma, G.W. and de Haan, A.B. (2008) Conceptual process design for aromatic/aliphatic separation with ionic liquids. *Chemical Engineering Research & Design*, **86** (7), 745–752.

76 Zimmermann, H. and Walzl, R., (2007) Ethylene, in *Ullmann's Encyclopedia of Industrial Chemistry*, Wiley-VCH Verlag GmbH, Weinheim., DOI:10.1002/14356007.a10_045.pub2.

77 Wasserscheid, P. and Welton, T. (2008) Outlook, in *Ionic Liquids in Synthesis*, 2nd edn (eds P. Wasserscheid and T. Welton), Wiley-VCH Verlag GmbH, Weinheim, pp. 689–704.

78 Maase, M., (2004) Ionic liquids on a large scale, how they can help to improve chemical processes, in *Ionic Liquids – A Road-Map to Commercialization* (CD ROM), Royal Society of Chemistry, Cambridge.

79 Maase, M. (2005) Cosi fan tutte ("They all can do it"): an improved way of doing it, in *Proceedings of the 1st International Congress on Ionic Liquids (COIL)*, Dechema, Salzburg, p. 37.

80 Johnson, R.D. (1986) The processing of biomacromolecules: a challenge for the eighties. *Fluid Phase Equilibrium*, **29**, 109–123.

81 Vernau, J. and Kula, M.R. (1990) Extraction of proteins from biological raw material using aqueous polyethylene glycol–citrate phase systems. *Biotechnology and Applied Biochemistry*, **12** (4), 397–404.

82 Creagh, A.L., Hasenack, B.B.E., Van der Padt, A., Sudhoelter, E.J.R. and Van't Riet, K. (1994) Separation of amino-acid enantiomers using micellar-enhanced ultrafiltration. *Biotechnology and Bioengineering*, **44** (6), 690–698.

83 Riedl, W. and Raiser, T. (2008) Membrane-supported extraction of biomolecules with aqueous two-phase systems. *Desalination*, **224** (1–3), 160–167.

84 Hatti-Kaul, R. (2000) Aqueous two-phase systems: a general overview. *Methods Biotechnology*, **11** (Aqueous Two-Phase Systems), 1–10.

85 Andrews, B.A., Schmidt, A.S. and Asenjo, J.A. (2005) Correlation for the partition behavior of proteins in aqueous two-phase systems: effect of surface hydrophobicity and charge. *Biotechnology and Bioengineering*, **90** (3), 380–390.

86 Costa, M.J.L., Cunha, M.T., Cabral, J.M.S. and Aires-Barros, M.R. (2000) Scale-up of recombinant cutinase recovery by whole broth extraction with PEG–phosphate aqueous two-phase. *Bioseparations*, **9** (4), 231–238.

87 Martínez-Aragón, M., Burghoff, S., Goetheer, E.L.V. and de Haan, A.B. (2009) Guidelines for solvent selection for carrier mediated extraction of proteins.

88 Pei, Y., Wang, J., Wu, K., Xuan, X. and Lu, X. (2009) Ionic liquid-based aqueous two-phase extraction of selected proteins. *Separation and Purification Technology*, **64** (3), 288–295.

89 Jiang, Y., Xia, H., Guo, C., Mahmood, I. and Liu, H. (2007) Phenomena and mechanism for separation and recovery of penicillin in ionic liquids aqueous solution. *Industrial & Engineering Chemistry Research*, **46** (19), 6303–6312.

90 Du, Z., Yu, Y.-L. and Wang, J.-H. (2007) Extraction of proteins from biological fluids by use of an ionic liquid/aqueous two-phase system. *Chemistry - A European Journal*, **13** (7), 2130–2137.

91 Dreyer, S. and Kragl, U. (2008) Ionic liquids for aqueous two-phase extraction and stabilization of enzymes. *Biotechnology and Bioengineering*, **99** (6), 1416–1424.

92 Tzeng, Y.-P., Shen, C.-W. and Yu, T. (2008) Liquid–liquid extraction of lysozyme using a dye-modified ionic liquid. *Journal of Chromatography. A*, **1193** (1–2), 1–6.

93 Lei, Z.G., Li, C.Y. and Chen, B.H. (2003) Extractive distillation: a review. *Separation Purification Review*, **32** (2), 121–213.

94 ED Sulfolane™ Process, UoP, http://www.uop.com/objects/ED%20Sulfolane.pdf. (Last accessed February 04, 2010).

95 Furter, W.F. and Cook, R.A. (1967) Salt effect in distillation: a literature review. *International Journal of Heat and Mass Transfer*, **10** (1), 23–36.

96 Pinto, R.T.P., Wolf-Maciel, M.R. and Lintomen, L. (2000) Saline extractive distillation process for ethanol purification. *ComputerS & Chemical Engineering*, **24** (2–7), 1689–1694.

97 Liao, B., Lei, Z., Xu, Z., Zhou, R. and Duan, Z. (2001) New process for separating propylene and propane by extractive distillation with aqueous acetonitrile. *Biochemical Engineering Journal*, **84** (3), 581–586.

98 Lei, Z., Wang, H., Zhou, R. and Duan, Z. (2002) Influence of salt added to solvent on extractive distillation. *Biochemical Engineering Journal*, **87** (2), 149–156.

99 Lei, Z., Zhou, R. and Duan, Z. (2002) Application of scaled particle theory in extractive distillation with salt. *Fluid Phase Equilibrium*, **200** (1), 187–201.

100 Ligero, E.L. and Ravagnani, T.M.K. (2003) Dehydration of ethanol with salt extractive distillation – a comparative analysis between processes with salt recovery. *Chemical Engineering and Processing*, **42** (7), 543–552.

101 Furter, W.F. (1992) Extractive distillation by salt effect. *Chemical Engineering Communications*, **116**, 35–40.

102 Beste, Y.A., Schoenmakers, H., Arlt, W., Seiler, M. and Jork, C. (2005) to BASF, Recycling of ionic liquids with extractive distillation. World Patent WO 2005016484.

103 Arlt, W., Seiler, M., Jork, C. and Schneider, T. (2002) to BASF Ionic liquids as selective additives for the separation of close-boiling or azeotropic mixtures. World Patent WO 200202074718.

104 Gmehling, J. and Krummen, M. (2003) Einsatz ionischer Flüssigkeiten als selektive Lösungsmittel für die Trennung aromatischer Kohlenwasserstoffe van nichtaromatischen Kohlenwasserstoffe durch extractieve Rektifikation und Extraktion. German Patent DE10154052.

105 Jork, C., Seiler, M., Beste, Y.A. and Arlt, W. (2004) Influence of ionic liquids on the phase behavior of aqueous azeotropic systems. *Journal of Chemical and Engineering Data*, **49** (4), 852–857.

106 Seiler, M., Jork, C., Kavarnou, A., Arlt, W. and Hirsch, R. (2004) Separation of azeotropic mixtures using hyperbranched polymers or ionic liquids. *AICHE Journal*, **50** (10), 2439–2454.

107 Seiler, M., Jork, C. and Arlt, W. (2004) Phasenverhalten von hochselektiven nichtflüchtigen Flüssigkeiten mit designbarem Eigenschaftsprofil und neue Anwendungen in der thermischen

Verfahrenstechnik. *Chemie Ingenieur Technik*, **76** (6), 735–744.
108 Y.A. Beste and H. Schoenmakers (2005) to BASF, Distillative method for separating narrow boiling or azeotropic mixtures using ionic liquids. World Patent WO 2005016483.
109 Beste, Y., Eggersmann, M. and Schoenmakers, H. (2005) Extraktivdestillation mit ionischen Flüssigkeiten. *Chemie Ingenieur Technik*, **77** (11), 1800–1808.
110 Beste, Y.A., Eggersmann, M. and Schoenmakers, H. (2005) Ionic liquids: breaking azeotropes efficiently by extractive distillation. In: *Sustainable (Bio) Chemical Process Technology – Incorporating the 6th Intenational Conference on Process Intensification*.
111 Lei, Z., Arlt, W. and Wasserscheid, P. (2006) Separation of 1-hexene and n-hexane with ionic liquids. *Fluid Phase Equilibrium*, **241** (1–2), 290–299.
112 Zhu, J., Chen, J., Li, C. and Fei, W., (2006) Study on the separation of 1-hexene and trans-3-hexene using ionic liquids. *Fluid Phase Equilibrium*, **247** (1–2), 102–106.
113 Beste, Y.A. and Schoenmakers, H. (2005) to BASF, Distllative method for separating narrow boiling or azeotropic mixtures using ionic liquids. World Patent WO 2005016483.
114 Jork, C., Seiler, M., Beste, Y.-A. and Arlt, W. (2004) Influence of ionic liquids on the phase behavior of aqueous azeotropic systems. *Journal of Chemical and Engineering Data*, **49** (4), 852–857.
115 Lei, Z., Arlt, W. and Wasserscheid, P. (2006) Separation of 1-hexene and n-hexane with ionic liquids. *Fluid Phase Equilibrium*, **241** (1–2), 290–299.
116 Lei, Z., Arlt, W. and Wasserscheid, P. (2007) Selection of entrainers in the 1-hexene/n-hexane system with a limited solubility. *Fluid Phase Equilibrium*, **260** (1), 29–35.
117 Liebert, V., Nebig, S. and Gmehling, J. (2008) Experimental and predicted phase equilibria and excess properties for systems with ionic liquids. *Fluid Phase Equilibrium*, **268** (1–2), 14–20.
118 Jork, C., Kristen, C., Pieraccini, D., Stark, A., Chiappe, C., Beste, Y.A. and Arlt, W. (2005) Tailor-made ionic liquids. *Journal of Chemical Thermodynamics*, **37** (6), 537–558.
119 Möllmann, C. and Gmehling, J. (1997) Measurement of activity coefficients at infinite dilution using gas–liquid chromatography. 5. Results for N-methylacetamide, N,N-dimethylacetamide, N,N-dibutylformamide and sulfolane as stationary phases. *Journal of Chemical and Engineering Data*, **42** (1), 35–40.
120 Weidlich, U., Roehm, H.J. and Gmehling, J. (1987) Measurement of γ^∞ using GLC. 2. Results for the stationary phases N-formylmorpholine and N-methylpyrrolidone. *Journal of Chemical and Engineering Data*, **32** (4), 450–453.
121 Santacesaria, E., Berlendis, D. and Carr,à S. (1979) Measurement of activity coefficients at infinite dilution by stripping and retention time methods. *Fluid Phase Equilibrium*, **3** (2–3), 167–176.
122 Maase, M. (2008) Industrial applications of ionic lquids, in *Ionic Liquids in Synthesis*, 2nd edn (eds P. Wasserscheid and T. Welton), Wiley-VCH Verlag GmbH, Weinheim, pp. 663–687.
123 Ito, E. and van Veen, J.A.R. (2006) On novel processes for removing sulfur from refinery streams. *Catalysis Today*, **116** (4), 446–460.
124 Bösmann, A., Datsevich, L., Jess, A., Lauter, A., Schmitz, C. and Wasserscheid, P. (2001) Deep desulfurization of diesel fuel by extraction with ionic liquids. *Chemical Communications*, (23), 2494–2495.
125 Schmidt, R. (2008) [bmim]AlCl$_4$ ionic liquid for deep desulfurization of real fuels. *Energy Fuels*, **22** (3), 1774–1778.
126 Huang, C., Chen, B., Zhang, J., Liu, Z. and Li, Y. (2004) Desulfurization of

gasoline by extraction with new ionic liquids. *Energy Fuels*, **18** (6), 1862–1864.
127 Xie, L.-L., Favre-Reguillon, A., Wang, X.-X., Fu, X., Pellet-Rostaing, S., Toussaint, G., Geantet, C., Vrinat, M. and Lemaire, M. (2008) Selective extraction of neutral nitrogen compounds found in diesel feed by 1-butyl-3-methylimidazolium chloride. *Green Chemistry*, **10** (5), 524–531.
128 Ko, N.H., Lee, J.S., Huh, E.S., Lee, H., Jung, K.D., Kim, H.S. and Cheong, M. (2008) Extractive desulfurization using Fe-containing ionic liquids. *Energy Fuels*, **22** (3), 1687–1690.
129 Zhang, S.G. and Zhang, Z.C. (2002) Novel properties of ionic liquids in selective sulfur removal from fuels at room temperature. *Green Chemistry*, **4** (4), 376–379.
130 Zhang, S., Zhang, Q. and Zhang, Z.C. (2004) Extractive desulfurization and denitrogenation of fuels using ionic liquids. *Industrial & Engineering Chemistry Research*, **43** (2), 614–622.
131 Alonso, L., Arce, A., Francisco, M., Rodriguez, O. and Soto, A. (2007) Gasoline desulfurization using extraction with [C_8mim][BF_4] ionic liquid. *AICHE Journal*, **53** (12), 3108–3115.
132 Liu, D., Gui, J., Song, L., Zhang, X. and Sun, Z. (2008) Deep desulfurization of diesel fuel by extraction with task-specific ionic liquids. *Petroleum Science and Technology*, **26** (9), 973–982.
133 Nie, Y., Li, C., Sun, A., Meng, H. and Wang, Z. (2006) Extractive desulfurization of gasoline using imidazolium-based phosphoric ionic liquids. *Energy Fuels*, **20** (5), 2083–2087.
134 Nie, Y., Li, C.-X. and Wang, Z.-H. (2007) Extractive desulfurization of fuel oil using alkylimidazole and its mixture with dialkylphosphate ionic liquids. *Industrial & Engineering Chemistry Research*, **46** (15), 5108–5112.
135 Jiang, X., Nie, Y., Li, C. and Wang, Z. (2008) Imidazolium-based alkylphosphate ionic liquids – a potential solvent for extractive desulfurization of fuel. *Fuel*, **87** (1), 79–84.
136 Nie, Y., Li, C., Meng, H. and Wang, Z. (2008) N,N-Dialkylimidazolium dialkylphosphate ionic liquids: their extractive performance for thiophene series compounds from fuel oils versus the length of alkyl group. *Fuel Processing Technology*, **89** (10), 978–983.
137 Wang, J.-l., Zhao, D.-s., Zhou, E.-p. and Dong, Z. (2007) Desulfurization of gasoline by extraction with N-alkylpyridinium-based ionic liquids. *Journal of Fuel Chemistry and Technology*, **35** (3), 293–296.
138 Holbrey, J.D., Lopez-Martin, I., Rothenberg, G., Seddon, K.R., Silvero, G. and Zheng, X. (2008) Desulfurization of oils using ionic liquids: selection of cationic and anionic components to enhance extraction efficiency. *Green Chemistry*, **10** (1), 87–92.
139 Esser, J., Wasserscheid, P. and Jess, A. (2004) Deep desulfurization of oil refinery streams by extraction with ionic liquids. *Green Chemistry*, **6** (7), 316–322.
140 Olivier-Bourbigou, H., Uzio, D., Diehl, F. and Magna, L. (2003) to Institut Français du Pétrole, Process for removal of sulfur and nitrogen compounds from hydrocarbon fractions. French Patent 2840916.
141 Lo, W.-H., Yang, H.-Y. and Wei, G.-T. (2003) One-pot desulfurization of light oils by chemical oxidation and solvent extraction with room temperature ionic liquids. *Green Chemistry*, **5** (5), 639–642.
142 Zhao, D., Wang, J. and Zhou, E. (2007) Oxidative desulfurization of diesel fuel using a Brønsted acid room temperature ionic liquid in the presence of H_2O_2. *Green Chemistry*, **9** (11), 1219–1222.
143 Zhao, D., Sun, Z., Li, F., Liu, R. and Shan, H. (2008) Oxidative desulfurization of thiophene catalyzed by $(C_4H_9)_4$NBr + $2C_6H_{11}$NO coordinated ionic liquid. *Energy Fuels*, **22** (5), 3065–3069.
144 Zhu, W., Li, H., Jiang, X., Yan, Y., Lu, J., He, L. and Xia, J. (2008) Commercially

available molybdic compound-catalyzed ultra-deep desulfurization of fuels in ionic liquids. *Green Chemistry*, **10** (6), 641–646.

145 Gargano, G.J. and Ruether, T. (2007) to Ultraclean Fuel Pty Ltd, Process for removing sulfur from liquid hydrocarbons. World Patent WO 2007106943.

146 Chakraborty, M. and Bart, H.-J. (2007) Highly selective and efficient transport of toluene in bulk ionic liquid membranes containing Ag^+ as carrier. *Fuel Processing Technology*, **88** (1), 43–49.

147 Branco, L.C., Crespo, J.G. and Afonso, C.A.M. (2008) Ionic liquids as an efficient bulk membrane for the selective transport of organic compounds. *Journal of Physical Organic Chemistry*, **21** (7–8), 718–723.

148 Matsumoto, M., Mikami, M. and Kondo, K. (2006) Separation of organic nitrogen compounds by supported liquid membranes based on ionic liquids. *J. Jpn. Pet. Inst.*, **49** (5), 256–261.

149 Branco, L.C., Crespo, J.G. and Afonso, C.A.M. (2002) Highly selective transport of organic compounds by using supported liquid membranes based on ionic liquids. *Angewandte Chemie (International Edition in English)*, **41** (15), 2771–2773.

150 Branco, L.C., Crespo, J.G. and Afonso, C.A.M. (2002) Studies on the selective transport of organic compounds by using ionic liquids as novel supported liquid membranes. *Chemistry - A European Journal*, **8** (17), 3865–3871.

151 Matsumoto, M., Inomoto, Y. and Kondo, K. (2005) Selective separation of aromatic hydrocarbons through supported liquid membranes based on ionic liquids. *The Journal of Membrane Science*, **246** (1), 77–81.

152 F. De Jong and J. De With (2005) to Shell Internationale Research Maatschappij, Process for the separation of olefins from paraffins using a supported ionic liquid membrane. World Patent Application WO 2005061422.

153 Schaefer, T., Branco, L.C., Fortunato, R., Izak, P., Rodrigues, C.M., Afonso, C.A.M. and Crespo, J.G., (2005) Opportunities for membrane separation processes using ionic liquids, in *Ionic Liquids IIIB: Fundamentals, Progress, Challenges and Opportunities*, (eds R.D. Rogers and K.R. Seddon), ACS Symposium Series, Vol. 902, American Chemical Society, Washington, DC, pp. 97–110.

154 Kröckel, J. and Kragl, U. (2003) Nanofiltration for the separation of nonvolatile products from solutions containing ionic liquids. *Chemical Engineering & Technology*, **26** (11), 1166–1168.

155 Han, S., Wong, H.T. and Livingston, A.G. (2005) Application of organic solvent nanofiltration to separation of ionic liquids and products from ionic liquid mediated reactions. *Chemical Engineering Research & Design*, **83** (A3), 309–316.

156 Wasserscheid, P., Kragl, U. and Kröckel, J. (2003) to Solvent Innovation GmbH, Method for separating substances from solutions containing ionic liquids by means of a membrane. World Patent WO 2003039719.

157 Schäfer, T. and Goulao Crespo, J.P.S. (2003) Removal and recovery of solutes present in ionic liquids by pervaporation. World Patent WO 2003013685.

158 David, W., Letcher, T.M., Ramjugernath, D. and Raal, D.J. (2003) Activity coefficients of hydrocarbon solutes at infinite dilution in the ionic liquid, 1-methyl-3-octylimidazolium chloride from gas–liquid chromatography. *Journal of Chemical Thermodynamics*, **35** (8), 1335–1341.

159 Won, J., Kim, D.B., Kang, Y.S., Choi, D.K., Kim, H.S., Kim, C.K. and Kim, C.K. (2005) An *ab initio* study of ionic liquid silver complexes as carriers in facilitated olefin transport membranes. *The Journal of Membrane Science*, **260** (1–2), 37–44.

160 Kang, S.W., Char, K., Kim, J.H. and Kang, Y.S. (2007) Ionic liquid as a solvent and the

long-term separation performance in a polymer/silver salt complex membrane. *Macromolecular Research*, **15** (2), 167–172.
161 Meisen, A. and Shuai, X. (1997) Research and development issues in CO_2 capture. *Energy Conversion Management*, **38** (Suppl., Proceedings of the Third International Conference on Carbon Dioxide Removal, 1996), S37–S42.
162 Rao, A.B. and Rubin, E.S. (2002) A technical, economic and environmental assessment of amine-based CO_2 capture technology for power plant greenhouse gas control. *Environmental Science & Technology*, **36** (20), 4467–4475.
163 Anthony, J.L., Maginn, E.J. and Brennecke, J.F., (2002) Gas solubilities in 1-*n*-butyl-3-methylimidazolium hexafluorophosphate, in *Ionic Liquids*, (eds R.D. Rogers and K.R. Seddon), ACS Symposium Series, Vol. 818, American Chemical Society, Washington, DC, pp. 260–269.
164 Anthony, J.L., Maginn, E.J. and Brennecke, J.F. (2002) Solubilities and thermodynamic properties of gases in the ionic liquid 1-*n*-butyl-3-methylimidazolium hexafluorophosphate. *The Journal of Physical Chemistry. B*, **106** (29), 7315–7320.
165 Pérez-Salado Kamps, Á., Tuma, D., Xia, J. and Maurer, G. (2003) Solubility of CO_2 in the ionic liquid [bmim][PF_6]. *Journal of Chemical and Engineering Data*, **48** (3), 746–749.
166 Aki, S.N.V.K., Mellein, B.R., Saurer, E.M. and Brennecke, J.F. (2004) High-pressure phase behavior of carbon dioxide with imidazolium-based ionic liquids. *The Journal of Physical Chemistry. B*, **108** (52), 20355–20365.
167 Cadena, C., Anthony, J.L., Shah, J.K., Morrow, T.I., Brennecke, J.F. and Maginn, E.J. (2004) Why is CO_2 so soluble in imidazolium-based ionic liquids? *Journal of the American Chemical Society*, **126** (16), 5300–5308.
168 Camper, D., Scovazzo, P., Koval, C. and Noble, R. (2004) Gas solubilities in room-temperature ionic liquids. *Industrial & Engineering Chemistry Research*, **43** (12), 3049–3054.
169 Baltus, R.E., Culbertson, B.H., Dai, S., Luo, H. and DePaoli, D.W. (2004) Low-pressure solubility of carbon dioxide in room-temperature ionic liquids measured with a quartz crystal microbalance. *The Journal of Physical Chemistry. B*, **108** (2), 721–727.
170 Scovazzo, P., Camper, D., Kieft, J., Poshusta, J., Koval, C. and Noble, R. (2004) Regular solution theory and CO_2 gas solubility in room-temperature ionic liquids. *Industrial & Engineering Chemistry Research*, **43** (21), 6855–6860.
171 Anthony, J.L., Anderson, J.L., Maginn, E.J. and Brennecke, J.F. (2005) Anion effects on gas solubility in ionic liquids. *The Journal of Physical Chemistry. B*, **109** (13), 6366–6374.
172 Shariati, A. and Peters, C.J. (2005) High-pressure phase equilibria of systems with ionic liquids. *Journal of Supercritical Fluids*, **34** (2), 171–176.
173 Shiflett, M.B. and Yokozeki, A. (2005) Solubilities and diffusivities of carbon dioxide in ionic liquids: [bmim][PF_6] and [bmim][BF_4]. *Industrial & Engineering Chemistry Research*, **44** (12), 4453–4464.
174 Anderson, J.L., Dixon, J.K. and Brennecke, J.F. (2007) Solubility of CO_2, CH_4, C_2H_6, C_2H_4, O_2 and N_2 in 1-hexyl-3-methylpyridinium bis(trifluoromethylsulfonyl)imide: comparison to other ionic liquids. *Accounts of Chemical Research*, **40** (11), 1208–1216.
175 Galán Sánchez, L.M., Meindersma, G.W. and de Haan, A.B. (2007) Solvent properties of functionalized ionic liquids for CO_2 absorption. *Chemical Engineering Research & Design*, **85** (1), 31–39.
176 Muldoon, M.J., Aki, S.N.V.K., Anderson, J.L., Dixon, J.K. and Brennecke, J.F. (2007) Improving carbon dioxide solubility in ionic liquids. *The Journal of Physical Chemistry. B*, **111** (30), 9001–9009.
177 Bates, E.D., Mayton, R.D., Ntai, I. and Davis, J.H., Jr. (2002) CO_2 capture by a

task-specific ionic liquid. *Journal of the American Chemical Society*, **124** (6), 926–927.

178 Davis, J.H., Jr. (2005) Task-specific ionic liquids for separations of petrochemical relevance: Reactive capture of CO_2 using amine-incorporating ions, in *Ionic Liquids IIIB: Fundamentals, Progress, Challenges and Opportunities*, (eds R.D. Rogers and K.R. Seddon), ACS Symposium Series, Vol. 902, American Chemical Society, Washington, DC, pp. 49–56.

179 Camper, D., Bara, J.E., Gin, D.L. and Noble, R.D. (2008) Room-Temperature ionic liquid–amine solutions: tunable solvents for efficient and reversible capture of CO_2. *Industrial & Engineering Chemistry Research*, **47** (21), 8496–8498.

180 Galán Sánchez, L.M., Meindersma, G.W. and de Haan, A.B., (2007) *Ionic Liquid Solvents for Gas Sweetening Operations in Greenhouse Gases*, CHEMRAWN-XVII and ICCDU-IX Conference on Green House Gases, 8-12 July 2007, Kingston, Canada, www.chem.queensu.ca/Conferences/CHEMRAWN/Meindersma_55.ppt.

181 Galán Sánchez, L.M. (2008) Functionalized ionic liquids absorption solvents for carbon dioxide and olefin separation. PhD thesis, Eindhoven University of Technology.

182 Bara, J.E., Gabriel, C.J., Lessmann, S., Carlisle, T.K., Finotello, A., Gin, D.L. and Noble, R.D. (2007) Enhanced CO_2 separation selectivity in oligo(ethylene glycol) functionalized room-temperature ionic liquids. *Industrial & Engineering Chemistry Research*, **46** (16), 5380–5386.

183 Bara, J.E., Gabriel, C.J., Hatakeyama, E.S., Carlisle, T.K., Lessmann, S., Noble, R.D. and Gin, D.L. (2008) Improving CO_2 selectivity in polymerized room-temperature ionic liquid gas separation membranes through incorporation of polar substituents. *The Journal of Membrane Science*, **321** (1), 3–7.

184 Hanioka, S., Maruyama, T., Sotani, T., Teramoto, M., Matsuyama, H., Nakashima, K., Hanaki, M., Kubota, F. and Goto, M. (2008) CO_2 separation facilitated by task-specific ionic liquids using a supported liquid membrane. *The Journal of Membrane Science*, **314** (1–2), 1–4.

185 Scovazzo, P., Kieft, J., Finan, D.A., Koval, C., DuBois, D. and Noble, R. (2004) Gas separations using non-hexafluorophosphate $[PF_6]^-$ anion supported ionic liquid membranes. *The Journal of Membrane Science*, **238** (1–2), 57–63.

186 Baltus, R.E., Counce, R.M., Culbertson, B.H., Luo, H.M., DePaoli, D.W., Dai, S. and Duckworth, D.C. (2005) Examination of the potential of ionic liquids for gas separations. *Separation Science Technology*, **40** (1–3), 525–541.

187 Kang, S.W., Char, K., Kim, J.H., Kim, C.K. and Kang, Y.S. (2006) Control of ionic interactions in silver salt–polymer complexes with ionic liquids: implications for facilitated olefin transport. *Chemistry of Materials*, **18** (7), 1789–1794.

188 Munson, C.L., Boudreau, L.C., Driver, M.S. and Schinski, W. (2002) to Chevron USA, Separation of olefins from paraffins using ionic liquid solutions. US Patent 6339182.

189 Munson, C.L., Boudreau, L.C., Driver, M.S. and Schinski, W. (2003) to Chevron USA, Separation of olefins from paraffins using ionic liquid solutions. US Patent 6623659.

190 Camper, D., Becker, C., Koval, C. and Noble, R. (2005) Low pressure hydrocarbon solubility in room temperature ionic liquids containing imidazolium rings interpreted using regular solution theory. *Industrial & Engineering Chemistry Research*, **44** (6), 1928–1933.

191 Morgan, D., Ferguson, L. and Scovazzo, P. (2005) Diffusivities of gases in room-temperature ionic liquids: data and correlations obtained using a lag-time technique. *Industrial & Engineering Chemistry Research*, **44** (13), 4815–4823.

192 Gutmann, M., Mueller, W. and Zeppenfeld, R. (2005) to Linde AG, Verfahren zur Olefinabtrannung aus Spaltgasen von Olefinanlagen mittels ionischer Flüssigkeiten. German Patent DE 10333546.

193 Camper, D., Becker, C., Koval, C. and Noble, R. (2006) Diffusion and solubility measurements in room temperature ionic liquids. *Industrial & Engineering Chemistry Research*, **45** (1), 445–450.

194 Ferguson, L. and Scovazzo, P. (2007) Solubility, Diffusivity and permeability of gases in phosphonium-based room temperature ionic liquids: data and correlations. *Industrial & Engineering Chemistry Research*, **46** (4), 1369–1374.

195 Ortiz, A., Ruiz, A., Gorri, D. and Ortiz, I. (2008) Room temperature ionic liquid with silver salt as efficient reaction media for propylene/propane separation: absorption equilibrium. *Separation and Purification Technology*, **63** (2), 311–318.

196 Huang, J.-F., Luo, H., Liang, C., Jiang, D.-e. and Dai, S. (2008) Advanced liquid membranes based on novel ionic liquids for selective separation of olefin/paraffin via olefin-facilitated transport. *Industrial & Engineering Chemistry Research*, **47** (3), 881–888.

197 Kang, Y.S., Jung, B., Kim, J.H., Won, J., Char, K.H. and Kang, S.W. (2005) Facilitated transport membranes for an alkene hydrocarbon separation. European Patent EP 1552875.

198 Kang, S.W., Hong, J., Char, K., Kim, J.H., Kim, J. and Kang, Y.S. (2008) Correlation between anions of ionic liquids and reduction of silver ions in facilitated olefin transport membranes. *Desalination*, **233** (1–3), 327–332.

199 Zhu, J.-Q., Chen, J., Li, C.-Y. and Fei, W.-Y. (2007) Centrifugal extraction for separation of ethylbenzene and octane using 1-butyl-3-methylimidazolium hexafluorophosphate ionic liquid as extractant. *Separation and Purification Technology*, **56** (2), 237–240.

200 S.A.F., Onink G.W. Meindersma and A.B. de Haan (2008) Ionic liquids in extraction operations: comparison of rotating disc contactor performance between [4-mebupy]BF_4 and sulfolane for aromatics extraction. in *Proceedings of ISEC 2008*, Tucson, AZ, pp. 1337–1342.

201 Godfrey, J.C. and Slater, M.J. (1994) *Liquid–Liquid Extraction Equipment*, John Wiley & Sons Inc., New York.

202 Lo, T.C., Baird, M.H.I. and Hanson, C. (1983) *Handbook of Solvent Extraction*, John Wiley & Sons Inc., New York.

203 Müller, E., Berger, R., Blass, E., Sluyts, D. and Pfennig, A. (2008) Liquid–liquid extraction. In: *Ullmann's Encyclopedia of Industrial Chemistry*, Wiley-VCH Verlag GmbH, Weinheim, DOI: 10.1002/14356007.b03_06.pub2.

204 Kamath, M.S., Rao, K.L., Jayabalou, R., Karanth, P.K. and Rau, M.G.S. (1976) Holdup studies in a rotary disc contactor. *Indian Journal of Technology*, **14** (1), 1–5.

205 Chiappe, C. and Pieraccini, D. (2005) Ionic liquids: solvent properties and organic reactivity. *Journal of Physical Organic Chemistry*, **18** (4), 275–297.

206 Klamt, A. and Eckert, F. (2000) COSMO-RS: a novel and efficient method for the a priori prediction of thermophysical data of liquids. *Fluid Phase Equilibrium*, **172** (1), 43–72.

207 Klamt, A. and Eckert, F. (2004) Prediction of vapor–liquid equilibria using COSMOtherm. *Fluid Phase Equilibrium*, **217** (1), 53–57.

208 Klamt, A. (2005) COSMO-RS, *From Quantum Chemistry to Fluid Phase Thermodynamics and Drug Design*, Elsevier, Amsterdam.

209 Diedenhofen, M., Eckert, F. and Klamt, A. (2003) Prediction of infinite dilution activity coefficients of organic compounds in ionic liquids using COSMO-RS. *Journal of Chemical and Engineering Data*, **48** (3), 475–479.

210 Putnam, R., Taylor, R., Klamt, A., Eckert, F. and Schiller, M. (2003) Prediction of infinite dilution activity coefficients using

COSMO-RS. *Industrial & Engineering Chemistry Research*, **42** (15), 3635–3641.
211 Banerjee, T. and Khanna, A. (2006) Infinite dilution activity coefficients for trihexyltetradecyl phosphonium ionic liquids: measurements and COSMO-RS prediction. *Journal of Chemical and Engineering Data*, **51** (6), 2170–2177.
212 Domanska, U., Pobudkowska, A. and Eckert, F. (2006) (Liquid + liquid) phase equilibria of 1-alkyl-3-methylimidazolium methylsulfate with alcohols or ethers or ketones. *Journal of Chemical Thermodynamics*, **38** (6), 685.
213 Domanska, U., Pobudkowska, A. and Eckert, F. (2006) Liquid–liquid equilibria in the binary systems (1,3-dimethylimidazolium or 1-butyl-3-methylimidazolium methylsulfate + hydrocarbons). *Green Chemistry*, **8** (3), 268–276.
214 Freire, M.G., Santos, L.M.N.B.F., Marrucho, I.M. and Coutinho, J.A.P. (2007) Evaluation of COSMO-RS for the prediction of LLE and VLE of alcohols + ionic liquids. *Fluid Phase Equilibrium*, **255** (2), 167–178.
215 Freire, M.G., Ventura, S.P.M., Santos, L.M.N.B.F., Marrucho, I.M. and Coutinho, J.A.P. (2008) Evaluation of COSMO-RS for the prediction of LLE and VLE of water and ionic liquids binary systems. *Fluid Phase Equilibrium*, **268** (1–2), 74–84.
216 Banerjee, T., Sahoo, R.K., Rath, S.S., Kumar, R. and Khanna, A. (2007) Multicomponent liquid–liquid equilibria prediction for aromatic extraction systems using COSMO-RS. *Industrial & Engineering Chemistry Research*, **46** (4), 1292–1304.
217 Banerjee, T., Singh, M.K. and Khanna, A. (2006) Prediction of binary VLE for imidazolium based ionic liquid systems using COSMO-RS. *Industrial & Engineering Chemistry Research*, **45** (9), 3207–3219.
218 Banerjee, T., Verma, K.K. and Khanna, A. (2008) Liquid–liquid equilibrium for ionic liquid systems using COSMO-RS: effect of cation and anion dissociation. *AICHE Journal*, **54** (7), 1874–1885.
219 Hansmeier, A.R., Broersen, A.C., Meindersma, G.W. and de Haan, A.B. (2006) COSMO-RS supported design of task specific ionic liquids for aromatic/aliphatic separations. in *22nd European Symposium on Applied Thermodynamics*, ESAT 2006, Elsinore, Denmark, pp. 38–41.
220 Diedenhofen, M., Klamt, A., Marsh, K. and Schaefer, A. (2007) Prediction of the vapor pressure and vaporization enthalpy of 1-n-alkyl-3-methylimidazolium-bis-(trifluoromethanesulfonyl)amide ionic liquids. *Physical Chemistry Chemical Physics*, **9** (33), 4653–4656.
221 Palomar, J., Ferro, V.R., Torrecilla, J.S. and Rodriguez, F. (2007) Density and molar volume predictions using COSMO-RS for ionic liquids. An approach to solvent design. *Industrial & Engineering Chemistry Research*, **46** (18), 6041–6048.
222 Freire, M.G., Neves, C.M.S.S., Carvalho, P.J., Gardas, R.L., Fernandes, A.M., Marrucho, I.M., Santos, L.M.N.B.F. and Coutinho, J.A.P. (2007) Mutual solubilities of water and hydrophobic ionic liquids. *The Journal of Physical Chemistry. B*, **111** (45), 13082–13089.
223 Freire, M.G., Carvalho, P.J. Gardas, R.L., Santos, L.M.N.B.F., Marrucho, I.M. and Coutinho, J.A.P. (2008) Solubility of water in tetradecyltrihexylphosphonium-based ionic liquids. *Journal of Chemical and Engineering Data*, **53** (10) 2378–2382, doi: 10.1021/je8002805.
224 Bosse, D. and Bart, H.J. (2005) Binary vapor–liquid equilibrium predictions with COSMOSPACE. *Industrial & Engineering Chemistry Research*, **44** (23), 8873–8882.
225 Kato, R. and Gmehling, J. (2005) Systems with ionic liquids: Measurement of VLE and γ^∞ data and prediction of their thermodynamic behavior using original UNIFAC, mod. UNIFAC(Do) and COSMO-RS(Ol). *Journal of Chemical Thermodynamics*, **37** (6), 603–619.

226 Sahandzhieva, K., Tuma, D., Breyer, S., Perez-Salado Kamps, A. and Maurer, G. (2006) Liquid–liquid equilibrium in mixtures of the ionic liquid 1-*n*-butyl-3-methylimidazolium hexafluorophosphate and an alkanol. *Journal of Chemical and Engineering Data*, **51** (5), 1516–1525.

227 Nebig, S., Bolts, R. and Gmehling, J. (2007) Measurement of vapor–liquid equilibria (VLE) and excess enthalpies (HE) of binary systems with 1-alkyl-3-methylimidazolium bis(trifluoromethylsulfonyl)imide and prediction of these properties and γ^∞ using modified UNIFAC (Dortmund). *Fluid Phase Equilibrium*, **258** (2), 168–178.

228 Zhang, X., Liu, Z. and Wang, W. (2008) Screening of ionic liquids to capture CO_2 by COSMO-RS and experiments. *AICHE Journal*, **54** (10), 2717–2728.

7
Applications of Ionic Liquids in Electrolyte Systems

William R. Pitner, Peer Kirsch, Kentaro Kawata, and Hiromi Shinohara

7.1
Introduction

Ionic liquids are a unique class of electrolytes because they combine in a single component the dual functions of ion conductor and solvent, functions normally carried out by, respectively, a supporting electrolyte and either water or a volatile organic solvent. The use of ionic liquids for their electrolytic properties has been an important consideration since their conception. Walden was looking for salts with low melting points and high conductivities when he combined nitric acid and ethylamine to form a conductive liquid [1]. In their search for an electrolyte bath for electroplating of aluminum, Hurley and Wier turned to low-melting analogues of $NaCl-AlCl_3$ mixtures, combining aluminum chloride with N-alkylpyridinium halides [2]. It was during the quest for novel battery electrolytes that researchers at the Naval Research Academy developed the dialkylimidazolium chloroaluminate systems [3]. It was not until Wilkes and Zawarotko combined dialkylimidazolium cations with air- and water-stable anions that the field of ionic liquids became attractive to a broad range of researchers with no interest in exploiting their electrolyte properties for traditional electrochemical applications [4]. However, their potential as components for a broad range of electrolyte systems has continued to attract active interest.

Today, research papers on the application of ionic liquids as components for batteries, electrochemical double-layer capacitors, dye-sensitized solar cells, fuel cells, electroplating baths and electrochemical sensors are abundant. This is despite the often noted property inherent in a solution of ions where strong coulombic interactions are expected to dominate transport properties: compared with molecular solvents and dilute electrolyte solutions, ionic liquids have relatively high viscosities and correspondingly low conductivities. For most electrochemical applications, high performance is often dependent upon maximizing current densities, a factor directly related to conductivity. This raises an obvious question: why choose an electrolyte

Handbook of Green Chemistry, Volume 6: Ionic Liquids
Edited by Peter Wasserscheid and Annegret Stark
Copyright © 2010 WILEY-VCH Verlag GmbH & Co. KGaA, Weinheim
ISBN: 978-3-527-32592-4

system where low conductivities are expected? Of the various answers which might be given, two dominate: safety and stability.

The concerns of designing an electrolyte system around a volatile, flammable organic solvent make the low vapor pressures and non-flammable nature of ionic liquids attractive. However, are ionic liquids really non-flammable? The safety claims often made about ionic liquids have sometimes been promoted without sufficient evidence. Recent demonstrations of the ability to design ionic liquids that are flammable and even explosive are a reminder not to make over-reaching claims about a broad class of materials. Also, as is true of many non-flammable materials, it has been demonstrated that ionic liquids when atomized can undergo combustion. However, the terms flammable and combustible, when used as classifications for indicating the relative hazard of materials, are based on the flash point of a liquid, but it is possible to select ionic liquids which would not be categorized as flammable on the basis of their flash points. This would justify the trade-off of higher current densities for the increased safety arising from not using a flammable organic solvent.

The second concern often addressed for ionic liquids is enhanced device stability which comes from the use of a non-volatile solvent. After all, it does not count for much that a device gives good performance when it comes off the assembly line if its lifetime is restricted from the slow but steady escape of a volatile electrolyte component. This is especially true for devices such as dye-sensitized solar cells, where lifetimes will be measured in years and decades and where exposure to the elements of sun, wind and rain are expected. Advances in stability brought about from the use of ionic liquid electrolytes as replacements for volatile organic solvents appear to be bringing this promising technology ever closer to the market, as demonstrated by the recent proliferation of start-up companies such as DyeSol and G24 Innovations active in this field.

Turning to new and untested materials as a solution to safety concerns and increased device stability naturally raises other concerns, specifically regarding their unknown toxicity and potential for biodegradability. Ionic liquids appropriate for use in electrolytes are often highly fluorinated materials or contain cyano groups and they tend to be hydrophobic. It must also be expected that materials designed to resist the extremes of oxidation and reduction occurring in an electrochemical device will also be resistant to the mechanisms of biodegradation. Luckily, as components of electronic equipment, the fate of ionic liquids used as electrolytes is already being given detailed consideration. Because electronic waste generally is composed of a mixture of highly valuable recyclable materials (gold, copper) and toxic materials (cadmium, tin), communities are already beginning to regulate their disposal. For example, the 25 Member States of the European Union (EU) are already required to comply with the Waste Electrical and Electronic Equipment Directive (WEEE Directive), which became EU law in February 2005, which sets targets for the collection, recycling and recovery of used electronic goods marked with an easily distinguished logo (Figure 7.1). As contained components of devices regulated by directives for their recovery and recycling, the risks involved in using ionic liquids as electrolytes are significantly minimized.

Figure 7.1 Since 13 August 2005, all electronic goods sold within the EU are required to bear the WEEE logo, indicating that the manufacturers of the goods are required to pay the costs of recovering, dismantling and disposing of the goods.

7.2 Electrolyte Properties of Ionic Liquids

The selection or design of an appropriate electrolyte is highly dependent upon the application area and a single ionic liquid is unlikely to be suitable as a universal electrolyte. The two most important factors looked for include ionic conductivity and electrochemical stability. As is common with all ionic liquids, these properties are determined by the choice of cation and anion. For the discussion of how the structure of an ionic liquid affects its chemicophysical properties (specifically its ionic conductivity and electrochemical stability), a simplified model of the ionic liquid will be assumed. In this model, an ionic liquid will be considered as having three basic parts: an anion, a cationic head group and a functional group attached to the head group; the functional group can be a simple straight-chain alkyl group or it can be more complex such as a hydroxy group, an alkoxy group or a branched-chain group. Variations of any of these three factors affect the chemicophysical properties of the resulting ionic liquid.

Although not much systematic information is readily available about the conductivity of ionic liquids, their viscosity has been systematically studied by a number of groups, allowing certain trends to be discussed. This information is useful to the electrochemist due to the well-understood relationship between conductivity and viscosity: factors in ionic liquid composition which tend to lower viscosity lead to materials with increased conductivity. Thus, one important strategy in ionic liquid electrolyte selection or design is minimization of viscosity in order to maximize conductivity.

One of the earliest investigations of the relationship between structure and chemicophysical properties was carried out by Hussey and co-workers in 1979 [5]. For a series of ionic liquids having the same head group and anion and varying only in the length of the straight alkyl chain functional group, it was observed that the viscosity increased and the conductivity decreased with increasing alkyl chain length. This trend has been well documented for a broad variety of head groups and anion. This trend is illustrated in Figure 7.2 with the results of viscosity and conductivity

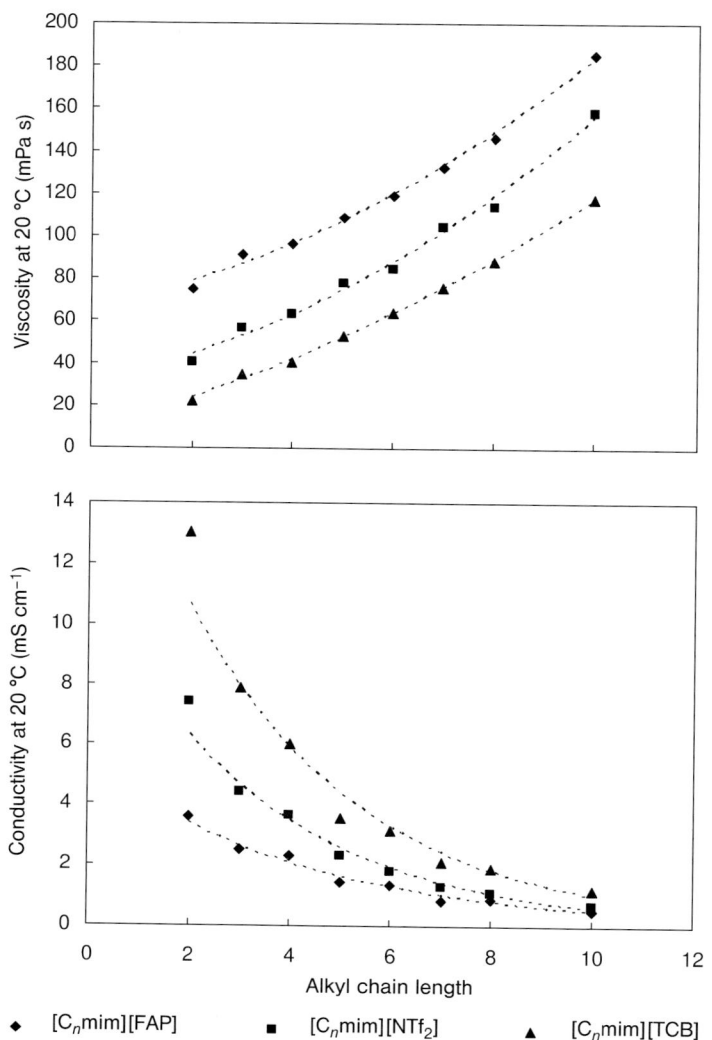

Figure 7.2 Effect of the alkyl chain length on the dynamic viscosity and specific conductivity of a series of 1-alkyl-3-methylimidazolium ionic liquids. Dashed lines are meant only as a guide to the eye.

Legend: ♦ [$C_n$mim][FAP]   ■ [$C_n$mim][NTf$_2$]   ▲ [$C_n$mim][TCB]

measurements for a series of ionic liquids based on 1-alkyl-3-methylimidazolium cations, [$C_n$mim]$^+$ (where C_n is ethyl, propyl, butyl, etc.), and three different anions, tetracyanoborate, [TCB]$^-$, bis(trifluoromethanesulfonyl)amide, [NTf$_2$]$^-$, and tris(pentafluoroethyl)trifluorophosphate, [FAP]$^-$. Functional groups on the alkyl chain also play a role, as illustrated by Table 7.1. Inclusion of a propylhydroxy (POH) functional group has the effect of increasing viscosity and density while decreasing conductivity. On the other hand, the methoxyethyl (MOE) group lead to decreased

Table 7.1 Effect of functional groups on the physical properties of a series of bis(trifluoromethanesulfonyl)imide ionic liquids: M_m = molar mass; ρ = density; η = dynamic viscosity; χ = specific conductivity.

Ionic Liquid	M_m/g mol^{-1}	ρ/g cm^{-3}	η/mPa s	χ/mS cm^{-1}
1-Alkyl-3-methylimidazolium				
[BMIM][NTF]	419.37	1.440	64	3.6
[MOEMIM][NTF]	421.34	1.510	59	3.8
[POHMIM][NTF]	421.34	1.538	142	1.9
N-Alkyl-pyridinium				
[BPYR][NTF]	416.37	1.453	77	2.8
[MOEPYR][NTF]	418.34	1.521	73	2.9
[POHPYR][NTF]	418.34	1.551	159	1.4
N-Alkyl-N-methylpyrrolidinium				
[BMPL][NTF]	422.41	1.399	98	2.2
[MOEMPL][NTF]	424.38	1.458	68	3.4
[POHMPL][NTF]	424.38	1.488	123	1.3
Alkyl-dimethyl-ethyl-ammonium				
[BDEA][NTF]	410.38	1.375	132	1.5
[MOEDEA][NTF]	412.37	1.435	77	2.7
[POHDEA][NTF]	412.37	1.465	224	1.0
N-Alkyl-N-methylpiperidinium				
[BMPP][NTF]	436.44	1.383	256	0.82
[MOEMPP][NTF]	438.41	1.447	148	1.41
[POHMPP][NTF]	438.41	1.465	455	0.49
N-Alkyl-N-methylmorpholinium				
[BMMO][NTF]	438.41	1.444	803	0.30
[MOEMMO][NTF]	440.38	1.512	470	0.53
[POHMMO][NTF]	440.38	1.538	1827	0.14

Alkyl groups: B = butyl; MOE = 2-methoxyethyl; POH = 3-hydroxypropyl

viscosities and increased conductivities, especially with aliphatic cationic head groups such as pyrrolidinium, piperidinium and tetraalkylammonium.

A second trend also becomes apparent from the two graphs in Figure 7.2: the viscosity and conductivity are dependent on the anion. This trend is further illustrated in Table 7.2, with the results of viscosity, density and conductivity measurements for a series of ionic liquids based on 1-ethyl-3-methylimidazolium cations, $[C_2\text{mim}]^+$.

Imidazolium is not the only organic cation which can be used to make an ionic liquid. Walden's ionic liquid contained an alkylammonium cation; the electroplating baths designed by Hurley and Wier were based on alkylpyridinium cations. Other common ionic liquid building blocks include tetraalkylphosphonium, dialkylpyrrolidinium, dialkylmorpholinium and dialkylpiperidinium. The effects of the cationic head group on the density, viscosity and conductivity of a family of related ionic liquids are illustrated in Table 7.1 and in Figure 7.3. Aromatic cations such as

Table 7.2 Effect of anion on the physical properties of a series of 1-ethyl-3-methylimidazolium ionic liquids: $M_m =$ molar mass; $\rho =$ density; $\eta =$ dynamic viscosity; $\chi =$ specific conductivity.

Ionic Liquid	M_m/g mol^{-1}	ρ/g cm^{-3}	η/mPa s	χ/mS cm^{-1}
[EMIM][TCM]	201.23	1.087	18.5	18
[EMIM][DCA]	177.22	1.113	20.8	18
[EMIM][TCB]	226.05	1.040	22.2	13
[EMIM][FSI]	291.30	1.447	22.4	15
[EMIM][ATF]	224.18	1.305	36.1	7.4
[EMIM][NTF]	393.32	1.524	40.5	7.4
[EMIM][SCN]	169.25	1.148	43.8	14
[EMIM][BF$_4$]	197.97	1.304	61.5	11
[EMIM][OTF]	260.24	1.381	65.4	3.4
[EMIM][FAP]	419.37	1.715	75.3	4.4

pyridinium and imidazolium tend to have the lowest viscosities and highest conductivities, due to their flat geometry and low charge density. The aliphatic cations, which are more dynamic in their assumption of geometry and suffer from higher charge density, tend to have increased viscosity and decreased conductivity relative to their aromatic analogues.

7.3
Electrochemical Stability

In most cases where ionic liquids are employed as electrolytes, the electrochemical stability of the ionic liquid is a key selection factor. The electrochemical window, a common expression of the electrochemical stability, is the range of electrode potentials over which the ionic liquid undergoes no faradaic processes and is bounded by the cathodic limit, where the ionic liquid is reduced, and by the anodic limit, where the ionic liquid is oxidized. In most, although not all, cases the cathodic limit of a given ionic liquid is determined by the reduction of the cation components, whereas the anodic limit is set by the oxidation of the anionic components. For many applications, such as batteries and electrochemical double-layer capacitors, it is advantageous to have as wide an electrochemical window as possible. For some applications, such as the electrodeposition of active metals at negative potentials or electropolymerization of monomers, one need only be concerned that the cathodic limit (electrodeposition) is negative enough or that the anodic limit (electropolymerization) is positive enough for the given process. For other applications, such as dye-sensitized solar cells, the electrochemical process of interest occurs over such a narrow potential range that the electrochemical window of the ionic liquid is almost an insignificant factor.

Measuring and reporting electrochemical windows of ionic liquids or, for that matter, any electrolyte system is a highly subjective activity. The anodic and cathodic

7.3 Electrochemical Stability | 197

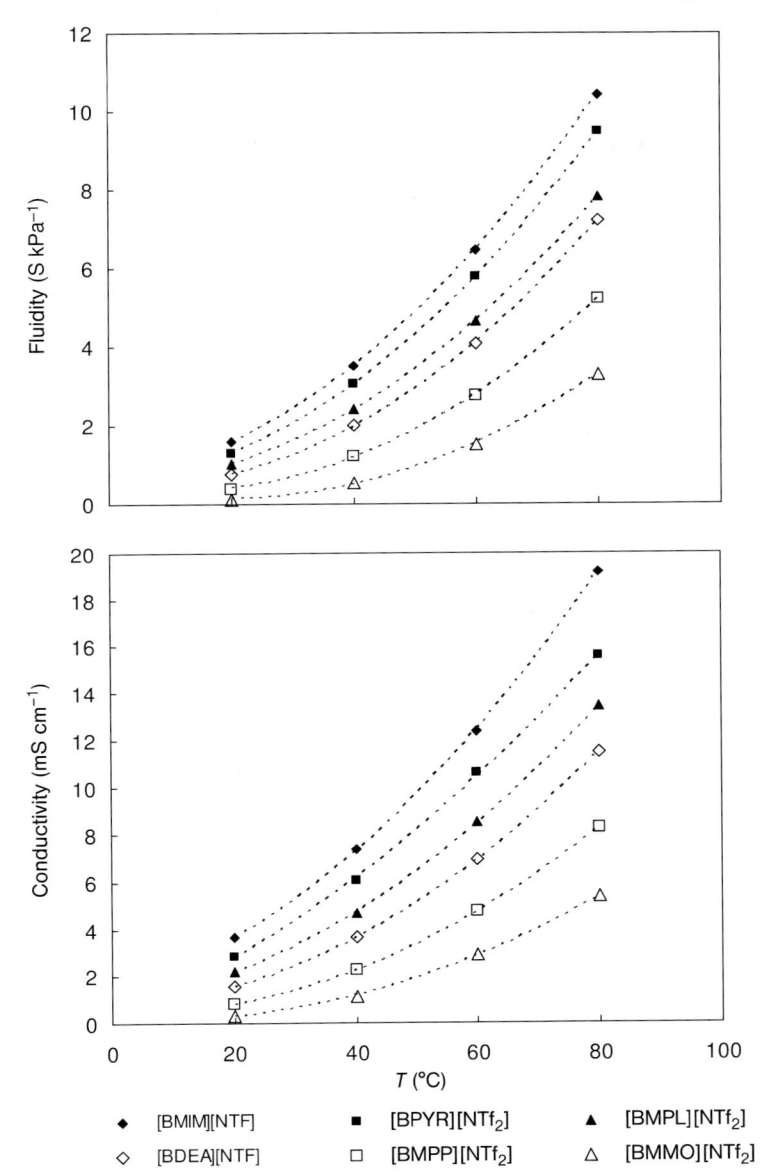

Figure 7.3 Effect of the cationic head group and temperature on the fluidity (i.e. the reciprocal of the dynamic viscosity) and specific conductivity of a series of bis(trifluoromethanesulfonyl)amide ionic liquids. Dashed lines are meant only as a guide to the eye.

♦ [BMIM][NTF]   ■ [BPYR][NTf$_2$]   ▲ [BMPL][NTf$_2$]
◇ [BDEA][NTF]   □ [BMPP][NTf$_2$]   △ [BMMO][NTf$_2$]

limits are controlled by both thermodynamic and kinetic factors and vary greatly with the choice of working electrode material, temperature and impurities. Defining what it means to be electrochemically inert is also highly subjective, with many arbitrary standards proposed. Recent work by Compton and co-workers has contributed greatly to our understanding of the electrochemical stability of ionic liquids [6].

7.4
Dye-sensitized Solar Cells

A recently emerging application for ionic liquids is dye-sensitized solar cells (DSSCs) [7, 8]. The principle of a DSSC (Figure 7.4) is based on a layer of sintered TiO_2 nanoparticles which is sensitized to a wide section of the solar spectrum by an adsorbed dye [9]. The electrons injected by the excited dye into the conduction band of the TiO_2 percolate to the cathode, which is typically made from glass, coated with a transparent, fluorine-doped tin oxide (FTO) electrode. The electron transfer process leaves the dye in an oxidized state. The charge of the dye cation is now transferred via an electrolyte solution containing an I^-/I_3^- redox couple towards the anode. At the anode, the reduction of I_3^- to I^- is catalyzed by a tiny amount of platinum catalyst on top of FTO glass. The highest efficiency so far achieved using a DSSC with a conventional solvent-based electrolyte is 11.2% [10], compared with around 15.5% for an amorphous silicon solar cell.

Although DSSCs have been around since 1991 and especially the economic side of this technology (US$0.7–1.0 W^{-1} for DSSC versus US$2–3 W^{-1} for amorphous silicon, or roughly one-third) [11] looks very attractive, there is so far no established market. One of the reasons for this surprising situation is technical difficulties, in particular the limited lifetime due to the leakage of the sealing allowing the

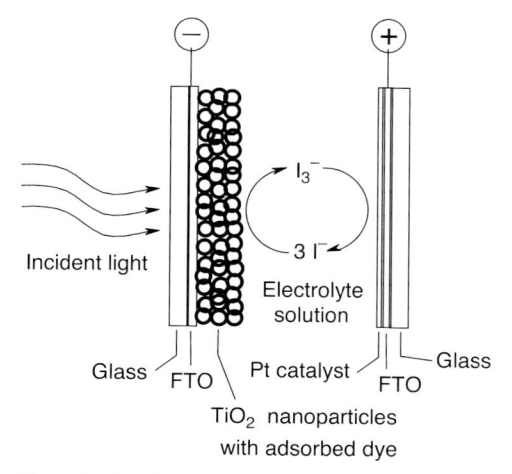

Figure 7.4 Working principle of a dye-sensitized solar cell (DSSC).

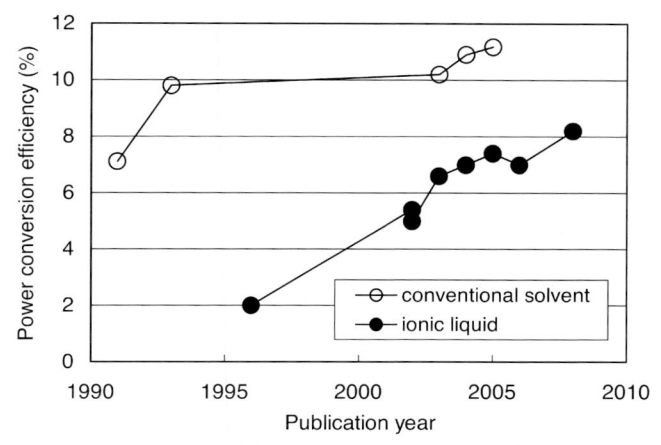

Figure 7.5 Comparison of the DSSC power conversion efficiency between conventional solvent and ionic liquid-based electrolytes.

evaporation of the electrolyte solvent out of the cell and the subsequent uptake of water into the cell.

Therefore, ionic liquids were recognized as the solvent of choice in the basic electrolyte solution [12]. Ionic liquids have many advantages compared with

Figure 7.6 Components of a typical ionic liquid-based DSSC electrolyte [12d] (a) and examples of sensitization dyes [9] (b).

conventional organic solvents. Their relatively small thermal expansion coefficient poses less stress on the sealing during day–night temperature cycles. They have negligible vapor pressure [13], i.e. there is no loss of solvent by evaporation and diffusion even during long-term operation under harsh outdoor conditions. Most ionic liquids are electrochemically, thermally and photochemically inert and they are excellent solvents for many ionic species.

A major disadvantage of ionic liquids compared with conventional solvents is their relatively high viscosity, which limits the mass transport of the I^-/I_3^- redox couple [12b]. This problem has recently been tackled by utilizing eutectic mixtures of ionic liquids containing cations with relatively small ionic radius and the tetracyanoborate anion [12d], some of which have melting temperatures well above room temperature. A so far unsurpassed efficiency of 8.2% under air-mass 1.5 global illumination has been achieved (Figure 7.5) [14].

Ionic liquid-based DSSC electrolytes (Figure 7.6) also contain some basic additives such as alkylbenzimidazoles. They are assumed to coordinate to free locations on the surface of the TiO_2 particles shifting the TiO_2 quasi-Fermi level towards the negative potential in addition to blocking the unwanted back-reduction of I_3^-.

The main development targets for the DSSC electrolyte beyond a wide liquid range are high conductivity and excellent long-term stability and reliability. As a further step towards commercialization, gelated and printable electrolyte formulations [15] are required to lower production prices, for example by enabling roll-to-roll processes for flexible DSSCs on plastic substrates [16].

In spite of many technical hurdles which still need to be overcome, most current developments are headed towards DSSC devices based on ionic liquids, mainly due to the excellent durability and reliability of this new generation of materials.

References

1 Walden, P. (1914) *Bull. Acad. Imp. Sci. St. - Pétersbourg*, 405.
2 Hurley, F.H. and Wier, T.P. (1951) *Journal of the Electrochemical Society*, **98**, 203.
3 Wilkes, J.S., Levisky, J.A., Wilson, R.A., and Hussey, C.L. (1982) *Inorganic Chemistry*, **21**, 1263.
4 Wilkes, J.S. and Zaworotko, M.J. (1992) *Chemical Communications*, 965.
5 Carpio, R.A., King, L.A., Lindstrom, R.E., Nardi, J.C. and Hussey, C.L. (1979) *Journal of the Electrochemical Society*, **126**, 1644.
6 O'Mahony, A.M., Silvester, D.S., Aldous, L., Hardacre, C. and Compton, R.G. (2008) *Journal of Chemical and Engineering Data*, **53**, 2884.
7 O'Regan, B. and Grätzel, M. (1991) *Nature*, **353**, 737.
8 (a) Grätzel, M. (2005) *Chemistry Letters*, **34**, 8; (b) Grätzel, M. (2005) *Inorganic Chemistry*, **44**, 6841; (c) Grätzel, M. (2007) *Philosophical Transactions of the Royal Society of London. Series B, Biological Sciences*, **365**, 993.
9 (a) Nazeeruddin, M.K., Kay, A., Rodicio, I., Humphry-Baker, R., Müller, E., Liska, P., Vlachopoulos, N. and Grätzel, M. (1993) *Journal of the American Chemical Society*, **115**, 115; (b) Wang, P., Zakeeruddin, S.M., Moser, J.-E., Zakeeruddin, M.K., Sekiguchi, T. and Grätzel, M. (2003) *Inorganic Materials*, **2**, 402; (c) Wang, P., Zakeeruddin, S.M., Humphry-Baker, R.,

Moser, J.-E. and Grätzel, M. (2003) *Advanced Materials*, **15**, 2101; (d) Hagberg, D.P., Yum, J.-H., Lee, H.-J., De Angelis, F., Marinado, T. Karlsson, K.M., Humphry-Baker, R., Sun, L., Hagfeldt, A., Grätzel, M. and Nazeeruddin, M.K. (2008) *Journal of the American Chemical Society*, **130**, 6259.

10 Chiba, Y., Islam, A., Watanabe, Y., Komiya, R., Koide, N. and Han, L. (2006) *Japanese Journal of Applied Physics Part 2*, **45**, L638.

11 Advanced PV Market, HiEdge, Tokyo, 2008.

12 (a) Singh, P., Rajeshwar, K., DuBow, J. and Job, R. (1980) *Journal of the American Chemical Society*, **102**, 4676; (b) Papageorgiou, N., Athanassov, Y., Armand, M., Bonhote, P., Pettersson, H., Azam, A. and Grätzel, M. (1996) *Journal of the Electrochemical Society*, **143**, 3099; (c) Kawata, K., Zakeeruddin, S.M., and Grätzel, M. (2002) *Electrochemical Society Proceedings*, **19**, 16; (d) Kuang, D., Wang, P., Ito, S., Zakeeruddin, S.M. and Grätzel, M. (2006) *Journal of the American Chemical Society*, **128**, 7732.

13 Wasserscheid, P. and Welton, T. (eds), (2008). *Ionic Liquids in Synthesis*, 2nd edn, Wiley-VCH Verlag GmbH, Weinheim.

14 Bai, Y., Cao, Y., Zhang, J., Wang, M., Li, R., Wang, P., Zakeeruddin, S.M. and Grätzel, M. (2008) *Inorganic Materials*, **7**, 626.

15 (a) Kato, T., Okazaki, A. and Hayase, S. (2006) *Journal of Photochemistry and Photobiology. A, Chemistry*, **A179**, 42; (b) Usui, H., Matsui, H., Tanabe, N., and Yanagida, S. (2004) *Journal of Photochemistry and Photobiology A: Chemistry*, **164**, 97.

16 Miyasaka, T., Kijitori, Y. and Ikegami, M. (2007) *Electrochemistry*, **75**, 2.

8
Ionic Liquids as Lubricants
Marc Uerdingen

8.1
Introduction

The combination of their unique properties, such as high thermal stability, wide liquidus range, extremely low vapor pressure and good flame resistance, makes ionic liquids interesting candidates for their use as lubricants. Liu and co-workers [1] were the first to propose in 2001 this ionic liquid application by reporting the results of their tribological investigations of several ionic liquids.

In the green chemistry context, it is evident that a lubricant with better lubrication properties leads to lower energy consumption in the corresponding mechanical application by just reducing energy losses by friction. Moreover, as all material protected by a better lubricant has a longer lifetime, advanced lubrication helps to save the energy and raw materials that are necessary to replace corroded parts of the machinery.

Since 2001, a remarkable number of publications on the use of ionic liquids as lubricants have appeared, many reporting on significant advantages of ionic liquids compared with conventional lubricants. A SciFinder search in January 2009 resulted in 144 hits.

Jimenez *et al.* [2] investigated several ionic liquids as neat lubricants and lubricant additives in steel–aluminum contact. Their study comprised different cation–anion combinations with a special focus on the effect of different alkyl chains at the imidazolium cation on the lubricating behavior. Another effect of structure–property interplay was disclosed in subsequent papers by Liu *et al.* demonstrating the effect of functional groups on the friction and wear behavior of aluminum alloy in lubricated aluminum–steel contact [3, 4]. In these publications, the ionic liquid was used as the base oil.

In contrast, Fox and Priest reported in a recent study the use of ionic liquids in group III hydroisomerized (high-quality) mineral base oil, mineral oils and different greases as an additive [5]. Of course, for their use in commercial lubricants, ionic liquid additive concepts need to be developed in order to obtain suitable ionic liquid formulations. However, a first technical application in which ionic liquid lubricant properties played a crucial role was reported by Linde AG [6] in 2006, namely the use of an ionic liquid as a liquid piston in a gas compressor.

8.2
Why Are Ionic Liquids Good Lubricants?

8.2.1
Wear and Friction Behavior

As Liu and several other authors have reported, ionic liquids show very good wear and friction behavior with various metal–metal combinations [1]. The method of choice to investigate the friction and wear behavior is the ball-on-disc tribometer, shown in Figure 8.1. A ball (upper specimen) is rotating with a given load on a disc (lower specimen) under defined temperature conditions.

Table 8.1 shows results of a steel–steel contact measurement using $[C_6mim][BF_4]$ as ionic liquid that were obtained using an SRV (OPTIMOL) tester [1].

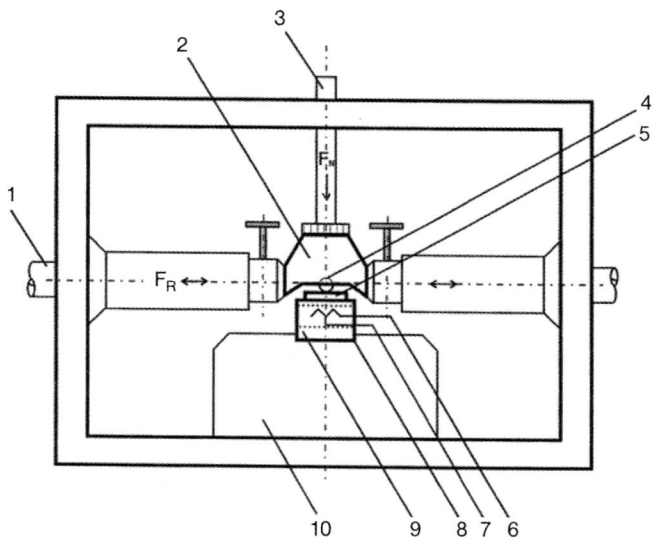

Figure 8.1 Ball-on-disc tribotester or SRV. 1, Drive axle; 2, upper specimen holder; 3, load axle; 4, specimen (upper); 5, specimen (lower); 6, temperature sensor; 7, heating; 8, lower specimen holder; 9, piezo measuring element; 10 receiving block.

8.2 Why Are Ionic Liquids Good Lubricants?

Table 8.1 Tribological properties of [C$_6$mim][BF$_4$].[a]

Load (N)	Friction coefficient			Wear volume ($\times 10^{-4}$ mm^3)		
	IL	X-1P	PFPE	IL	X-1P	PFPE
200	0.060	0.070	0.120	0.05	0.07	0.60
300	0.055	0.065	0.110	0.22	2.21	1.90
400	0.050	—[b]	0.10	0.39	—[b]	5.03
500	0.045	—[b]	—[b]	0.45	—[b]	—[b]
600	0.045	—[b]	—[b]	0.53	—[b]	—[b]

[a]Conditions: steel–steel contact, SRV tester, frequency 25 Hz, amplitude 1 mm, duration 30 min.
[b]Lubrication failure.
Adapted from [1].

As one of the most remarkable results of these studies, Liu et al. found that [C$_6$mim][BF$_4$] and [C$_2$C$_6$im][BF$_4$] show with steel–steel contact lower friction coefficients than even phosphazene and perfluoropolyether lubricants which are known as high-performance lubricants. The same group [7] has also reported excellent tribological properties for [C$_2$C$_8$im][BF$_4$]. In another study, they reported [8] lower friction coefficients for three 1,3-dialkylimidazolium hexafluorophosphates ([C$_6$mim][PF$_6$], [C$_2$C$_6$im][PF$_6$] and [C$_3$C$_6$im][PF$_6$]) compared with a liquid paraffin lubricant activated with ZDDP (zinc dithiophosphates) under vacuum conditions. Kamimura et al. [9] demonstrated that [C$_4$mim][BF$_4$] and [(C$_2$)$_2$C$_1$(C$_2$OC$_1$)N]BF$_4$] (diethylmethylmethoxyethylammonium tetrafluoroborate) show lower friction coefficients than a PAO (poly-alpha-olefin), a POE (polyolester) and PFPE (perfluoropolyether) base oil (Figures 8.2 and 8.3 and Table 8.2).

Kamimura et al. also demonstrated the effect of carboxylic acids as additives to ionic liquid base oils [10]. With added decanoic acid, the wear and friction performance of [C$_2$mim]][NTf$_2$] was enhanced significantly.

Liu et al. [11] demonstrated that asymmetrically substituted tetraalkylphosphonium ionic liquids of the type [P$_{4,4,4,n}$][BF$_4$] (with $n = 8$, 10, 12) showed for various contacts superior properties compared with [C$_2$C$_6$im][PF$_6$] in terms of anti-wear performance and load-carrying capacity. The same group also reported [12] that tetraalkylphosphonium tetrafluoroborates show excellent tribological performance as lubrication oils.

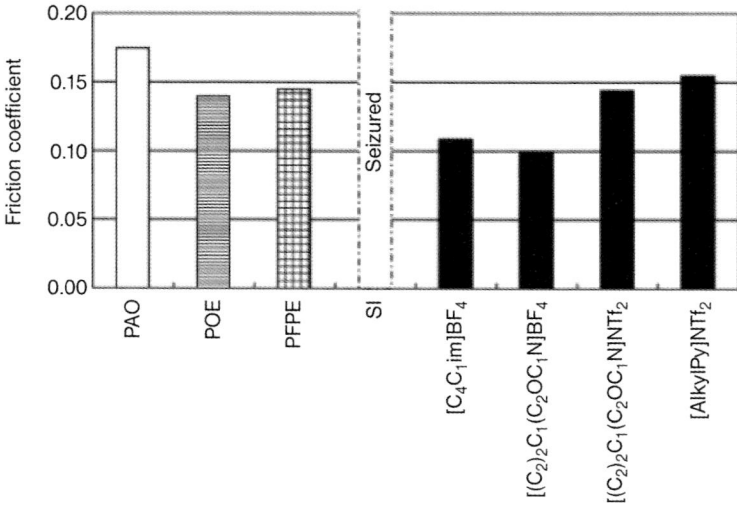

Figure 8.2 Friction coefficients for several ILs compared with conventional (PAO, POE and PFPE lubricants. Adapted from [9].

In a similar manner, ionic liquids based on tetraalkylammonium cations were found to show excellent tribological performance as lubricating oils. Obviously, [$HN_{8,8,8}$][NTf_2] offers a 30% reduction in friction compared with a standard mineral oil. Interestingly, this performance seems to be a strong function of the cation type, as imidazolium-based ILs proved to be less favorable than conventional mineral oils [13, 14].

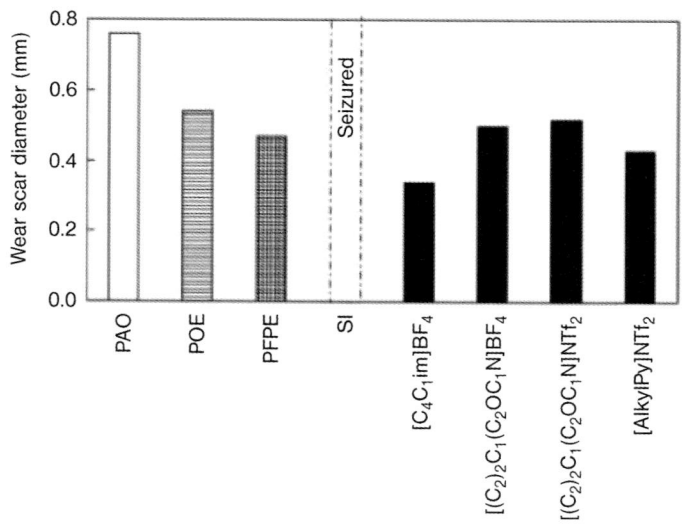

Figure 8.3 Wear scar diameters for several ILs compared with conventional PAO, POE)and PFPE lubricants. Adapted from [9].

Table 8.2 Test conditions of the tribological tests displayed in Figures 8.2 and 8.3.

Parameter	Value
Operation parameters	
Applied load (N)	196
Hertz contract stress (GPa)	2.47
Rotation (rpm)	1200
Sliding velocity (m s^{-1})	0.46
Oil temperature (°C)	75
Test duration (min)	60
Test ball	
Ball material	SIJ2(115)
Diameter (mm)	12.7
Hardness (HRC)	62

Adapted from [9].

In general, the tetrafluoroborate derivatives of imidazolium cations with longer chains give lower friction and wear values than the corresponding hexafluorophosphate salts [2].

As another general trend, longer *n*-alkyl chains in the imidazolium cation reduce the wear and friction and increase the viscosity of the ionic liquid. This was demonstrated by Jimenez and Bermudez [15] for [C$_6$mim][BF$_4$] and [C$_8$mim][BF$_4$]. Their study demonstrated that these ionic liquids exhibit better thermal stabilities and lubricating performances than mineral and synthetic oils at extreme temperatures. [C$_8$mim][BF$_4$] proved to be a highly effective lubricant in the full temperature range explored in this study.

It is remarkable that in all these studies, ionic liquids were applied that contain the hydrolytically labile [PF$_6$]$^-$ or [BF$_4$]$^-$ anions. Hydrolysis of these anions is known to liberate highly corrosive HF with the decomposition process being triggered by the presence of water, high temperatures and acidity. In fact, Phillips *et al.* reported in 2007 that under tribocorrosion conditions Fe–oxide and Fe–fluoride species are formed at steel surfaces when lubricated with tetrafluoroborate or hexafluorophosphate salts [16]. They studied steel surfaces after sliding contact with [C$_2$mim][BF$_4$] and [C$_4$mim][PF$_6$] at 300 °C by Mössbauer spectroscopy (Figure 8.4) and X-ray induced photoelectron spectroscopy (XPS). These measurements confirmed the presence of mainly FeF$_2$ on the steel surface. It can be expected that such Fe fluorides act as a passivating layer, which explains the very promising wear and friction results, even though the surface reaction of ionic liquid decomposition products does in fact represent a kind of corrosion process.

The same authors also oxidized the steel sample before contacting it with the ionic liquid. Also in this latter case they observed a reaction of the fluorinated ionic liquid with the surface, although to a lesser extent compared with bare metal.

In another report by Liu and co-workers [17], hydrolysis-stable bis(trifluoromethanesulfonyl)amide ionic liquids were applied. Under tribocorrosion conditions on steel–steel contact, they authors observed by XPS and by scanning electron

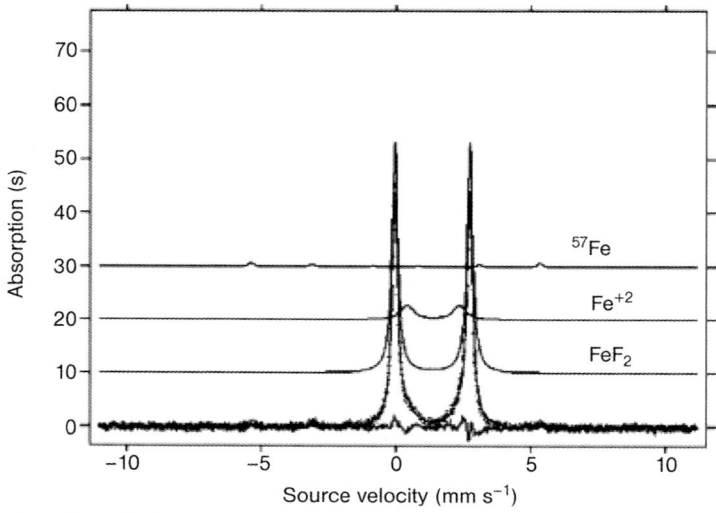

Figure 8.4 Mössbauer spectrum of a steel surface after treatment with a hexafluorophosphate ionic liquid. Reproduced from [16].

microscopy (SEM/EDS) measurements a boundary film of complex chemical nature including FeS and organic fluoride species on the steel surface.

In our own work, we took a different approach and focused on the investigation of non-fluorinated ionic liquids in lubrication applications. 1-Ethyl-3-methylimidazolium ethylsulfate ($[C_2mim][C_2H_5SO_4]$) and Ammoeng 102 [ethyl bis(polyethoxyethanol) tallow ammonium ethylsulfate] were tested and compared with fluorinated ionic liquids and standard lube oils. Table 8.3 summarizes the results of our SRV, reciprocating friction and wear results [18]. This comparison highlights impressively the superior friction properties of the halide-free ionic liquids under investigation.

$$m+n = 14\text{-}25$$
$$R = C_{14}\text{-}C_{18}$$

Ammoeng 120

Jimenez and co-workers also investigated fluoride-free imidazolium-based ionic liquids using tosylate as the counterion [2, 15, 19]. Compared with hexafluorophosphate and tetrafluoroborate ionic liquids, reduced tribocorrosion and consequently reduced friction and wear for ionic liquids with short alkyl chains at the cation were observed.

These results indicate clearly that the very attractive tribological properties of ionic liquids not only are due to the formation of fluorinated surface species (as seen with fluorinated ionic liquids) but also have to be attributed to the nature of the Coulombic interactions in the lubricating film.

8.2 Why Are Ionic Liquids Good Lubricants?

Table 8.3 Wear and friction coefficients of the fluoride-free ionic liquids [C$_2$mim][C$_2$H$_5$SO$_4$] and Ammoeng 102 compared with fluorinated ionic liquids and the standard lube oil ISO VG 32.

Ionic liquid	Friction coefficient		Wear scar (mm)	Measurement of friction behavior (SRV) according to DIN 51834-02 V:
	f_{min}	f_{max}		Ball: 100Cr6, $D = 10$ mm Ring: $d = 18.6$ mm, $h = 2.6$ mm
[C$_4$mim][BF$_4$]	0.070	0.090	0.85	**Testing parameters:** Temperature: 90 °C
[C$_4$mim][PF$_6$]	0.080	0.095	0.55	Initial running: 30 s
[C$_6$mim][PF$_6$]	0.085	0.085	0.50	Duration: 120 min
Ammoeng 102	0.087	0.095	0.57	Amplitude of oscillation: 1 mm
[N$_{1888}$][Tf$_2$N]	0.090	0.095	0.65	**Normal force:**
[C$_2$mim][C$_2$H$_5$SO$_4$]	0.047	0.085	0.92	Initially: 30 N
[1-C$_2$,3-C$_1$py][C$_4$F$_9$SO$_3$]	0.067	0.079	0.52	During testing: 100 N Frequency: 50 Hz
ISO VG 32	0.130	0.150	0.80	**Determination of:** f_{min} and f_{max} (friction number) after 15, 30 and 90 min Amount of wear, wear scar WK, is determined by analysis under a microscope

According to the literature, the tribology of two friction partners separated by a lubricant can be described by the Stribeck curve (Figure 8.5) [20].

In particular, at the far left side of the Stribeck curve (area of the boundary friction), where only a few molecules form the lubricating contact, inner-molecular interaction of the lubricant itself is of very high influence. Due to the inner Coulombic forces of ionic liquids and their ability to form extended hydrogen bond networks, the formation of more stable double layers and similar aggregates becomes more favorable. Also, the clearly higher polarity of ionic liquids compared with standard lube oils is of great relevance as it leads to enhanced surface wettability and stronger

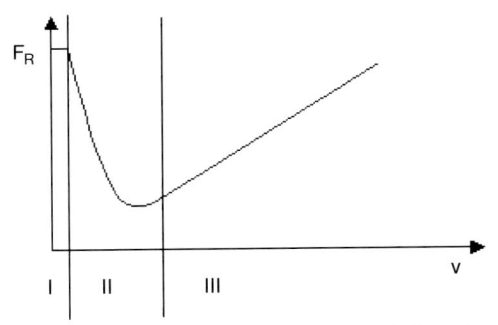

Figure 8.5 Stribeck curve to describe the tribology of two friction partners separated by a lubricant. F_R, friction; v, velocity. I, boundary friction; II, mixed film friction (elastohydrodynamic); III, hydrodynamic lubrication. Adapted from [20].

liquid adsorption to the metallic contact surface [21]. All these particular ionic liquid interaction potentials contribute to the fact that more ionic liquid ion pairs are in metal contact and contribute to a significant friction and wear reduction compared with conventional, non-polar and non-ionic lubricants.

At higher rotation or agitation, however, a thin lubricating film is formed (elastohydrodynamic lubricating) and the two friction partners are separated by this film. Innerlubricant interactions have a less pronounced influence in this elastohydrodynamic regime. At even higher velocity, more and more layers of the lubricant film are formed and in this hydrodynamic status the friction is directly dependent on the velocity.

8.2.2
Pressure Behavior

A further consequence of the strong Coulombic and hydrogen bonding interactions in ionic liquids is their very advantageous pressure behavior. The pressure behavior can be tested in the so-called Shell Four-Ball Wear Test. Even without using any pressure additives, very high pressure values of >8500 N can be reached with greases based on ionic liquids (for a typical PFPE grease the value is 7000 N and for mineral oil-based greases the loads are 4000–5000 N) [18]. Improvements in pressure behavior can also be obtained by adding ionic liquids to commercial lubricants and greases. Fox and Priest, for example, added 1 wt% of $[C_6mim][PF_6]$ to a ready-to-use formulated "high-temperature grease" and were able to increase the load by about 60% in the Shell Four-Ball Wear Test [22, 53].

The American Gear Manufacture Association relation for the film thickness with the pressure coefficient can be used to estimate the film-generating capability of a lubricant, which is an important value for evaluating its elastohydrodynamic lubrication (EHL). Fernandez and co-workers showed that imidazolium bis(trifluoromethanesulfonyl)amide-based ILs are very suitable candidates for EHL applications such as roller-bearing and gear boxes [23].

The compressibility of ionic liquids was quantified in investigations by Predel and Schlücker [24]. They reported that the ionic liquid $[C_2mim][C_2H_5SO_4]$ shows a volume loss of only 0.34% on going from 1 to 100 bar compared with approximately 1% for the commonly used hydraulic oil Shell Tellus 22. This ionic liquid was even found to be remarkably less compressible than water.

In general, it can be concluded from the existing literature reports that ionic liquids are characterized by a remarkable pressure behavior. The latter can be understood in terms of the strong ionic and hydrogen bonding forces that form these liquids. Consequently, ionic liquids show very promising performance (as pure liquids or as additives) in all engineering applications where high loads have to be handled by a lubrication system or by a working fluid.

8.2.3
Thermal Stability

In addition to the tribological behavior, thermal stability under the application conditions is a key selection criterion for a suitable lubricant. Consequently, a

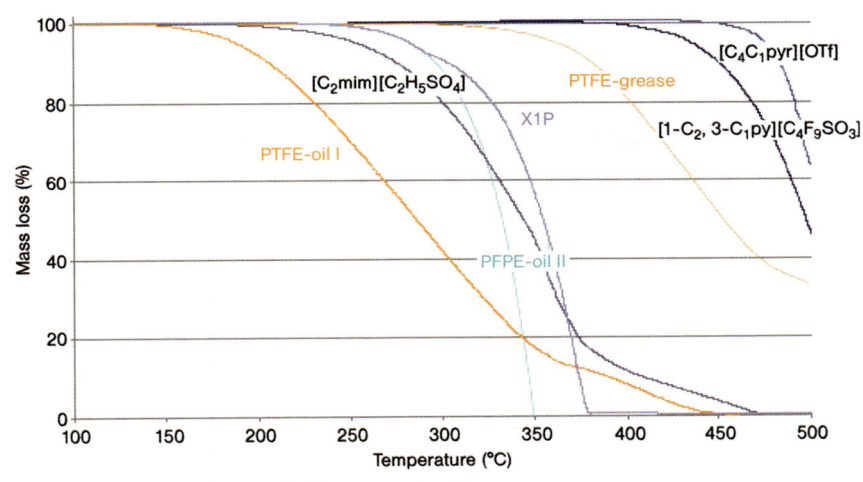

Figure 8.6 Thermal stability of different ionic liquids in comparison with the commercial lubricants F-oil, X1P, PFPE oil II and a PTFE grease. TGA measurement at 5 K min^{-1}.

number of research groups have focused their efforts on the optimization of the already fairly high thermal stability of ionic liquids.

We investigated the thermal decomposition of several ionic liquids in comparison with commercially applied polytetrafluoroethylene (PTFE) greases or PFPE oil. As shown in Figure 8.6, several ionic liquid base oils clearly outperform the commonly applied standard synthetic oil in a standard thermogravimetric analysis (TGA) using a 5 K min^{-1} heating ramp.

It should be clearly noted in this context that the onset temperature of a TGA can only give a first hint at the potential thermal stability limit and possible service temperature. Long-term isothermal stability tests are essential to determine the application range of an ionic liquid lubricant. Depending on the required lifetime of the lubricant at a given temperature, the thermal analysis has to be adapted. In Figure 8.7, the results of an isothermal stability test at 275 °C over 250 h are given for different ionic liquids.

According to Zhang and co-workers [25], the thermal stability of ionic liquids decreases with the anion trend $[PF_6]^- > [NTf_2]^- > [OTf]^- > [BF_4]^- >$ non fluorinated anions. Compared with conventional hydrocarbon-based lubricants, Minami et al. [26] reported not only an enhanced temperature stability of ionic liquids but also a higher thermo-oxidative stability. Whereas the former was determined in an isothermal test at 200 °C over 1000 h, the latter was studied in the Rotating Bomb Oxidation Test (RBOT). For all $[NTf_2]^-$ salts tested, the isothermal test showed a highly reduced weight loss over time compared with classical lubricants. Whereas for a standard PAO the weight loss was >50% after 48 h at 200 °C, the weight loss with the best ionic liquid, $[(C_2)_2C_1(C_2OC_1)N][NTf_2]$, was only 0.13% after 1000 h at 200 °C. The result of the RBOT is shown in Figure 8.8. From this comparison, it becomes very

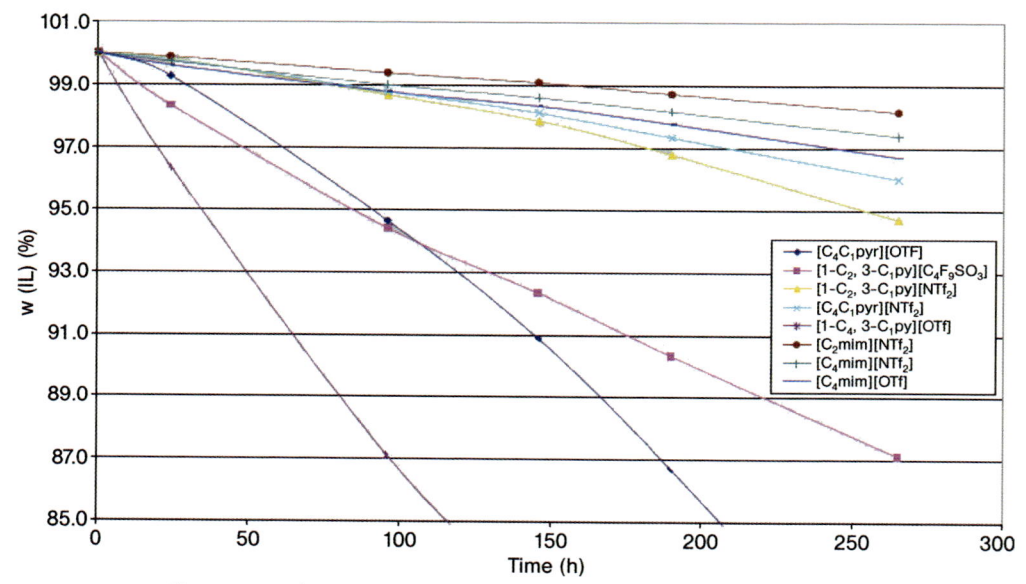

Figure 8.7 Isothermal temperature stability test of various ionic liquids. Temperature stability at 275 °C.

evident that ionic liquids show a higher oxidative stability than classical lubricants. It is an obvious but important fact that the negligible vapor pressures of the ionic liquids under investigation play a crucial role in preventing oxidative decomposition in the RBOT.

For the $[NTf_2]^-$ salts, the observed stability trend, as established from the temperature stability test, the store aging test, the viscosity/performance test and

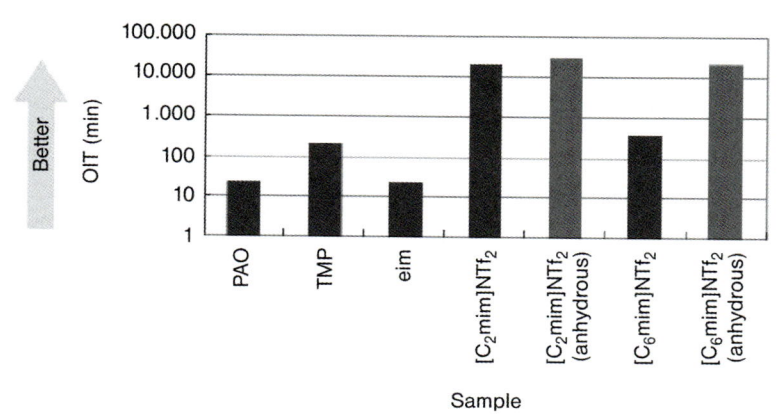

Figure 8.8 Comparison of the standard lubricants PAO and TMP with different ionic liquids and ethylimidazole (C_2im), the precursor amine for the ionic liquid synthesis, in the RBOT (150 °C, 620 kPa O_2). Adapted from [26].

the RBOT, was found to follow the order $[C_2mim][NTf_2] > [C_6mim][NTf_2] > [C_{12}MIM][NTf_2]$.

To enhance further the stability of the ionic liquid cation, Shreeve and co-workers [27] reported the synthesis and application of dicationic ionic liquids with a bridging polyfluoroalkyl moiety between two substituted imidazolium rings. For these geminal dicationic ionic liquids, they reported higher onset temperatures of up to 415 °C in comparison with the classical imidazolium representatives. Moreover, very promising tribological data with steel–steel contact were reported for these dicationic systems at 300 °C.

8.2.4
Viscosity Index and Pour Point

For many mechanical applications, the viscosity and, in particular, the viscosity index (VI) are very important properties. In many technical specifications, the required viscosity is given in $mm^2 s^{-1}$ instead of mPa s.

The temperature dependence of the viscosity of a lubricant is given by the VI, which contains information on the viscosity at 40 and 100 °C.

The pour point is the temperature at which the first crystallite formation occurs when the lubricant is cooled. Consequently, the pour point is a critical property to define the lowest handling and storage temperature of the lubricant.

Both characteristics, the viscosity/viscosity index and the pour point, determine to a large extent the selection of a suitable lubricant for a given application. Table 8.4 summarizes viscosity and melting point data for ionic liquids in comparison with common synthetic oils [26], and it is obvious that the ionic liquids provide VIs comparable to those of commercial synthetic lubricants and suitable liquid ranges.

Of course, lubricant properties have to vary over a broad range depending on the intended application. In the case of mineral oils, modifications can be introduced by the proper choice of the distillation cut, leading to different mineral oil fractions. In the case of synthetic lubricants, property modifications are introduced by varying the alkyl chain length and the functional groups. In the case of ionic liquid-based lubricants, it is mainly the variation of the hydrophobic character at the cation that has been explored in detail so far. From all these studies, it is evident that longer alkyl chains in the cation lead to a higher viscosity of the ionic liquid (for several examples, see Figure 8.9). Figure 8.9 also shows the significant increase in IL viscosities at lower temperature, which is still a drawback compared with common oils. Low-viscosity ionic liquids, such as $[C_2mim][NTf_2]$, show very attractive viscosities at room temperature. However, at 100 °C, the viscosity is usually below $10 \, mm^2 s^{-1}$, which is too low for some applications.

The use of viscosity-modifying additives or ionic liquid mixtures can be expected to optimize the viscosity index further. However, no results on the successful application of commercial viscosity modifiers in ionic liquids have been reported. For binary mixtures, the viscosity can be estimated with a method reported by Romani and co-workers [28] To simplify, an ideal behavior can be adopted and followed by a linear

Table 8.4 Viscosity/viscosity index and melting point data for ionic liquids in comparison with different synthetic lubricants.

Compound/description	Density at 15 °C (g cm^{-3})	Viscosity (mm^2 s^{-1})		Viscosity index	Melting point (°C)
		at 40 °C	at 100 °C		
1-Ethyl-3-methylimidazolium bis(trifluoromethanesulfonyl)amide	1.53	13.4	3.75	185	−15.3
1-Ethyl-3-methylimidazolium tetrafluoroborate	1.29	16.7	4.65	224	14.6
1-Hexyl-3-methylimidazolium bis(trifluoromethanesulfonyl)amide	1.38	27.4	5.59	148	<−50
1-Dodecyl-3-methylimidazolium bis(trifluoromethanesulfonyl)amide	1.25	61.8	9.54	136	21.9
N,N-Diethyl-N-methyl-N-(2-meth- oxyethyl)ammonium bis(trifluorometha- nesulfonyl)amide	1.42	20.4	4.80	168	<−50
N,N-Diethyl-N-methyl-N-(2-meth- oxyethyl)ammonium tetrafluoroborate	1.18	146.8	17.0	125	9
Poly-alpha-olefin-type synthetic hydrocarbon	–	16.8	3.90	120	–
Trimethylolpropane esters of octanoic and decanoic acids	–	19.0	4.50	139	–
Polydimethylsiloxane	–	22.9	9.66	450	–
Perfluoropolyether	–	18.4	5.00	235	–

Adapted from [26].

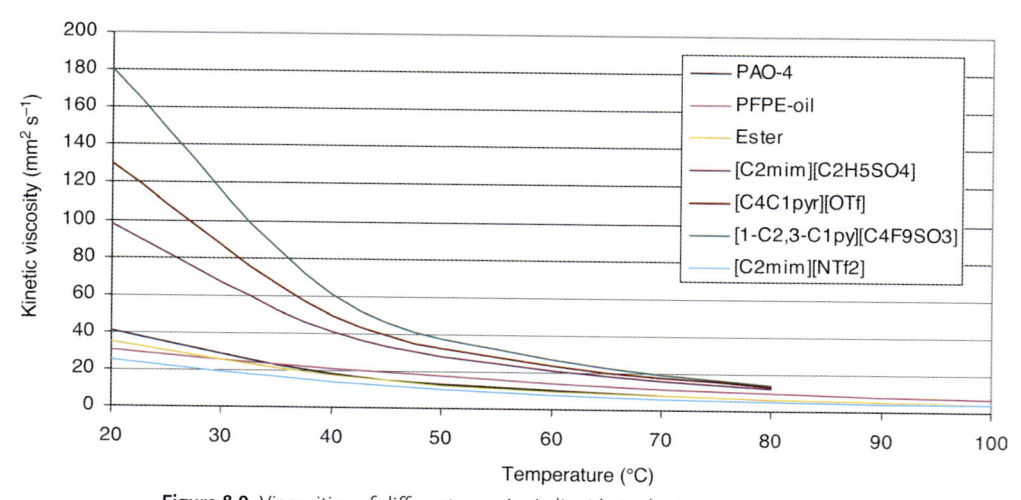

Figure 8.9 Viscosities of different pure ionic liquids in the temperature range 20–100 °C.

approach the viscosity can be fine-tuned in the range between the limits of the pure ionic liquids.

8.2.5
Corrosion

It is well known that ionic liquids can be hydrophobic or hydrophilic in nature depending on the character of their ions. Their corrosion behavior is very different according to this classification. Hydrophilic lubricants, in general, cannot be used in water-containing environments, because their water uptake would lead to a loss of lubricating properties.

It has been reported that some ionic liquid structures, namely alkyl sulfates and chlorides, can be corrosive, in particular towards brass and copper [29]. However, the first successful attempts to moderate this effect using suitable corrosion inhibitors have been demonstrated. In 2004 we reported for the first time that ionic liquids can be doped with 2000 ppm of $1H$-benzotriazole to reduce the corrosion rate for copper from >0.4 to <0.1 mm per year[29]. Liu and co-workers also reported a similar effect by using benzotriazole to reduce the corrosiveness of hexafluorophosphate ionic liquids [30].

In our own work, we demonstrated that hydrophobic ionic liquid lubricants can be formulated to be stable against the most typically employed construction metals by using corrosion additives, which had to be adapted to the ionic nature of the base fluid. Figure 8.10 shows the test procedure used in this work and some selected results.

From the results in Figure 8.10, it is obvious that for the hydrophobic ionic liquid $[C_4mim][NTf_2]$ the right additive can prevent corrosion on ST 1203 steel over 7 days at 60 °C. In prolonged tests, also over 600 h, no corrosion was detected [18]. From these early studies, it can be concluded that the technical application of ionic liquid lubricants will require a well-designed additive package. This is not surprising as every conventional commercial lubricant also comprises a sophisticated additive package to adjust the viscosity and the corrosion behavior.

8.2.6
Electric Conductivity

Due to their ionic nature, ionic liquids are intrinsically conductive. Consequently, the use of ionic liquids as a base oil or as an additive to conventional base oils requires no further conductive additives (which may affect the tribological behavior) to obtain a lubricant with electric conductivity.

Conventional technical solutions for conductive lubricants include mostly graphite as additive. Graphite is a particulate material and for some applications, for example in bearings, it is not suitable to bring small particles into the lubricating regime. The particles can lead to failure of the lubricant and thus of the bearing. Working without the conductive additive, however, can cause fretting.

→ **Test procedure for screening for hydrophobic ILs**

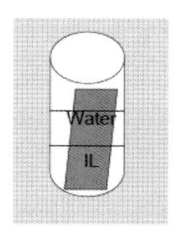

Conditions:

7 days at 60°C

Tested metal:
Steel (ST1203)
Stainless Steel (V2A),
Copper,
CuSN6

Corrosion test at 60°C for 7 days

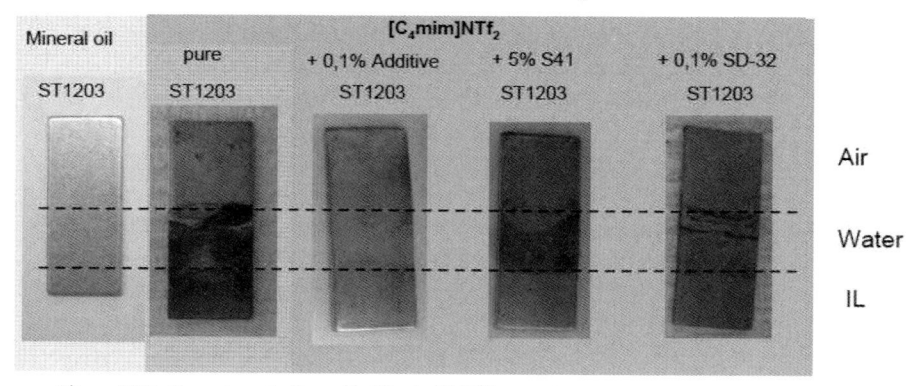

Figure 8.10 Corrosion studies with [C$_4$mim][NTf$_2$].

By using an ionic liquid as a conductive additive, this problem can be solved. A typical graphite grease has an electric resistance of 2 kΩ cm, whereas an ionic liquid-based grease based on [C$_2$mim][C$_2$H$_5$SO$_4$] has a resistance of only 0.2 kΩ cm [18]. Without doubt the intrinsic electric conductivity is a very attractive feature of ionic lubricants in some applications.

8.2.7
Ionic Greases

A classical lubricant grease consists of 70% base oil and 30% thickener. Typical thickeners are alkali metal stearates, such as potassium, calcium or lithium hydroxystearates. The calcium or lithium stearates are known to show higher hydrophobicity and thermal stability. These stearates are typically produced by an *in situ* neutralization or by saponification in the base oil. However, handling of strong bases in ionic liquids is often problematic as many ionic liquids contain acidic protons or eliminate alkenes in the presence of strong base [31].

A commonly applied thickener is polyurethane. The synthesis of polyurethane is possible in ionic liquids if the ionic liquid does not contain functionalized side chains such as hydroxyl end groups, which are prone to react in the polymerization themselves.

Another version of making greases is to use particulate thickeners such as silica, PTFE or polyurethane, but again the presence of particles in the grease formulation is critical for some applications. Another disadvantage of these particle grease formulations is a higher oil separation rate.

8.3 Applications, Conclusion and Future Challenges

Today in industry, a lot of energy is lost by friction. In the transport sector alone, 10–15% of the energy generated in automotive engines is converted into heat by friction and is lost for the initial purpose of transport [23].

The lubricants that are mainly used today are mineral oil based and therefore restricted in their chemical structures and properties. Modern synthetic lubricants can overcome some of these barriers as they allow higher temperature stability, higher loads, better tribological behavior, reduced noise, applicability in vacuum and better chemical inertness.

Without doubt, ionic liquids are highly attractive candidates for existing and future lubrication problems. Their physico-chemical and tribological nature makes them a very promising class of new synthetic ionic lubricants with a number of entirely new features such as electric conductivity, extremely low compressibility, negligible vapor pressure, chemical inertness, strong inner-lubricant forces and superior thermal stability.

There is a need for high thermal stability lubricants in aircraft, spacecraft [32, 33], industrial baking ovens and other fields where the performance of commonly used PFPE-based lubricants is not good enough to satisfy all industrial needs.

There is also a need for lubricants with extremely low vapor pressures and very low compressibilities. These unique properties open new horizons for working fluids in machinery, compressors, vacuum pumps and hydraulic systems [18].

Another obviously strong point of ionic liquid-based lubricants is thin-film lubrication. In microelectrochemical–mechanical systems (MEMS) [34], for example, ionic liquids can be expected to be excellent lubricants as their strong inner-lubricant forces allow for better lubrication in extremely thin films.

Of course, with all these very promising opportunities, there are still some needs and challenges to be addressed. The registration and systematic toxicological testing of ionic liquids is essential prior to any open-access use of ionic liquid lubricants. In the lubricant industry there are some restrictions on the use of chemicals, such as the European Eco-label, which have to be considered along with all the technical requirements.

In this context, it is very encouraging that the first technical application of an ionic liquid lubricant has already been commercialized. Linde AG announced in 2006 the lubrication of several commercial compressors with ionic liquids [6]. Moreover, the first ionic liquid producers have started specific research in ionic liquid lubricant formulation, an effort from which commercial ionic liquid lubricant formulations can be expected soon [35]. It is clear from today's point of view that ionic liquid lubricants have a huge potential to contribute to substantial savings of energy and raw materials in the future. Hence ionic liquid-based lubricants can be considered as an important green chemistry aspect of ionic liquid technology.

References

1 Ye, C., Liu, W. and Chen, Y. (2001) *Chemical Communications*, **21**, 2144.
2 Jimenez, A., Bermudez, M., Iglesias, P., Carrion, F. and Martinez-Nicolas, G. (2006) *Wear*, **260**, 766.
3 Mu, Z., Zhou F., Zang S., Liang, Y. and Liu, W. (2005) *Tribology International*, **38**, 725–731.
4 Jin, C., Ye, C., Phililips, B., Zabinski, J., Liu, X., Liu, W. and Shreeve, J. (2006) *Journal of Materials Chemistry*, **16**, 1529–1535.
5 Fox, M.F. and Priest, M. (2008) *Journal of Engineering Tribology*, **222**, 291–303.
6 Linde Reports on Science and Technology, Linde AG, Wiesbaden, January (2006).
7 Liu, W., Ye, C., Gong, Q., Wang, H. and Wang, P. (2002) *Tribology Letters*, **13**, 81.
8 Wang, H., Lu, Q., Ye, C. and Liu, W. (2006) *Wear*, **261**, 1174.
9 Kamimura, H., Kubo, T., Minami, I. and Mori, S. (2007) *Tribology International*, **40**, 620.
10 Kamimura, H., Chiba, T., Watanabe, N., Kubo, T., Nanao, H., Minami, I. and Mori, S. (2006) *Tribology Online*, 40–43.
11 Liu, X., Zhou, F., Liang, Y. and Liu, W. (2006) *Wear*, **261**, 1174.
12 Weng, L., Liu, X., Liang, Y. and Xue, Q. (2007) *Tribology Letters*, **26**, 11.
13 Qu, J., Truhan, J.J., Dai, S., Luo, H. and Blau, P.J. (2006) *Tribology Letters*, **22**, 207.
14 Qu, J., Truhan, J.J., Dai, S., Luo, H. and Blau, P.J. (2006) *Tribology Letters* **22**, 207–214.
15 Jimenez, A. and Bermudez, M. (2007) *Tribology Letters*, **26**, 23.
16 Phillips, B.S., John, G. and Zabinski, J.S. (2007) *Tribology Letters*, **26**, 85–91.
17 Lu, Q., Wang, H., Ye, C., Liu, W. and Xue, Q. (2004) *Tribology International*, **37**, 547–552.
18 Uerdingen, M. (2008) presentation at the 16th International Colloquium on Tribology, Esslingen.
19 Jimenez, A., Bermudez, M. Carrion, F. and Martinez-Nicolas, G. (2006) *Wear*, **261**, 347.
20 Stribeck, R. (2002) *Z. Verein. Dtsch. Ing.*, **46**, 1341–1348.
21 Qu, J., Truhan, J.J., Dai, S., Luo, H. and Blau, P.J. (2006) *Tribology Letters*, **22**, 207–214.
22 Mistry, K., Fox, M.F. and Priest, M. (2009) *Journal of Engineering Tribology*, **223**, 563–569.
23 Pensoda, A.S., Comunas, M.J.P. and Fernandez, J. (2008) *Tribology Letters*, **31**, 107–118.
24 Predel, T. and Schlücker, E. (2007) ProcessNet-Jahrestagung, Oral Presentation, Aachen.
25 Reddy, R.G., Zhang, Z., Arenas, M.F. and Blake, D.M. (2003) *High Temperature Material Processes*, **22**, 87.
26 Minami, I., Kamimura, H. and Mori, S. (2007) *Journal of Synthetic Lubrication*, **24**, 135–147.
27 Zeng, Z., Philipps, B.S., Xiao, J.-C. and Shreeve, J.M. (2008) *Chemistry of Materials*, **20**, 2719–2726.
28 Navia, P., Troncoso, J. and Romaní, L. (2008) *Journal of Solution Chemistry*, **37**, 677–688.

29 Uerdingen, M., Treber, C., Balser, M., Schmitt, G. and Werner, C. (2005) *Green Chemistry*, **7**, 321–325.
30 Liu, X., Zhou, F., Liang, Y. and Liu, W. (2006) *Tribology Letters*, **23**, 191–196.
31 Clement, N.D. and Cavell, K.J. (2004) *Angewandte Chemie (International Edition in English)*, **43**, 3845–3847.
32 Jin, C.M., Ye, C., Phillips, B.S. and Zabisnki, J.S. (2006) *Journal of Materials Chemistry*, **16**, 1529.
33 Merstallinger, A., Liedtke, V., Wendrinsky, J., Baca, L., Fastner, U., Doerr, N. and Gaillard, L. (2006) 10th International Symposium on Materials in a Space Environment, European Space Agency, Collioure, France.
34 Bhushan, B., Palacio, M. and Kinzig, B. (2008) *Journal of Colloid and Interface Science*, **317**, 275–287.
35 Scott, A. (2002) *Chemical Week*, **164**, 35.

9
New Working Pairs for Absorption Chillers
Matthias Seiler and Peter Schwab

9.1
Introduction

Ionic liquids [1, 2] are most commonly defined as substances composed entirely of ions with melting points below 100 °C. More specifically, ionic liquids are organic salts that are fluid at (or close to) ambient temperature. Being a laboratory curiosity just a decade ago, ionic liquids are now available on a large scale and some industrial applications are in the early stages of commercialization. Several more uses are currently being discussed and developed and much work remains to uncover fully the high potential of this novel class of liquids. Their scientific and technical appeal is predominantly based on their variability and versatility. By selecting a suitable combination of cation and anion, the properties of ionic liquids can be adjusted over a relatively wide range. Among the most interesting and characteristic properties of ionic liquids are their extremely low volatility, their high thermal and electrochemical stability, their wide liquid ranges, their high dissolution power for a large variety of organic, inorganic and organometallic compounds and the adjustability of their polarity as a function of different variables, e.g. composition [3–8]. Although a lot of effort in past years was put into obtaining pure compounds consisting of only one cation and one anion, it has become more evident recently that binary, ternary and higher mixtures of ionic liquids might be advantageous for finding suitable compositions in particular applications. Furthermore, ionic liquids might have many advantages and might open up new possibilities even when used in mixtures with other chemicals or materials. In fact, while early focus was put on developing ionic liquids as solvents for synthesis, the perception that ionic liquids are limited to replacing organic solvents in synthesis has changed as it has been shown that they can be employed as process aids, e.g. in the hydrosilylation of olefins (immobilization of otherwise homogeneous catalysts) and in the specific absorption of gases. Furthermore, due to their unique properties, ionic liquids create many interesting possibilities as far apart as performance additives and functional fluids in various materials and applications.

Handbook of Green Chemistry, Volume 6: Ionic Liquids
Edited by Peter Wasserscheid and Annegret Stark
Copyright © 2010 WILEY-VCH Verlag GmbH & Co. KGaA, Weinheim
ISBN: 978-3-527-32592-4

Based on these qualities, the consideration of ionic liquids in the search for improved chiller absorbents, e.g. for the replacement of LiBr, appears rather self-evident. In the green chemistry and green engineering context, the use of ionic liquids as new absorbents in absorption chillers is of great potential relevance. This is due to the fact that heat pump/absorption chiller technology is able to make use of so far unused, low-temperature heat streams that are wasted today as their transformation into electrical energy is not efficient enough. Estimations for the European chemical industry alone calculate the amount of these waste heat streams to be several hundred gigawatts. These waste heat streams can be converted by absorption chillers and/or heat pumps either to provide cooling or to be transformed to a higher temperature level [9]. To realize this enormous potential for energy savings and thus CO_2 emission reductions, more efficient absorption chiller technologies and heat storage technologies are required.

Ionic liquids are highly interesting absorbents for absorption chillers. In the following, the requirements and challenges for the use of ionic liquids in absorption chillers are discussed. After presenting a brief state of the art summary, the progress in the development of ionic liquids as potential absorbents for refrigerants such as water is reported.

9.2
Absorption Chillers

As described by Ziegler [9], sorption systems, such as those used for chiller applications, refrigeration and industrial waste heat recovery, transform the latent heat of the liquid–vapor or solid–vapor phase change between different temperature levels. Absorption chillers use two different working fluids, the refrigerant and the absorbent. In Figure 9.1, a basic absorption chiller cycle is illustrated. The four most important components are the evaporator E, where cooling is provided, the generator G, to where the driving heat is conveyed, and the absorber A and condenser C, from which the pumped heat is disposed away. In the generator, vapor is desorbed

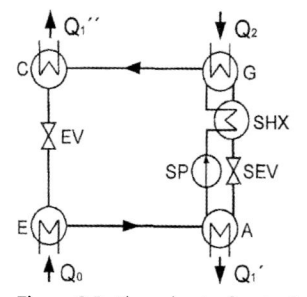

Figure 9.1 Flow sheet of a single-effect absorption chiller [9].
A, absorber; C, condenser; E, evaporator; EV, expansion valve;
G, generator; SEV, solution expansion valve; SHX, solution heat exchanger; SP, solution pump.

from the solution because of the heat input. The vapor is condensed, throttled and evaporated as in compression systems. After evaporation, the vapor is absorbed in the solution which is cooled in the absorber. The solution is pumped to the generator to be regenerated and throttled back to the absorber. To improve the efficiency, a solution heat exchanger SHX is introduced into the solution circuit. For cooling applications, the efficiency (coefficient of performance, COP) is defined as the ratio of cooling output, Q_0, to driving heat input, Q_2 [9].

Absorption chillers are generally classified as direct- or indirect-fired and as single-, double- or triple-effect absorption chillers. In direct-fired units, the heat source can be a gas or some other fuel that is burned in the unit. Indirect-fired units use steam or some other transfer fluid that brings in heat from a separate source, such as a boiler or heat recovered from an industrial process. Hybrid systems, which are relatively common with absorption chillers, combine gas systems and electric systems for load optimization and flexibility [10].

The main freedom in designing absorption systems lies in the choice of the working pair and in the principle of multi-staging. Thus, a crucial fraction of fixed and variable costs for absorption processes is defined by absorbent and refrigerant properties. Although many innovative working pairs have been suggested over the last decades [9], only two meet the prevailing industrial requirements: (a) H_2O–LiBr for water chillers and (b) NH_3–H_2O for refrigeration.

However, these state-of-the-art working pairs show a number of drawbacks, including corrosiveness, a restricted temperature lift due to crystallization (H_2O–LiBr), toxicity, high working pressure and the need for rectification (NH_3–H_2O). Especially for multi-effect cycles with their potential for primary energy savings, absorbents combining a high selectivity and capacity with thermal and chemical stability and a moderate corrosiveness at high process temperatures could lead to new industrial applications.

This chapter focuses on the H_2O–LiBr-driven absorption chiller and the replacement of the LiBr absorbent by suitable ionic liquids.

9.3
Requirements and Challenges

When analyzing the state of the art concerning the potential use of ionic liquid absorbents (see below), it becomes obvious that most of the requirements for using ionic liquids in absorption chillers have either not been evaluated at all or not evaluated sufficiently. Therefore, in this section, the most important ionic liquid requirements are presented:

In order to consider an ionic liquid as a genuine candidate for utilization in an absorption chiller, several prerequisites have to be met. First and foremost, water should be readily soluble in the ionic liquid without any liquid–liquid miscibility gap or crystallization border (thermodynamics). Second, the ionic liquid should be hygroscopic and absorb water with an adequate capacity. Third, heat and mass transfer in the absorber, the generator and the heat exchangers is essential. Moreover, the ionic liquid should wet the surface of the heat exchanging areas in the chiller well over the relevant concentration levels and retain a low viscosity at both low water content and low system temperatures. Furthermore, the ionic liquids should not be corrosive and needs to be compatible with standard chiller materials. High long-term thermal and chemical stability is required, especially when considering double- and triple-effect chillers.

9.3.1
Thermodynamics, Heat and Mass Transfer

Apart from meeting essential heat and mass transfer requirements [9], the ionic liquid–refrigerant phase behavior is one of the most important aspects when evaluating the suitability of ionic liquids for chiller applications. Complete miscibility, the absence of crystal formation under operating conditions and a decrease in the partial pressure of water (P_{H_2O}) of comparable magnitude as for LiBr are crucial for optimizing the coefficient of performance. In the case of ionic liquid absorbents, decreasing the partial pressure of the refrigerant is driven by the anion–cation combination. Especially with respect to water, a small activity coefficient γ_{H_2O} mainly depends on the nature of the anion [11, 12]. Further details are discussed below.

9.3.2
Crystallization Behavior

Ionic liquids are characterized by a rather complex crystallization behavior. In fact, the existence of room temperature ionic liquids is a result of their inherent difficulty to crystallize. In many cases, the crystallization process is kinetically inhibited, so that supercooling can be observed over extended periods of time (up to weeks). Measurements should therefore be checked carefully by differential scanning calorimetric (DSC) analyses before reporting a melting point. Sometimes, pour points offer good indications for the behavior under operating conditions. Furthermore, some ionic liquids completely lack a melting point, and show a glass transition temperature instead [13].

The melting point is a function of the substitution pattern and symmetry on the cation and the choice of the anion [14]. For any given anion, the melting point of the ionic liquid decreases with increasing linear alkyl chain length before reaching a minimum usually between six and eight carbon atoms [15]. This observation can be explained by a decrease in the effective Coulombic force between the ions and an impediment to efficient crystal packing. For longer chain lengths, the intermolecular van der Waals forces increase in relevance and lead to higher melting points. Liquid crystalline compounds can be obtained with chain lengths greater than 14 carbon atoms.

Not surprisingly, any branching of the alkyl chains or the introduction of functional groups which can form hydrogen bonds raises the melting point. For any given cation, the melting point depends very strongly on the anion. As might be expected, increasing electron delocalization in the anion lowers the melting point.

For example, the melting point of [C$_2$mim]Cl is reported as 87 °C, whereas the corresponding dicyanamide salt melts below −21 °C [16].

9.3.3
Corrosion Behavior

The corrosion behavior of pure ionic liquids towards different metals and alloys is currently under intense investigation, as also discussed elsewhere in this book [14, 17]. The anion plays a prominent role in the corrosion process and therefore predominantly determines the corrosiveness of the compound towards ferrous metals. In the case of non-ferrous metals, the complexation ability of both anion and the cation has to be considered. As far as the anions are concerned, halides are known to be problematic and should be avoided. Furthermore, halide impurities are commonplace due to current synthetic processes and possible contamination should be taken into account. At elevated temperatures, decomposition processes can liberate species that are more aggressive towards the metal or alloy. Generally, corrosiveness of ionic liquids towards ferrous metals is less problematic than towards non-ferrous metals.

9.3.4
Viscosity

The correct determination of the viscosities of pure ionic liquids is challenging because the viscosity is very sensitive to even small amounts of contaminants. However, some general observations are worth reporting. Viscosities of ionic liquids vary over a wide range [14, 15]. Many ionic liquids behave more like honey at room temperature and there is really no upper limit. However, the lowest reported viscosities are in the range of ethylene glycol and therefore considerably higher than those of water and some conventional organic solvents. Importantly, the viscosities are very strongly temperature dependent and even ionic liquids that are highly viscous at room temperature become easy to handle at moderately elevated temperatures well below their thermal decomposition temperature [18]. The temperature dependence is less pronounced for low-viscosity ionic liquids, posing a limit on the viscosities attainable even at high temperatures.

As far as the selection of cations and anions is concerned, the main criteria to obtain lower viscosities are the size and the delocalization of the ions, which limit their attractive intramolecular forces. Interestingly, functionalization in the side chain can be an important factor. Ether, hydroxyethyl and nitrile groups often reduce the viscosity significantly compared with their non-derivatized analogues of similar size [19].

9.3.5
Thermal Stability

The issue of thermal stability has been the subject of controversy in the literature, mainly due to the application of different methodologies and the interpretation of data [20]. Most commonly, degradation temperatures are derived from the weight loss measured during fast scan thermogravimetric analysis (TGA) and/or differential thermal analysis (DTA). The correlation is justifiable due to the very low volatility of ionic liquids below the degradation temperature. However, even though the analytical methods are cheap and easily reproducible and deliver data that are generally comparable, reported values for thermal stability vary widely. Therefore, it is important to look beyond the reported T_{onset} values and consider the key parameters and assumptions under which the values were obtained. Important parameters are the rate of heating, the purity of the compound, the atmosphere and the material of the pan in which the compound is heated. However, even the determination of the T_{onset} value from the measurement is open to interpretation. Some report T_{onset} as the onset point of weight loss, some report it after 5% weight loss, and others use the step tangent method or the minimum in the DTA curve method. These methods, however, reveal only the short-term thermal stability and sometimes severely overestimate the maximum operating temperatures [21–23].

Therefore, these values should be used for comparison relative to other compounds. Another useful method to obtain onset temperatures of degradation is gas chromatography [24]. When employing ionic liquids as the stationary phase, the detection of the phase bleeding provides excellent comparative data [25].

The results published in the literature reveal several noteworthy general rules for the thermal stability of ionic liquids. The decomposition temperature depends very strongly on the nucleophilicity of the anion [13, 14, 26, 27] because one of the main degradation pathways is the nucleophilic attack of the anion on the cation. Among the cations, phosphonium salts are the most stable, followed by imidazolium salts and ammonium salts. The thermal stability of imidazolium and similar cations can be improved by increasing the linear, short-chain alkyl substitution on the ring. Also, it is well documented that halide-based anions exhibit poor thermal stability whereas anions with fluorinated alkyl chains show the highest reported decomposition temperature [14, 15].

9.4
State of the Art and Selected Results

Several research groups have studied the phase behavior of ionic liquid–water systems [12, 14, 28–39]. In this context, a wide variety of ionic liquids have been investigated, such as [C$_2$mim][PF$_6$], [C$_4$mim][PF$_6$], [C$_8$mim][PF$_6$], [C$_4$mim][BF$_4$], [C$_8$mim][BF$_4$], [C$_4$mim]Cl, [C$_2$mim][Tf$_2$N], [C$_4$mim][Tf$_2$N], [C$_1$mim][DMP], [C$_4$mim]Br, [C$_2$OHmim][BF$_4$], [C$_4$mim]I, [C$_2$mim][diethylphosphate] and [C$_4$mim][dibutylphosphate].

In addition, other potential ionic liquid–refrigerant systems have been studied: Cai *et al.* developed a dynamic model for a single-effect absorption chiller cycle using ionic liquids as absorbent [35]. Thermodynamic properties were obtained from an equation of state for the refrigerant–absorbent mixtures. The coefficient of performance (COP) of a system using CO$_2$–[C$_4$mim][PF$_6$] was much lower than that of a traditional system using NH$_3$–H$_2$O but questions remain regarding the accuracy of the thermodynamic approach chosen. Some design and operation parameters that affect the cycle performance were identified and the transient response of the cycle was investigated.

Kim *et al.* investigated [C$_4$mim]Br–trifluoroethanol and [C$_4$mim][BF$_4$]–trifluoroethanol as potential working pairs for absorption heat pumps [36]. Refractive indices and heat capacities were determined in the temperature range 298.2–323.2 K. The partial pressure of trifluoroethanol (TFE) was measured using the boiling point method in the concentration range 40.0–90.0 wt% of ionic liquid and were successfully correlated with an Antoine-type equation. The partial pressure results indicate that the absorption performance of [C$_4$mim]Br is better than that of the ionic liquid [C$_4$mim][BF$_4$].

Ionic liquid–CO$_2$-working pairs such as [C$_4$mim][PF$_6$]–CO$_2$ combination to be used for absorption refrigeration were proposed by Sen and Paolucci [37]. Based on a promising absorption performance and a good thermal stability, it was concluded that the ionic liquid proposed represents a suitable absorbent for absorption refrigeration. Key properties defining the ideal ionic liquid for absorption refrigeration systems were also discussed. However, it was noted that most thermodynamic and thermophysical ionic liquid properties, which are required to discuss the ionic liquid potential for absorption refrigeration, are not available in the literature.

Yokozeki and Shiflett [38, 39] determined the solubilities of ammonia in [C$_2$mim][acetate], [C$_2$mim][SCN], [C$_2$mim][C$_2$H$_5$SO$_4$], [dimethylethylammonium][acetate] under isothermal conditions between 283 and 373 K and in a concentration range between 30 and 85 mol% of ammonia. For the investigated ionic liquid, very high NH$_3$ solubilities were found. The pressure–temperature data obtained were successfully described using an equation of state. All the excess properties (enthalpy, entropy and Gibbs energy) showed negative values, reflecting strong intermolecular complex formation. It was concluded that the ionic liquid–NH$_3$ systems investigated represent promising working pairs for the field of absorption refrigeration. Further valuable results on the physiochemical and thermophysical properties of ionic liquids can be found elsewhere [44–55].

In Figure 9.2, vapor–liquid equilibria (VLE) of different ionic liquid–water-systems at 35 °C are depicted. The ionic liquids used are commercially available from Evonik Degussa. The VLE measurements were carried out according to the procedure described by Evonik Degussa [11, 12, 40]. Unlike LiBr, at 35 °C no crystal formation occurs over the entire concentration range.

The ionic liquids investigated have alkylimidazolium cations. As demonstrated elsewhere, the cation has a negligible influence on the partial pressure of water [3, 11, 12, 40]. However, the choice of the anion is essential when aiming at lowering P_{H_2O}. Whereas the ethylsulfate anion of Tego ionic liquid IMES ([C$_2$mim][C$_2$H$_5$SO$_4$]) corresponds to a moderate decrease in P_{H_2O}, the dimethylphosphate anion of the ionic liquid [C$_1$mim][dimethylphosphate] leads to a stronger attraction of the water molecules, corresponding to a smaller activity coefficient γ_{H_2O}. However, with increasing compactness of the anion, the partial pressure can be decreased further. This becomes obvious for [C$_2$mim][acetate], an ionic liquid that even allows the minimum pressure of the LiBr–H$_2$O system to be reached at high ionic liquid concentration.

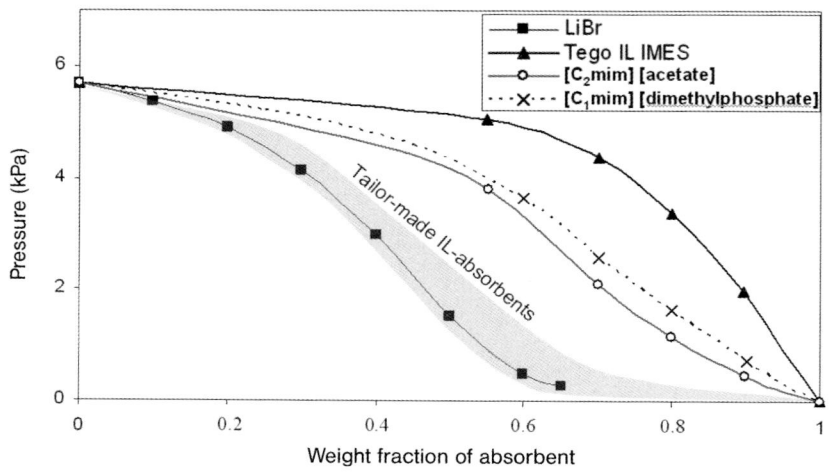

Figure 9.2 Experimental VLE results for selected ionic liquid–water systems at 308.15 K.

Therefore, the absorption performance for alkylimidazolium-based ionic liquids improves with increasing compactness of the anion according to our preliminary conclusions, i.e. γ_{H_2O} ([acetate]$^-$) < γ_{H_2O} ([dimethylphosphate(DMP)]$^-$) < γ_{H_2O} ([ethylsulfate]$^-$).

Based on these results, Evonik Degussa currently synthesizes and evaluates tailor-made ionic liquid absorbents. The absorption performance of the latter candidates is indicated by the gray shaded area in Figure 9.2. In addition, adjusting the property profile of an ionic liquid-based working pair to the aforementioned requirements by the use of high performance additives can help leveraging the full potential of ionic liquids in this field. Further state-of-the-art information on the potential use of ionic liquids in absorptions chillers can be found elsewhere [11, 12, 40–43].

9.5
Abbreviations

P_{H_2O} partial pressure of water (kPa)
γ_{H_2O} activity coefficient
VLE   vapor–liquid equilibrium
TFE   trifluoroethanol
COP   coefficient of performance

References

1 Rogers, R.D. and Seddon, K.R. (2003) Ionic liquids – solvents of the future. *Science*, **31**, 792–793.

2 Wasserscheid, P. (2001) Ionische Flüssigkeiten –innovative Lösungsmittel. *Nachrichten Chemie*, **49**, 12–16.

3 Seiler, M., Jork, C., Kavarnou, A., Arlt, W. and Hirsch, R. (2004) Separation of azeotropic mixtures using hyperbranched polymers or ionic liquids. *AICHE Journal*, **50**, 2439–2454.

4 Wasserscheid, P. and Keim, W. (2000) Ionische Flüssigkeiten – neue Lösungen für die Übergangsmetallkatalyse. *Angewandte Chemie*, **112**, 3926–3945.

5 Wasserscheid, P. (2003) Potential to apply ionic liquids in industry, In: *Green Industrial Applications of Ionic Liquids*, (eds Rogers, R.D. and Seddon, K.R.) NATO Science Series, II: Mathematics, Physics and Chemistry, Vol. 92, Kluwer, Dordrecht, pp. 29–47.

6 Davis, J.H. (2002) Synthesis of task-specific ionic liquids, In: *Ionic Liquids in Synthesis*, (eds Wasserscheid, P. and Welton, T.) VCH Verlag GmbH, Weinheim, pp. 33.

7 Jork, C., Kristen, C., Pieraccini, D., Stark, A., Chiappe, C., Beste, Y.A. and Arlt, W. (2005) Tailor-made ionic liquids. *Journal of Chemical Thermodynamics*, **37**, 537–558.

8 Plechkova, N.V. and Seddon, K.R. (2008) Applications of ionic liquids in the chemical industry. *Chemical Society Reviews*, **37**, 123–150.

9 Ziegler, F. (1999) Recent developments and future prospects of sorption heat pump systems. *International Journal of Thermal Science*, **38**, 191–208.

10 Tang, T., Villarreal, L. and Green, J. *Advanced Design Guideline by the New Buildings Institute for the Southern California Gas Company*, Contract P13311, part of SoCalGas' Third Party Initiative (1998).

11 Jork, C., Seiler, M. and Weyershausen, B. World Patent, WO 2006/134015, 2006.

12 Seiler, M., Schwab, P. and Ziegler, F. Sorption systems using ionic liquids, In: Proceedings of the International Sorption Heat Pump Conference, 23–26 September 2008, Seoul, Korea.

13 Huddleston, J.G., Visser, A.E., Reichert, W.M., Willauer, H.D., Broker, G.A. and Rogers, R.D. (2001) *Green Chemistry*, **3**, 156–163.

14 Wasserscheid, P. and Welton, T. (2008) *Ionic Liquids in Synthesis* 2nd edn., Wiley-VCH Verlag GmbH, Weinheim.

15 Handy, S. (2005) *Current Organic Chemistry*, **9**, 959–988.

16 MacFarlane, D.R., Forsyth, S.A., Golding, J. and Deacon, G.B. (2002) *Green Chemistry*, **4**, 444–449.

17 Endres, F., Abbott, A., MacFarlane, D.R., (eds) (2008) *Electrodeposition in Ionic Liquids*, Wiley-VCH Verlag GmbH, Weinheim.

18 Wasserscheid, P., van Hal, R. and Boesmann, A. (2002) *Green Chemistry*, **4**, 400–404.

19 Welton, T. (2002) Polarity, In: *Ionic Liquids in Synthesis*, (eds Wasserscheid, P. and Welton, T.) Wiley-VCH Verlag GmbH, Weinheim, pp. 94–103.

20 Scammells, P.J., Scott, J.L. and Singer, R.D. (2005) *Australian Journal of Chemistry*, **58**, 155–169.

21 Fredlake, C.P., Crosthwaite, J.M., Hert, D.G., Aki, S.N.V. and Brennecke, J.F. (2004) *Journal of Chemical and Engineering Data*, **49**, 954.

22 Baranyai, K., Deacon, G.B., MacFarlane, D.R., Pringle, J.M. and Scott, J.L. (2004) *Australian Journal of Chemistry*, **57**, 145.

23 Fox, D.M., Awad, W.H., Gilman, J.W., Maupin, P.H., DeLong, H.C. and Trulove, P.C. (2003) *Green Chemistry*, **5**, 724.

24 Poole, C.F., Furton, K.G. and Kersten, B.R. (1986) *Journal of Chromatographic Science*, **24**, 400–405.

25 Poole, S.K. and Poole, C.F. (1988) *Journal of Chromatography*, **17**, 435–439.

26 Ngo, H.L., LeCompte, K., Hargens, L. and McEwen, A.B. (2000) *Thermochimica Acta*, **357–358**, 97.

27 Stuff, J.R. (1989) *Thermochim. Acta*, **152**, 421–425.

28 Anthony, J.L., Maginn, E.J. and Brennecke, J.F. (2001) Solutions thermodynamics of imidazolium-based ionic liquids and water. *The Journal of Physical Chemistry. B*, **105**, 10942–10949.

29 Calvar, N., González, B., Gómez, E. and Domininguez, Á. (2006) Vapor–liquid equilibria for the ternary system ethanol

+ water + 1-butyl-3-methylimidazolium chloride and the corresponding binary systems at 101.3 kPa. *Journal of Chemical and Engineering Data* **51**, 2178–2181.

30 Kato, R. and Gmehling, J. (2005) Measurement and correlation of vapor–liquid equilibria of binary systems containing the ionic liquids [EMIM] [(CF$_3$SO$_2$)$_2$N], [BMIM][(CF$_3$SO$_2$)$_2$N], [MMIM][(CH$_3$)PO$_4$] and oxygenated organic compounds respectively water. *Fluid Phase Equilibria*, **231**, 38–43.

31 Kim, K., Park, S.Y., Choi, S. and Lee, H. (2004) Vapor pressures of the 1-butyl-3-methylimidazolium bromide + water, 1-butyl-3-methylimidazolium tetrafluoroborate + water and 1-(2-hydroxyethyl)-3-methylimidazolium tetrafluoroborate + water systems. *Journal of Chemical and Engineering Data*, **49**, 1550–1553.

32 Kim, K., Demberlnyamba, D., Shin, B., Yeon, S., Choi, S., Cha, J., Lee, H., Lee, C. and Shim, J. (2006) Surface tension and viscosity of l-butyl-3-methylimidazolium iodide and l-butyl-3-methylimidazolium tetrafluoroborate and solubility of lithium bromide + 1-butyl-3-methylimidazolium bromide in water. *Korean Journal of Chemical Engineering*, **2**, (31), 113–116.

33 Zhao, J., Jiang, X., Li, C. and Wang, Z. (2006) Vapor pressure measurement for binary and ternary systems containing a phosphoric ionic liquid. *Fluid Phase Equilibria*, **247**, 190–198.

34 Wong, D.S.H., Chen, J.P., Chang, J.M. and Chou, C.H. (2002) Phase equilibria of water and ionic liquids [emim][PF$_6$] and [bmim][PF$_6$]. *Fluid Phase Equilibria*, **194–197**, 1089–1095.

35 Cai, W., Sen, M. and Paolucci, S. Dynamic modelling of an absorption refrigeration system using ionic liquids. In: *Proceedings of 2007 International Mechanical Engineering Congress and Exposition*, Seattle, WA, 2007.

36 Kim, K., Shin, B., Lee, H. and Ziegler, F. (2004) Refractive index and heat capacity of 1-butyl-3-methylimidazolium bromide and 1-butyl-3-methylimidazolium tetrafluoroborate and vapor pressure of binary systems for 1-butyl-3-methylimidazolium bromide + trifluoroethanol and 1-butyl-3-methylimidazolium tetrafluoroborate + trifluoroethanol. *Fluid Phase Equilibria*, **218**, 215–220.

37 Sen, M. and Paolucci, S. Using carbon dioxide and ionic liquids for absorption refrigeration. In: *7th IIR Gustav Lorentzen Conference on Natural Working Fluids*, Trondheim, Norway, 2006.

38 Yokozeki, A. and Shiflett, M.B. (2007) Vapor–liquid equilibria of ammonia + ionic liquid mixtures. *Applied Energy*, **84**, 1258–1273.

39 Shiflett, M. and Yokozeki, A. (2006) Solubility and diffusivity of hydrofluorocarbons in room-temperature ionic liquids. *AICHE Journal*, **52**, 1205–1219.

40 Seiler, M. Kälteprozesse mit ionischen Flüssigkeiten. In: *High Pressure Meets Advanced Fluids*, ProcessNet Jahrestreffen, Aachen, Germany, 2008.

41 Shiflett, M. and Yokozeki, A. (2006) Vapor–liquid–liquid equilibria of hydrofluorocarbons + 1-butyl-3-methylimidazolium hexafluorophosphate. *Journal of Chemical and Engineering Data*, **51**, 1931–1939.

42 Boesmann, A. and Schubert, F.T. World Patent, WO 2005/113702, 2005.

43 Shiflett, M. and Yokozeki, A. World Patent, WO 2006/084262, 2006.

44 Tokuda, H., Hayamizu, K. Ishii, K., Susan, Md.A.B.H., and Watanabe, M. (2004) Physicochemical properties and structures of room temperature ionic liquids. 1. Variation of anionic species. *The Journal of Physical Chemistry. B*, **108**, 16593–16600.

45 Huddleston, J.G., Visser, A.E., Reichert, W.M., Willauer, H.D., Broker, G.A. and Rogers, R.D. (2001) Characterization and comparison of hydrophilic and hydrophobic room-temperature ionic liquids incorporating the imidazolium cation. *Green Chemistry*, **3**, 156–164.

46 Reddy, R., Zhang, Z., Arenas, M.F. and Blake, D.M. (2003) Thermal stability and corrosivity evaluations of ionic liquids as thermal energy storage media. *High-Temp. Mater. Processes*, **22**, (2), 87–94.

47 Bonhôte, P., Dias, A.P., Papageorgiou, N., Kalyanasundaram, K. and Graetzel, M. (1996) Hydrophobic, highly conductive ambient-temperature molten salts. *Inorganic Chemistry*, **35**, 1168–1178.

48 Branco, L.C., Rosa, J.N., Ramos, J.J.M. and Alfonso, A.M. (2002) Preparation and characterization of new room temperature ionic liquids. *Chemistry - A European Journal*, **8**, (16), 3671–3677.

49 Tokuda, H., Hayamizu, K. and Ishii, K., Susan, Md.A.B.H., and Watanabe, M. (2005) Physicochemical properties and structures of room temperature ionic liquids. 2. Variation of the alkyl chain length in imidazolium cation. *The Journal of Physical Chemistry. B*, **109**, 6103–611.

50 Crosthwaite, J.M., Muldoon, M.J., Dixon, K.J., Anderson, J.L. and Brennecke, J.F. (2005) Phase transition and decomposition temperatures, heat capacities and viscosities of pyridinium ionic liquids. *Journal of Chemical Thermodynamics*, **37**, 9–68.

51 Fredlake, C.P., Crosthwaite, J.M., Hert, D.G., Aki, S.N.V.K. and Brennecke, J.F. (2004) Thermophysical properties of imidazolium-based ionic liquids. *Journal of Chemical and Engineering Data*, **49**, 954–964.

52 Awad, W.H., Gilman, J.W., Nyden, M., Harris, R.H., Sutto, T.E., Callahan, J., Trulove, P.C., Delong, H.C. and Fox, D.M. (2004) Thermal degradation studies of alkyl-imidazolium salts and their application in nanocomposites. *Thermochimica Acta*, **409**, 3–11.

53 Baltus, R.E., Robert, M.C., Culhertson, B.H., Luo, H., Depaoli, D.W., Dai, S. and Duckworth, D.C. (2005) *Separation Science Technology*, **40**, (1–3), 525–541.

54 Gordon, C.M. (2001) New developments in catalysis using ionic liquids. *Applied Catalysis A-General*, **222**, 101–117.

55 Seddon, K., Stark, A. and Torres, M..-J. (2002) in Clean Solvents: Alternative Media for Chemical Reactions and Processing, *ACS Symposium Series*, Vol. 819, American Chemical Society, pp. Washington, DC, 34–49.

Part IV
Ionic Liquids and the Environment

10
Design of Inherently Safer Ionic Liquids: Toxicology and Biodegradation

Marianne Matzke, Jürgen Arning, Johannes Ranke, Bernd Jastorff, and Stefan Stolte

10.1
Introduction

Green chemistry and its 12 principles, defined and described by Paul Anastas and John C. Warner in 1998 [1], represent an essential prerequisite for a sustainable chemistry, sustainable chemical products and a sustainable and hence responsible chemical industry. "Inherently safer chemical products", the "design" of new and "environmentally benign substances", all these keywords are closely linked to green chemistry and they are gaining more and more importance following the Agenda 21 process and – probably more important – in the light of the new European chemical legislation for the Registration, Evaluation, Authorization and Restriction of Chemicals (REACH).

But how can we reach the goal of a real sustainable chemical, which has to be on the one hand non-toxic, readily biodegradable and with its synthesis leaving a small ecological footprint – the key issues with respect to green chemistry – but on the other hand meets all technological and economic needs? Using the case of ionic liquids as an example of a promising group of industrial chemicals, which bear a high potential for a sustainable design process, the above-mentioned issues and related questions will be discussed in this chapter. Since ionic liquids represent a highly diverse group of chemicals, they serve as an ideal model to demonstrate the problems and challenges that chemical and biological complexity poses on the way towards sustainable chemical substances. However, the case of ionic liquids also shows how these problems can be addressed and how subsequently the goal of tailor-made sustainable substances can be realized by an integrated approach based on an interdisciplinary and tiered strategy, in which academic and industrial research closely cooperate. Thus, by considering not only economic and technological features but also ecological and toxicological issues right from the beginning of the development process of ionic liquids, environmental disasters as in the case of dichlorodiphenyltrichloroethane (DDT) – which started as "the compound of the century" in 1939 and ended up as an internationally banned toxicant in 1972 – can be avoided.

More recent examples of industrial chemicals that only fulfill the technological and economic issues of a sustainable chemical and fail in the ecotoxicological part are perfluorooctanesulfonate (PFOS), nonylphenols and brominated flame retardants. They have been found to be highly persistent in the environment, bioaccumulative or exhibit endocrine-disrupting effects in different species [2–8].

Regarding these latter issues and the overall context of the present book – what does green chemistry mean for ionic liquids? – this chapter focuses on the safety aspects of ionic liquids. Using the "Thinking in terms of Structure–Activity Relationships" (T-SAR) concept, introduced by Jastorff et al. [9], in the first part of the chapter, the available toxicological and ecotoxicological data for ionic liquids are systematically analyzed and summarized. It will be discussed how the different substructural elements – cationic head group, side chain and anion – influence the (eco)toxicity of ionic liquids in various cellular, aquatic and terrestrial test systems. Based on these results, the potential of each of these substructural elements to be used for the design of more inherently safer ionic liquid structures is considered.

The second part of this chapter deals with the biodegradability of ionic liquids. The present literature is reviewed concerning aerobic degradation of ionic liquids and their metabolites and degradation pathways and also the toxicity of ionic liquids towards microorganisms, for example in sewage sludge. Again, the three substructural elements head group, side chain and anion will be analyzed regarding their potential for the design process of readily biodegradable ionic liquids.

In the following subsections, the theoretical T-SAR approach and the test kit concept together with the tiered strategy for the design of inherently safer ionic liquids will be introduced in more detail.

10.1.1
The T-SAR Approach and the "Test Kit" Concept

The theoretical T-SAR analysis of a chemical entity is based on its three-dimensional structural formula representing the "identity card" of each single substance. From this starting point, the most important molecular interaction potentials, the stereochemistry and functional groups causing a certain reactivity can be identified. For example, the potential of a chemical to act as hydrogen bond donor or acceptor can easily be revealed, and also its hydrophobic interaction potential. Furthermore, the flexibility, the three-dimensional space filling and the various bonding angles of a molecule – describing the spatial orientation of the interaction potentials – can be estimated. Especially this spatial pattern of molecular interaction potentials is of high relevance for the interactions of substances with biomolecules such as enzymes or receptor proteins. Jastorff et al. [10] have given a detailed description of the algorithm for the theoretical analysis of the structural formula of a chemical entity. In Figure 10.1, the T-SAR analysis of a chemical entity is presented schematically, including the color-coding scheme used to highlight the most important interaction potentials. This theoretical approach forms the basis of the analysis of the data presented within this chapter and it was applied as a useful tool to identify and group the key structural elements determining ionic liquid (eco)toxicity and biodegradability.

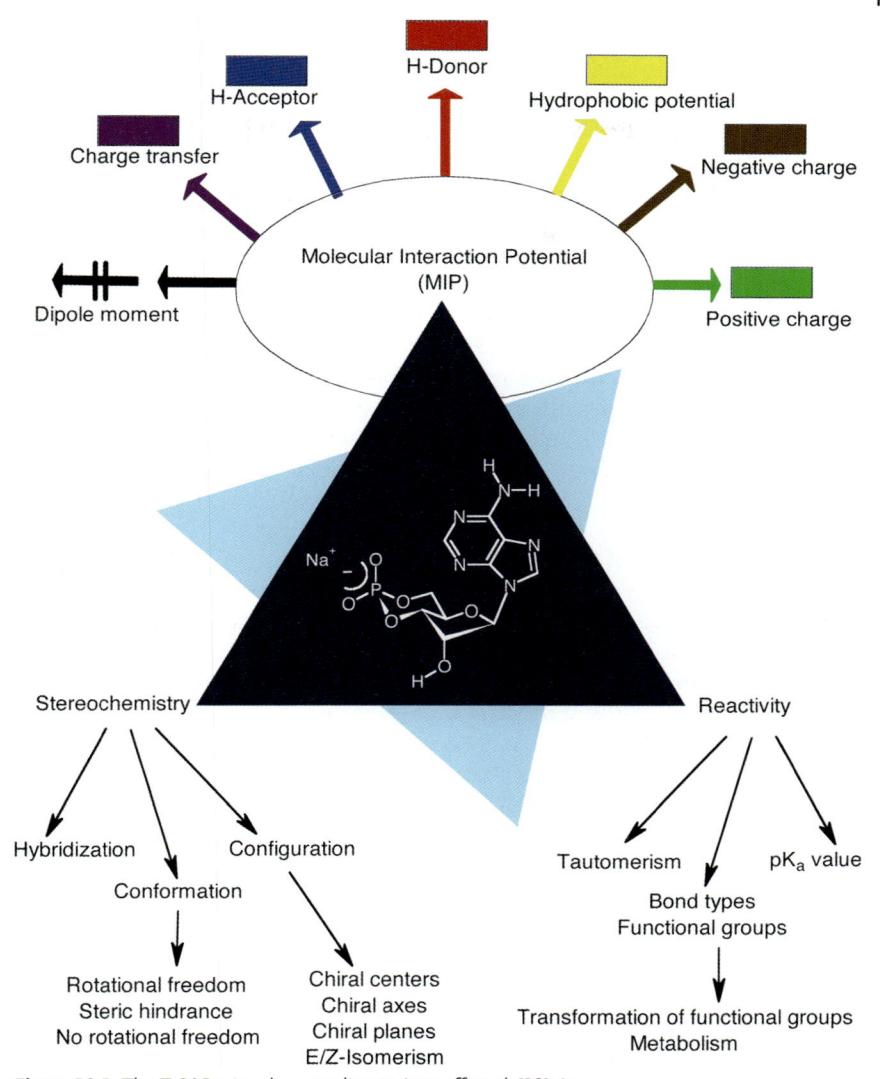

Figure 10.1 The T-SAR triangle according to Jastorff et al. [10]. It provides an overview of the T-SAR analysis of a chemical structure. Additionally, the color-coding scheme to highlight molecular interaction potentials is presented.

Subsequently, this T-SAR analysis allows the formulation of working hypotheses regarding physicochemical properties (for instance, partition coefficients, vapor pressure, solubilities), possible biotic and abiotic metabolization pathways and (eco)toxicological impacts. Especially the analysis of biotic and abiotic metabolites provides valuable knowledge on the way towards readily biodegradable and non-persistent chemical products. These working hypotheses can then be verified in toxicity and ecotoxicity assays selected from a flexible test battery [11–13].

Another, more practical, benefit of the theoretical T-SAR approach arises from the fact that it is capable of revealing general mechanisms and trends of harmful impacts of chemicals by simply selecting a few "lead structures" from a large pool of substances. The selection of such lead structures – representing the structural and substructural key elements that characterize a certain substance class – forms the basis of the so-called "test kit concept" [14]. The knowledge generated from testing these test kits can then be carefully applied to compounds that have not yet been explored. Thereby, it supports the implementation of qualitative and quantitative structure–activity relationships (SARs and QSARs). SARs and QSARs represent a powerful and highly efficient tool in the comparative hazard assessment of chemicals, since they help to reduce the number of necessary tests and costs dramatically.

10.1.2
Strategy for the Design of Sustainable Ionic Liquids

The strategy proposed here aims at considering the technological needs of chemical substances – in our case ionic liquids – while avoiding hazardous structures in the early steps of the research and development stage of new industrial chemical products. Thus, it must be a major goal of chemists, biologists and technologists to identify the most promising candidates for real sustainable products out of the enormous pool of technologically accessible substances. Striving for sustainability is – especially in the field of chemistry – a transdisciplinary (life sciences together with engineering sciences) and an interdisciplinary (chemistry, biology, toxicology, physics, etc.) task. This implies that a fundamental cornerstone of the strategic approach must be the close cooperation of academic scientists and the producing industry to cover all needs for sustainable chemical products. In the case of ionic liquids, such a collaboration has been successfully applied in the eco-design of inherently safer ionic liquid structures [15].

Additionally, the development of sustainable substances is always an iterative process. With respect to this aspect, we propose a tiered strategy to theoretically analyze chemical structures using the above-introduced T-SAR approach, to test preselected "lead structures" in a flexible test battery and finally to use the data generated for further synthesizing a first set of already optimized structures that subsequently will be fed into this cycle again [14].

The crucial advantage of such a tiered and differentiated approach compared with the routinely applied testing procedures is that the T-SAR guided analysis yields test systems that are able to provide valuable data, which can be used to accept or reject certain working hypotheses. Thus, this testing strategy can efficiently increase not only the quantity but also in particular the quality of knowledge within the framework of ecotoxicology. A schematic overview on the tiered testing strategy starting from the theoretical T-SAR analysis is given in Figure 10.2.

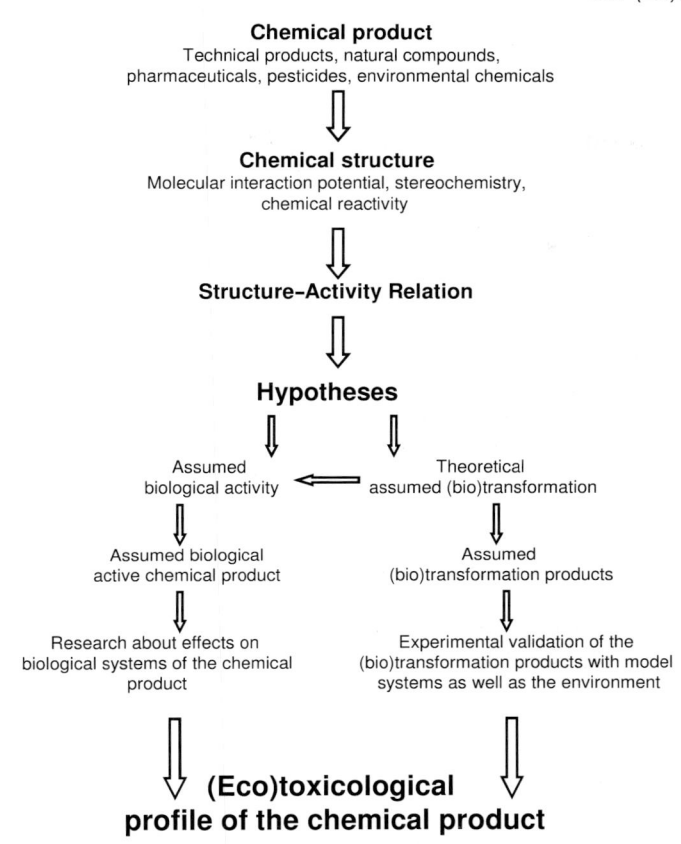

```
         Chemical product
   Technical products, natural compounds,
pharmaceuticals, pesticides, environmental chemicals
                    ⇩
          Chemical structure
  Molecular interaction potential, stereochemistry,
              chemical reactivity
                    ⇩
        Structure–Activity Relation
                    ⇩
              Hypotheses
              ⇩        ⇩
      Assumed            Theoretical
  biological activity  ⇐  assumed (bio)transformation
         ⇩                      ⇩
  Assumed biological        Assumed
active chemical product  (bio)transformation products
         ⇩                      ⇩
 Research about effects on  Experimental validation of the
biological systems of the chemical  (bio)transformation products with model
         product          systems as well as the environment
              ⇩        ⇩
         (Eco)toxicological
      profile of the chemical product
```

Figure 10.2 Approach for the development of (eco)toxicological profiles of chemical compounds according to Jastorff et al. [9].

Integrating such (eco)toxicological profiles into early steps of the research and development processes of new chemical products is a necessary milestone on the way towards inherently safer and hence "greener" chemicals.

The following parts of this chapter systematically analyze – based on the T-SAR approach and the test kit concept – ionic liquid structures along with recent literature data on the (eco)toxicological profiles of cationic head groups, side chains and anions of ionic liquids. Special emphasis will be placed on the prospective designing potential that each of these substructural elements bears with respect to the goals of green chemistry.

10.2
(Eco)toxicity of Ionic Liquids

In the past few years, the image of ionic liquids has changed. In the beginning, the whole substance class was discussed in an undifferentiated way as being "green" and "environmentally benign". Those statements were justified with the negligible vapor pressure, which results in reduced air emission, non-flammability and non-explosiveness. Indeed, these pronounced advantages for the operational safety includes a high "green" potential, but without a sound knowledge of the (eco) toxicological behavior, no justification for this classification can be given.

Additionally, based on the high structural diversity of this substance class, general statements such as "ionic liquids are green" or "ionic liquids are toxic" have to be avoided. These suggestive generalizations are counter-productive for the development of products and processes containing ionic liquids.

However, in the meantime, a paradigm shift has appeared owing to an increased knowledge based on a growing number of studies analyzing the hazard potential of many ionic liquids in different biological test systems. From these studies, it is now known that there are ionic liquids with a low and a high hazard potential and that the "greenness" depends strongly on the structure.

An overview on the existing (eco)toxicological studies and structures investigated is given later in Schemes 10.1 and 10.2, an overview of investigated organisms is given in Table 10.1 and an overview of tested ionic liquids is given in Table 10.2. According to the T-SAR approach presented, a systematic analysis of the different structural elements based on the experimental and theoretical data collected will be performed to contribute to the structural design of ionic liquids with a reduced hazard for humans and the environment.

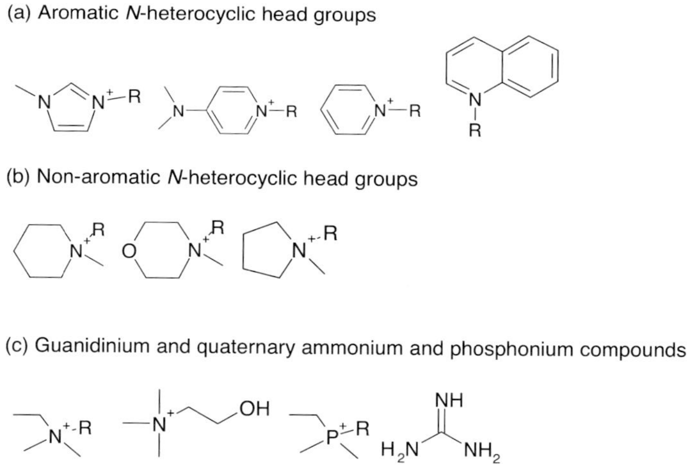

(a) Aromatic *N*-heterocyclic head groups

(b) Non-aromatic *N*-heterocyclic head groups

(c) Guanidinium and quaternary ammonium and phosphonium compounds

Scheme 10.1 Tested cationic head group structures of ionic liquids. The side chain is replaced by R. Side chains: linear alkyl side chains (methyl to octadecyl) and side chains containing functional groups such as ether (in different positions), terminal hydroxyl, carboxyl, ester and nitrile functions.

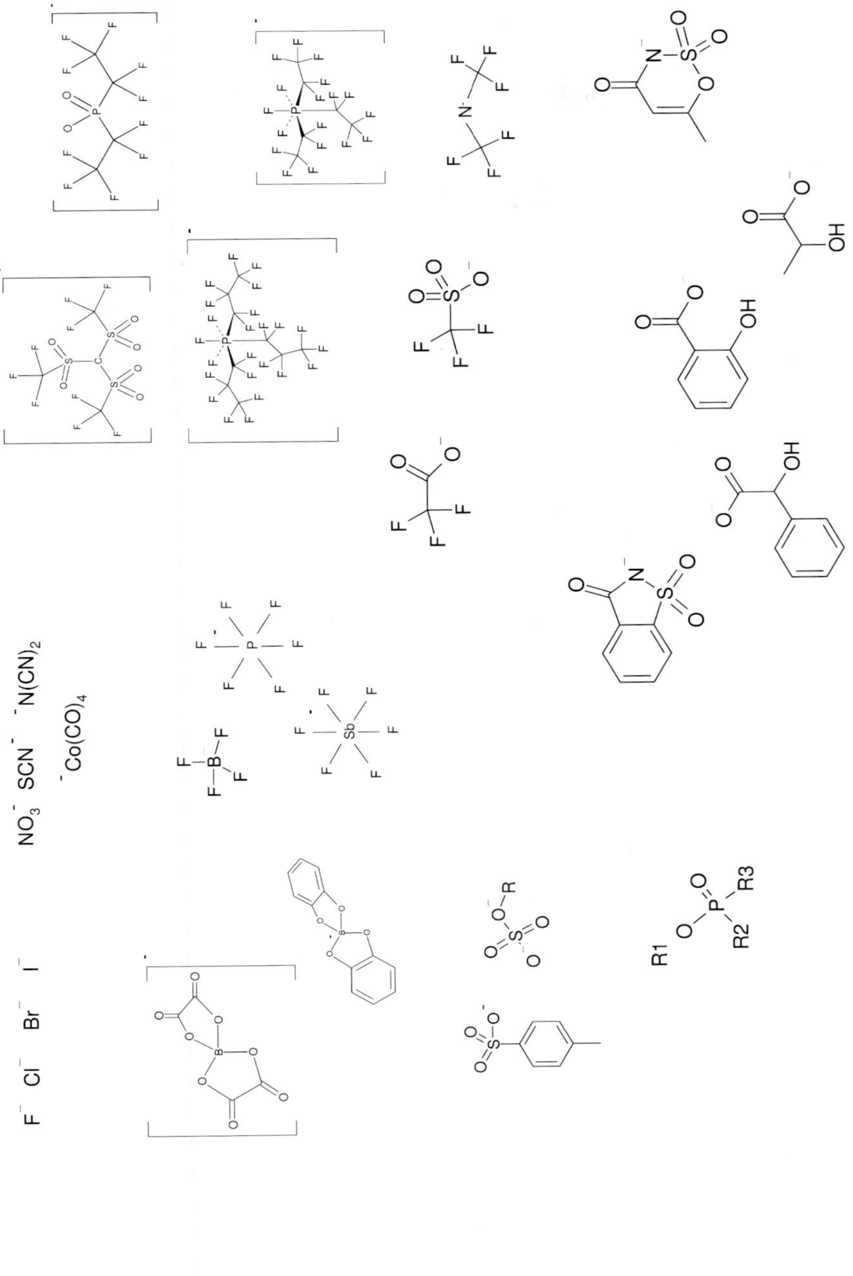

Scheme 10.2 Overview of the most important ionic liquid anions tested.

Table 10.1 Overview of test organisms investigated.

Enzymes	Acetylcholinesterase [11, 12, 15, 62, 64] AMP deaminase [61] Monooxygenase P450 [69]
Cell cultures	**Human** HT-29 (colon carcinoma) [29],CaCo-2 (colon carcinoma) [29, 31], HeLa (cervical carcinoma) [63, 70] **Animal** IPC-81(promyelotic leukemia, rat) [12, 15, 55, 56, 58, 62, 65, 66], C6 (glioma cells, rat) [56]
Microorganisms	**Bacteria – cocci** *Micrococcus luteus* [23, 47–51], *Staphylococcus aureus* [23, 27, 32, 47–51] (*S. aureus* 209 KCTC 1916 [26]), *Enterococcus faecalis* [23, 47, 48], *Moraxella catarrhalis* [23, 47, 48], *Escherichia coli* [23, 24, 47, 48] (*E. coli* KCTC 1924 [26]), *Enterococcus hirae* [23, 49, 50], *Proteus vulgaris* [23, 47–51], *Klebsiella pneumoniae* [23, 47–51], *Staphylococcus epidermis* [23, 49, 50], *Pediococcus pentosaceus* NRIC 0099 [42] **Bacteria – rods** *Pseudomonas aeruginosa* [23, 47–51], *Salmonella typhimurium* KCTC 1926 [26], *Salmonella typhimurium* TA98/TA100 [28], *Photobacterium phosphoreum* [32], *Serratia marcescens* [23, 50], *Vibrio fischeri* [12, 13, 25, 27, 35, 41, 55, 59, 60], *Pseudomonas fluorescens* [27], *Lactobacillus homohiochi* NRIC 0119 [42], *Lactobacillus homohiochi* NRIC 1815 [42], *Lactobacillus fructivorans* NRIC 0224 [42], *Lactobacillus fructivorans* NRIC 1814 [42], *Lactobacillus delbruekii* subsp. *lactis* NRIC 1683 [42], *Leuconostoc fallax* NRIC 0210 [42], *Lactobacillus rhamnosus* NBRC 3863 [43] **Bacteria – bacilli** *Bacillus subtilis* [23, 26, 27, 47, 48] (*B. subtilis* KCTC 1914 [26]), *Bacillus cereus* [30], *Bacillus coagulans* NBRC 12583 [42] **Fungi** *Saccharomyces cerevisiae* [27, 30], *Pichia pastoris* [30], *Candida albicans* [23, 26, 47–50] (*C. albicans* KCTC 1940 [26]), *Rhodotorula rubra* [23, 47–51], *Sclerophoma pityophila* [51]
Plants	**Limnic algae** *Scenedesmus quadricauda* (unicellular green alga) [34], *Chlamydomonas reinhardtii* (unicellular green alga) [34], *Selenastrum capricornutum* (unicellular green alga) [19–22, 52, 71], *Scenedesmus vacuolatus* (unicellular green alga) [12, 13, 46], *Chlorella regularis* [26] **Marine algae** *Oocystis submarina* (unicellular green alga) [38], *Cyclotella meneghiniana* (diatom) [38] **Aquatic plant** *Lemna minor* (duckweed) [12, 13, 15, 37] **Terrestrial plants** *Hordeum vulgare* (barley) [16], *Raphanus sativus* (radish) [16], *Lepidium sativum* (garden cress) [12, 15, 44], *Triticum aestivum* (wheat) [12, 15, 44–46]

Table 10.1 (Continued)

Invertebrates	**Aquatic**
	Daphnia magna (water flea) [17, 25, 32, 60, 71], *Physa acuta* (freshwater snail) [18]
	Terrestrial
	Folsomia candida (springtail) [12], *Caenorhabditis elegans* (roundworm) [68]
Vertebrates	**Amphibians**
	Rana nigromaculata (frog) [40]
	Fish
	Danio rerio (zebrafish) [53], *Lebistes reticulatus* (guppy) [33]
	Mammals
	Rattus norvegicus Fischer 344 (rat) [36], white New Zealand rabbit [36], mouse (BALC/c mice) [36]

10.2.1
Influence of the Side Chain

The majority of the existing ecotoxicological data deal with the influence of the alkyl side chain on the toxicity of ionic liquids. Within these studies, different head groups (for example imidazolium, pyridinium or phosphonium) were analyzed with alkyl side chains ranging from C_1 up to C_{18} combined with different anions (mainly halides). A broad range of (biological) test systems ranging from isolated enzymes (e.g. acetylcholinesterase), different cell lines (e.g. IPC-81, HeLa, C6, HT-29 and CaCo-2), different bacteria strains (e.g. *Vibrio fischeri*, *Escherichia coli*, *Staphylococcus aureus*, *Bacillus subtilis*), yeast (*Saccharomyces cerevisiae*), algae (e.g. *Chlamydomonas reinhardtii*, *Scenedesmus vacuolatus*, *Oocystis submarina*, *Cyclotella meneghiniana*, *Pseudokirchneriella subcapitata*), plants (e.g. *Lemna minor*, *Triticum aestivum*, *Lepidium sativum*), invertebrates (e.g. *Daphnia magna*, *Caenorhabditis elegans*, *Folsomia candida*) and even some tests with vertebrates (mice, rats, frogs or rabbits) were used for these investigations (Table 10.1). In these tests, it was consistently found that an increasing alkyl side chain length led to increased toxic effects. This effect is known as the "side chain effect".

Depending on the test system used, the observed side chain effect differs in its intensity. Swatlowski *et al.* [68] investigated toxicities for 1-butyl-3-methylimidazolium chloride, 1-methyl-3-octylimidazolium chloride and 1-methyl-3-tetradecylimidazolium chloride with the roundworm *Caenorhabditis elegans*. Differences in toxicity were observed for all three compounds but, according to a classification system for narrative descriptions of acute toxicities 1-butyl-3-methylimidazolium chloride and 1-methyl-3-octylimidazolium chloride were "not acute toxic" and 1-methyl-3-tetradecylimidazolium chloride was only "slightly toxic".

A much more pronounced side chain effect was found in an aquatic test system using the limnic green algae *Scenedesmus vacuolatus* [13]. Within this test system, moderate EC_{50} values were determined for 1-ethyl-3-methylimidazolium chloride (600 µM) and an intensely increasing toxicity for 1-butyl-3-methylimidazolium

Table 10.2 Overview of tested ionic liquids.

Cation	Anion	Ref.
1-Ethyl-3-methylimidazolium	Bis[1,2-benzenediolato(2−)-O1,O2-borate	[12, 66]
	Bis[oxalato(2−)-borate	[66]
	Bis(pentafluoroethyl)phosphinate	[66]
	Ethylsulfate	[59, 66]
	Methylsulfate	[39]
	Tetrafluoroborate	[38, 39, 44, 46, 66, 69, 70]
	Chloride	[11, 13, 41, 66, 69]
	Bromide	[70]
	Trifluoroacetate	[69]
	Bis(trifluoromethanesulfonyl)amide	[70]
	Trifluoromethanesulfonate	[39, 69]
	Hexafluorophosphate	[66]
	Trifluorotris(pentafluoroethyl)phosphate	[66]
1-Propyl-3-methylimidazolium	Tetrafluoroborate	[13]
	Bromide	[19, 52]
	Chloride	[11, 58]
	Hexafluorophosphate	[58]
	Tetrafluoroborate	[55, 58]
1-Butyl-3-methylimidazolium	Chloride	[11–13, 17, 19, 22, 27, 31, 36, 39, 45, 58, 59, 61, 63–66, 68–71]
	Hexafluorophosphate	[17, 18, 20, 22, 29, 30, 42, 43, 53, 55, 58, 61, 63, 64, 66, 70, 71]
	Tetrafluoroborate	[12, 15, 17, 22, 29, 30, 38, 39, 44–46, 53, 55, 56, 58, 60, 63, 64, 66, 70]
	Bromide	[17, 18, 21, 22, 27, 28, 34, 37, 52, 55, 58, 61, 64, 66, 70]
	p-Toluenesulfonate	[55, 66]
	Octylsulfate	[12, 22, 64, 66]
	Methylsulfate	[31, 66]
	Bis(trifluoromethanesulfonyl)amide	[12, 13, 25, 29, 39, 45, 53, 65, 66, 70]
	2-(2-Methoxyethoxy)ethylsulfate	[64, 66]
	Trifluoromethanesulfonate	[22, 53, 64, 66]
	Dicyanamide	[27, 29, 53, 64, 66]
	Nitrate	[53]
	Hexafluoroantimonate	[22, 66]
	Bis(trifluoromethyl)amide	[12, 66]
	Hydrogensulfate	[45, 66]
	Methanesulfonate	[66]

Table 10.2 (Continued)

Cation	Anion	Ref.
	Thiocyanate	[66]
	Iodide	[66]
	Tetracarbonylcobaltate	[66]
	Trifluorotris(pentafluoroethyl)phosphate	[66]
	p-Toluenesulfonate	[61]
	Acesulfamate	[29]
1-Methyl-3-pentylimidazolium	Chloride	[11, 13, 55, 58]
	Hexafluorophosphate	[55, 58]
	Tetrafluoroborate	[13, 55, 58]
1-Hexyl-3-methylimidazolium	Chloride	[11, 13, 31, 41, 55, 58, 59, 66, 69, 70]
	Hexafluorophosphate	[31, 42, 43, 55, 58, 59, 66, 70]
	Tetrafluoroborate	[38, 55, 56, 58, 63, 66, 70]
	Bis(trifluoromethanesulfonyl)amide	[39, 66]
	Trifluorotris(pentafluoroethyl)phosphate	[66]
	Tris(trifluoromethanesulfonyl)methide	[66]
	Trifluorotris(heptafluoropropyl)phosphate	[66]
	Hexafluoroantimonate	[39]
	Bromide	[18, 19, 25, 27, 28, 34, 37, 52, 54]
1-Heptyl-3-methylimidazolium	Chloride	[11, 13, 55, 58]
	Hexafluorophosphate	[55, 58]
	Tetrafluoroborate	[13, 55, 58]
1-Methyl-3-octylimidazolium	Chloride	[11, 13, 31, 55, 58, 59, 68, 69]
	Bis(trifluoromethanesulfonyl)amide	[70]
	Hexafluorophosphate	[8, 31, 39, 42, 43, 55, 58, 59]
	Tetrafluoroborate	[12, 15, 29, 44, 46, 55, 56, 58, 64, 70]
	Bromide	[18, 19, 25–28, 34, 37, 40, 52, 54, 70]
	Methylsulfate	[39]
1-Methyl-3-nonylimidazolium	Chloride	[11, 55, 58]
	Hexafluorophosphate	[55, 58]
	Tetrafluoroborate	[13, 55, 58]
1-Decyl-3-methylimidazolium	Chloride	[11, 13, 26, 31, 55, 58, 70]
	Hexafluorophosphate	[55, 58, 70]
	Tetrafluoroborate	[29, 55, 56, 58, 63, 64, 70]
1-Dodecyl-3-methylimidazolium	Bromide	[13, 26]
	Chloride	[71]

(Continued)

Table 10.2 (Continued)

Cation	Anion	Ref.
1-Methyl-3-tetradecylimidazolium	Chloride Bromide	[13, 26, 68] [13, 26]
1-Hexadecyl-3-methylimidazolium	Chloride Bromide	[11, 71] [26]
1-Octadecyl-3-methylimidazolium	Chloride	[11, 13, 71]
1-Butyl-3-ethylimidazolium	Tetrafluoroborate	[55, 58, 63, 64]
1-Hexyl-3-ethylimidazolium	Bromide Tetrafluoroborate	[55, 58] [55, 58]
1-Decyl-3-ethylimidazolium	Bromide	[55, 58]
1-Ethyl-3-propylimidazolium	Bromide	[11, 55]
1,3-Diethylimidazolium	Ethylsulfate Bromide	[31] [11, 58]
1,3-Dimethylimidazolium	Methylsulfate Chloride Tetrafluoroborate Bis(trifluoromethanesulfonyl)amide Bromide	[31, 59] [70] [70] [70] [70]
1-Butyl-2,3-dimethylimidazolium	Tetrafluoroborate Hexafluorophosphate	[29] [53]
1-Hexyl-2,3-dimethylimidazolium	Chloride Tetrafluoroborate Bromide	[31] [58] [25, 54]
1-(2-Methoxyethyl)-3-methylimidazolium	Chloride Tetrafluoroborate Dicyanamide Bis(trifluoromethanesulfonyl)amide	[11, 13, 65] [60] [60] [13, 65]
1-(2-Ethoxyethyl)-3-methylimidazolium	Bis(trifluoromethanesulfonyl)amide	[13, 65]
1-(Ethoxymethyl)-3-methylimidazolium	Chloride Bis(trifluoromethanesulfonyl)amide	[11, 13] [13]
1-(2-Ethoxyethyl)-3-methylimidazolium	Bromide	[11, 13, 65]
1-(2-Hydroxyethyl)-3-methylimidazolium	Iodide Bis(trifluoromethanesulfonyl)amide Hexafluorophosphate Tetrafluoroborate Acesulfamate Saccharinate	[11, 13, 65] [13, 65] [29] [29] [29] [29]
1-Hydroxyethyl-2-methyl-3-tetradecylimidazolium	Chloride	[26]

Table 10.2 (Continued)

Cation	Anion	Ref.
1-Hexadecyl-3-hydroxyethyl-2-methylimidazolium	Chloride	[26]
1-(3-Carboxypropyl)-3-methylimidazolium	Chloride Bis(trifluoromethanesulfonyl)amide Chloride	[15, 58] [13, 65] [11, 13, 65]
1-[2-(2-Methoxyethoxy)-ethyl]-3-methylimidazolium	Hexafluorophosphate Bis(trifluoromethanesulfonyl)amide	[29] [29]
1-(2-Methoxypropyl)-3-methylimidazolium	Bis(trifluoromethanesulfonyl)amide Chloride	[13, 65] [11, 13, 65]
1-(7-Carboxyheptyl)-3-methylimidazolium	Bromide	[15, 58]
1-(8-Hydroxyoctyl)-3-methylimidazolium	Bromide	[11, 15, 58]
1-Benzyl-3-methylimidazolium	Methylsulfate Chloride	[31] [31]
1-Benzyl-3-ethylimidazolium	Ethylsulfate	[31]
1-p-Fluorobenzyl-3-methylimidazolium	Chloride	[31]
1-p-Chlorobenzyl-3-methylimidazolium	Chloride	[31]
1-Methyl-3-(phenylmethyl)imidazolium	Chloride Bromide Hexafluorophosphate Tetrafluoroborate Bis(trifluoromethanesulfonyl)amide	[11, 58] [70] [58] [58, 64, 70] [70]
1-Methyl-3-[(4-methylphenyl)methyl]imidazolium	Chloride Hexafluorophosphate Tetrafluoroborate	[11, 58] [58] [58]
1-Methyl-3-(2-phenylethyl)imidazolium	Chloride Hexafluorophosphate	[11, 58] [58]
1-Phenylpropyl-3-methylimidazolium	Trifluoromethanesulfonate Hexafluoroantimonate	[39] [39]
1-Methyl-3-(2-phenylethyl)imidazolium	Tetrafluoroborate	[64]
1-Methyl-3-(3-oxobutyl)imidazolium	Bromide	[11]
1-(Cyanomethyl)-3-methylimidazolium	Chloride Bis(trifluoromethanesulfonyl)amide	[11, 13, 69] [13]

(Continued)

Table 10.2 (Continued)

Cation	Anion	Ref.
1-Benzyl-3-methylimidazolium	Tetrafluoroborate	[38]
1-Butyl-3-ethyl-imidazolium	Ethylsulfate	[31]
1,3-Dialkoxymethylimidazolium (chain length C_3–C_{12})	Chloride, Tetrafluoroborate, Hexafluorophosphate, Bis(trifluoromethanesulfonyl)amide	[49]
1-Alkyl- and 1-Alkoxymethylimidazolium (chain length C_3–C_{12})	L-Lactates	[50]
1-Alkyl- and 1-Alkoxymethylimidazolium (chain length C_3–C_{12})	DL-Lactates	[50]
1-Alkyl- and 1-Alkoxymethylimidazolium (chain length C_3–C_{12})	DL-Lactates, L-lactates, salicylate	[51]
3-Alkoxymethyl-1-methylimidazolium salts (chain length C_3–C_{12}, C_{14}, C_{16})	Chloride, Tetrafluoroborate, Hexafluorophosphate	[49]
1-Alkoxymethyl-3-(nicotinylaminomethyl)benzimidazolium (chain length C_2–C_{12})	Chloride	[47]
1-Ethylpyridinium	Chloride	[11]
	Bromide	[70]
	Bis(trifluoromethanesulfonyl)amide	[70]
1-Propylpyridinium	Bromide	[11, 52]
1-Butylpyridinium	Tetrafluoroborate	[58]
	Bis(trifluoromethanesulfonyl)amide	[53]
	Chloride	[11, 27, 69, 71]
	Bromide	[13, 27, 58, 65, 70]
1-Pentylpyridinium	Bromide	[11]
1-Hexylpyridinium	Chloride	[11]
	Bromide	[25, 54]
1-Octylpyridinium	Chloride	[11]
	Bromide	[18, 70]
1-Propyl-3-methylpyridinium	Bromide	[52]
1-Butyl-2-methylpyridinium	Tetrafluoroborate	[58]
	Chloride	[11]
1-Butyl-3-methylpyridinium	Tetrafluoroborate	[58, 64]
	Chloride	[11]

Table 10.2 (Continued)

Cation	Anion	Ref.
	Dicyanamide	[27]
	Bromide	[18, 21, 25, 27, 28, 37, 52, 54]
	Hexafluorophosphate	[64]
1-Hexyl-3-methylpyridinium	Bromide	[18, 25, 27, 28, 52, 54]
	Chloride	[31]
1-Methyl-3-octylpyridinium	Bromide	[25, 27, 28, 52, 54]
1-Butyl-4-methylpyridinium	Hexafluorophosphate	[58]
	Tetrafluoroborate	[58]
	Chloride	[11]
1-Butyl-3,4-dimethylpyridinium	Tetrafluoroborate	[58]
1-Butyl-3,5-dimethylpyridinium	Tetrafluoroborate	[58]
	Bromide	[25, 27]
4-(Dimethylamino)-1-ethylpyridinium	Bromide	[11]
4-(Dimethylamino)-1-butylpyridinium	Bis(trifluoromethanesulfonyl)amide	[65]
	Chloride	[11, 13, 65]
	Bromide	[25]
4-(Dimethylamino)-1-hexylpyridinium	Chloride	[11]
	Bromide	[25, 54]
1-(2-Ethoxyethyl)-pyridinium	Bromide	[11]
1-(Ethoxymethyl)-pyridinium	Chloride	[11]
1-(2-Hydroxyethyl)-pyridinium	Iodide	[11]
1-(2-Methoxyethyl)-pyridinium	Chloride	[11]
1-(3-Methoxypropyl)-pyridinium	Chloride	[11]
1-(3-Hydroxypropyl)-pyridinium	Chloride	[11]
1-(Cyanomethyl)-pyridinium	Chloride	[11]
1-Hexyl-4-piperidinopyridinium	Bromide	[25, 54]
1-Butyl-3,5-dimethylpyridinium	Bromide	[54]

(Continued)

Table 10.2 (Continued)

Cation	Anion	Ref.
1-Propyloxymethyl-3-hydroxypyridinium	Chloride Acesulfamate Saccharinate	[62] [62] [62]
1-Butyloxymethyl-3-hydroxypyridinium	Chloride Acesulfamate Saccharinate	[62] [62] [62]
1-Hexyloxymethyl-3-hydroxypyridinium	Chloride Acesulfamate Saccharinate	[62] [62] [62]
1-Heptyloxymethyl-3-hydroxypyridinium	Chloride Acesulfamate Saccharinate	[62] [62] [62]
1-Undecanyloxymethyl-3-hydroxypyridinium	Chloride Acesulfamate Saccharinate	[62] [62] [62]
1-Alkoxymethyl-3-(1-benzotriazol-1-ylmethylamino)pyridinium (chain length C_3–C_{12})	Chloride	[47]
1-Alkoxymethyl-3-(1-benzimidazolmethylamino)-pyridinium (chain length C_3–C_{12})	Chloride	[47]
N,N'-Bis[3-(1-alkoxymethyl)pyridinium (chain length C_1–C_{12})	Chloride	[47]
1-Alkoxymethyl-3-carbamoylpyridinium	Chlorides, Iodides, Bromides, Tetrafluoroborates	[48]
1-Hexyl-1-methylpyrrolidinium	Chloride	[41, 58]
1-Methyl-1-octylpyrrolidinium	Chloride Tetrafluoroborate	[11, 58] [58]
1-(2-Methoxyethyl)-1-methylpyrrolidinium	Chloride Bis(trifluoromethanesulfonyl)amide	[11, 65] [65]
1-(3-Methoxyethyl)-1-methylpyrrolidinium	Chloride	[11]
1-(Ethoxymethyl)-1-methylpyrrolidinium	Chloride	[11, 65]
1-(2-Ethoxyethyl)-1-methylpyrrolidinium	Bis(trifluoromethanesulfonyl)amide Bromide	[65] [11, 65]
1-(2-Ethoxymethyl)-1-methylpyrrolidinium	Bis(trifluoromethanesulfonyl)amide	[65]

Table 10.2 (Continued)

Cation	Anion	Ref.
1-(2-Hydroxyethyl)-1-methylpyrrolidinium	Iodide Bis(trifluoromethanesulfonyl)amide	[11, 65] [65]
1-(3-Methoxypropyl)-1-methylpyrrolidinium	Chloride Bis(trifluoromethanesulfonyl)amide	[65] [65]
1-(3-Hydroxypropyl)-1-methylpyrrolidinium	Bis(trifluoromethanesulfonyl)amide Chloride	[65] [11, 65]
1-(Cyanomethyl)-1-methylpyrrolidinium	Bis(trifluoromethanesulfonyl)amide Chloride	[65] [11, 65]
1,1-Dihexylpyrrolidinium	Tetrafluoroborate	[58]
n-Dodecyl-N-hydroxyethylpyrrolidinium	Chloride	[26]
1-Butylquinolinium	Tetrafluoroborate Bromide	[58] [11, 58]
1-Hexylquinolinium	Tetrafluoroborate Tetrafluoroborate	[11] [58]
1-Octylquinolinium	Tetrafluoroborate Bromide	[58] [11, 58]
Tetramethylammonium	Bromide	[25, 28]
Tetraethylammonium	Chloride Bromide	[11] [25, 28]
Tetrabutylammonium	Bromide	[11, 18, 21, 25, 28, 37, 54]
	Bromide	[28]
Butylethyl dimethylammonium	Chloride Bis(trifluoromethanesulfonyl)amide	[11, 65] [65]
Decylbenzyldimethyl-ammonium	Chloride	[11, 58]
Dodecylbenzyldimethyl-ammonium	Chloride	[58]
Tetradecylbenzyldimethyl-ammonium	Chloride	[58]
Ethyl(2-methoxyethyl) dimethylammonium	Chloride	[65]
(Ethoxymethyl) ethyldimethylammonium	Chloride Bis(trifluoromethanesulfonyl)amide	[65] [65]
Ethyl(2-methoxyethyl) dimethylammonium	Bis(trifluoromethanesulfonyl)amide	[65]
Ethyl(2-ethoxyethyl) dimethylammonium	Bis(trifluoromethanesulfonyl)amide Chloride	[65] [65]
Ethyl(2-hydroxyethyl) dimethylammonium	Iodide Bis(trifluoromethanesulfonyl)amide	[65] [65]

(Continued)

Table 10.2 (Continued)

Cation	Anion	Ref.
Ethyl(3-hydroxypropyl)dimethylammonium	Bis(trifluoromethanesulfonyl)amide	[65]
Cyanomethylethyl-dimethylammonium	Chloride Bis(trifluoromethanesulfonyl)amide	[65] [65]
Hexyltriethylammonium	Bromide	[25]
PEG-5 cocomonium	Methylsulfate	[71]
4-Butyl-4-methylmorpholinium	Bromide Bis(trifluoromethanesulfonyl)amide	[11, 13, 65] [65]
4-(2-Methoxyethyl)-4-methylmorpholinium	Chloride Bis(trifluoromethanesulfonyl)amide	[11, 65] [65]
4-(Ethoxymethyl)-4-methylmorpholinium	Chloride Bis(trifluoromethanesulfonyl)amide	[11, 65] [65]
4-(2-Ethoxyethyl)-4-methylmorpholinium	Bromide Bis(trifluoromethanesulfonyl)amide	[11, 65] [65]
4-(2-Hydroxyethyl)-4-methylmorpholinium	Iodide Bis(trifluoromethanesulfonyl)amide	[11, 65] [65]
4-(3-Methoxypropyl)-4-methylmorpholinium	Bis(trifluoromethanesulfonyl)amide Chloride	[65] [11, 65]
4-(3-Hydroxypropyl)-4-methylmorpholinium	Chloride Bis(trifluoromethanesulfonyl)amide	[11, 65] [65]
4-(Cyanomethyl)-4-methylmorpholinium	Bis(trifluoromethanesulfonyl)amide Chloride	[65] [11, 65]
1-Butyl-1-methylpiperidinium	Bromide Bis(trifluoromethanesulfonyl)amide	[11, 13, 65] [65]
1-(2-Hydroxyethyl)-1-methylpiperidinium	Bis(trifluoromethanesulfonyl)amide Iodide	[65] [11, 65]
1-(3-Hydroxypropyl)-1-methylpiperidinium	Bis(trifluoromethanesulfonyl)amide Chloride	[65] [11, 65]
1-(Ethoxymethyl)-1-methylpiperidinium	Bis(trifluoromethanesulfonyl)amide Chloride	[65] [11, 65]
1-(2-Ethoxyethyl)-1-methylpiperidinium	Bis(trifluoromethanesulfonyl)amide Bromide	[65] [11, 65]
1-(2-Methoxyethyl)-1-methylpiperidinium	Bromide Bis(trifluoromethanesulfonyl)amide	[11, 65] [65]
1-(3-Methoxypropyl)-1-methylpiperidinium	Chloride Bis(trifluoromethanesulfonyl)amide	[11, 65] [65]
1-(Cyanomethyl)-1-methylpiperidinium	Bis(trifluoromethanesulfonyl)amide Chloride	[65] [11, 65]
Tetrabutylphosphonium	Bromide	[11, 18, 21, 25, 54, 58]

Table 10.2 (Continued)

Cation	Anion	Ref.
Ethyltrihexylphosphonium	Bromide	[23]
Tributylethylphosphonium	Diethylphosphate	[25]
Trihexyl(tetradecyl)phosphonium	Bis(2,4,4-trimethylpentyl)phosphinate	[23, 64]
	Dicyanamide	[23, 64]
	Hexafluorophosphate	[23, 64]
	Tetrafluoroborate	[23, 58, 64]
	Bromide	[25]
	Chloride	[23, 24, 71]
	Diisobutylphosphinate	[23]
	Dicyclohexylphosphinate	[23]
	Diisobutyldithiobutylphosphinate	[23]
	Decanoate	[23]
	Trifluoromethanesulfonate	[23]
	Bis(trifluoromethanesulfonyl)amide	[23, 24, 29]
	Diethylphosphate	[71]
Methyltrioctylammonium	Bis(trifluoromethanesulfonyl)amide	[24]
	Chloride	[24]
Choline	Chloride	[25]
	Bis(trifluoromethanesulfonyl)amide	[25]
	Acesulfamate	[29]
	Saccharinate	[29]
Tetra-n-hexyldimethyl-guanidinium	Hexafluorophosphate	[29]
	Bis(trifluoromethanesulfonyl)amide	[29]
	Dicyanoamide	[29]
Ammoeng 100		[53]
Ammoeng 110		[53]
Ammoeng 112		[53]
Ammoeng 130		[53]

chloride (130 μM), 1-methyl-3-octylimidazolium chloride (0.0017 μM) up to 1-decyl-3-methylimidazolium chloride (0.0003 μM), which was approximately 150 times higher than that for a toxic reference compound (atrazine) [12].

The dependence between increasing alkyl side chain length and increasing toxicity does not hold true any longer for very long alkyl side chains. At a certain chain length, the toxicity cannot be increased any further. This phenomenon is well known from the literature for highly lipophilic substances (log K_{ow} > 5) as the "cut-off" effect. For this phenomenon, different explanations are discussed based either on insufficient solubility (nominal concentration deviating from real test concentration) or on kinetic aspects (uptake is slowed down because of steric effects for compounds with a large molecular size). This "cut-off" effect has been observed for ionic liquids in several test systems: for *Vibrio fischeri* the toxicity decreased from 1-dodecyl-3-methylimidazolium chloride to 1-tetradecyl-3-methylimidazolium chloride, for *Scenedesmus vacuolatus* this effect occurred from C_{10} to C_{12} and for *Selenastrum capricornutum* it was observable from C_{12} to C_{16}. Even though the toxicity decreased, the effects were still at a high level.

Several studies have investigated the biological effects of side chains containing functional groups in comparison with alkyl side chains. Jastorff et al. [15] determined the cytotoxicity towards rat leukemia cells (IPC-81) for 1-butyl-3-methylimidazolium in comparison with hydroxybutyl, 3-oxobutyl and 3-carboxypropyl. It was found that for all three compounds tested, the toxicity clearly decreased compared with the butyl side chain. The same trend was found for 1-octyl-3-methylimidazolium when polar functional groups were introduced. Those results formed the basis to investigate the biological effects to acetylcholinesterase, IPC-81 cells, *Vibrio fischeri*, *Scenedesmus vacuolatus* and *Lemna minor* for a large set of ionic liquids containing functionalized side chains (ether groups in different positions, terminal hydroxyl and nitrile functions) combined with different head groups and chloride as counter ion [13, 65]. The results from those tests also showed trends for lower biological activities for the ionic liquids investigated with functionalized side chains. Samori et al. [60] confirmed these observations in assays with *Daphnia magna* and *Vibrio fischeri* by testing two imidazolium-based ionic liquids containing ether groups showing EC_{50} values one order of magnitude higher (less toxic) than 1-butyl-3-methylimidazolium. The antimicrobial properties of ionic liquids with long alkyl and alkoxy side chains were investigated by Pernak et al. [48] and Demberelnyamba et al. [26]. Both studies showed that the trend towards reduced toxicities caused by the incorporation of functional groups into the side chain (partly or totally) disappeared when the functionalized side chain length was increased. For ammonium compounds with long polyethylene glycol chains, clear toxic effects were found for *Daphnia magna* and *Selenastrum capricornutum* [54]. Pretti et al. [53] found toxicity towards fish (*Danio rerio*) for ammonium salts with long alkyl substituents (stearyl or cocosyl).

10.2.2
Influence of the Head Group

In contrast to the side chain of ionic liquids, the influence of the head group on (eco)toxicity was found to be of minor importance in most cases. The toxicity depends on the organisms investigated and is diverse: for some cases the influence on (eco)toxicity was weaker than the side chain effect, vague or even not present. In the following, some basic results will be pointed out. Scheme 10.1 shows the most important head groups investigated.

Stolte et al. [13] investigated the influence of seven head groups on IPC-81 cells, *Vibrio fischeri*, *Scenedesmus vacuolatus* and *Lemna minor*. In these studies, three aromatic head groups [4-(dimethylamino)pyridinium, pyridinium and 1-methylimidazolium], three non-aromatic heterocycles (4-methylmorpholinium, 1-methylpiperidinium and 1-methylpyrrolidinium) and one non-cyclic quaternary ammonium compound (*N,N,N*-dimethylethylammonium) were investigated, all substituted with a butyl side chain. The 1-butyl-4-(dimethylamino)pyridinium compound drastically influenced the toxicity towards all test organisms and the drastic toxicities can most likely be explained by the higher hydrophobicity of the compound in comparison with the other butyl-substituted head groups. In contrast, very low toxicities were found for the morpholinium compound, which is probably based on its low hydrophobicity. The ammonium compound showed low toxicities to the algae and the bacteria, but the toxicities observed for *Lemna minor* were distinct ($EC_{50} = 6.8\,\mu M$) [13].

These results are supported by the studies of Arning *et al.* [11] who tested the influence of different head groups on electric eel acetylcholinesterase. Here, three molecular key interaction potentials were identified (the positively charged nitrogen atom, a broad delocalized aromatic ring system and a certain lipophilicity) as being responsible for the acetylcholinesterase inhibition potential. The 4-(dimethylamino) pyridinium and the quinolinium head groups showed the strongest inhibition potential. In contrast, the morpholinium head group was found to be only weakly inhibiting or even inactive.

Wang *et al.* [70] investigated the effects of different head groups (imidazolium, pyridinium, ammonium, phosphonium and choline derivatives) on HeLa cells. It can be concluded that the choline derivatives and the ammonium compounds are less toxic than the pyridinium- and imidazolium-based derivatives. These results are also supported by data from Couling *et al.* [25], who also found that pyridinium and imidazolium compounds are more toxic than choline or quaternary ammonium structures. In contrast, Bernot *et al.* [18] analyzed the influence of four different head groups (imidazolium, pyridinium, phosphonium and ammonium) on the behavior (grazing rates and movement) of the freshwater snail *Physa acuta* and they concluded that no clear influence of the head group was detectable.

10.2.3
Influence of the Anion

The observed effects of different anionic moieties on the toxicity were heterogeneous and varied between the different studies. Scheme 10.2 shows the most important anions investigated.

Two studies analyzed the influence of the anionic moiety on isolated enzymes (electric eel acetylcholinesterase and AMP deaminase). The enzyme inhibition assay with acetylcholinesterase proved to be unsuitable for examining the influence of the anionic moiety on the toxicity owing to the strongly limited interactions of anions in general with the active binding site of the acetylcholinesterase [12]. Here, inhibiting effects were observed only for fluoride and fluoride-containing anions which are vulnerable to hydrolyses. In contrast, for the AMP deaminase, which is important for maintaining the ATP pool of cells, fluoride-containing anions had stronger inhibitory effects than chlorides [61].

Stolte *et al.* [66] investigated the influence of 30 anions (including borates, methides, amides, imides, phosphates, phosphinates, antimonates, sulfates and sulfonates) on the cytotoxicity of the rat leukemia cell line IPC-81. Especially for highly fluorinated anions or anions that are sensitive to hydrolytic processes an increased cytotoxicity was observable. Some anions, e.g. bis[1,2-benzdiolato(2–)]borate, reached EC_{50} values comparable to those of highly reactive and toxic biocides such as N-methylisothiazol-3-one. The authors suggested that the mode of toxic action is based on their hydrophobicity and chemical reactivity. Based on these results, Matzke et al. [12] investigated a selection of six lipophilic and hydrolytically sensitive, structurally very heterogeneous anions by employing a flexible (eco)toxicological test battery. This test battery included an isolated enzyme, IPC-81 cells, marine bacteria, limnic green algae, duckweed, wheat, cress and a soil invertebrate. The anion test kit contained pure inorganic anions (chloride and tetrafluoroborate), an organic sulfate (octylsulfate), fluorinated anions $\{[NTf_2]^-$ and $[(CF_3)_2N]^-\}$ and an inorganic borate complex. Based on the various structures, different modes of toxic action were expected, expressed by a diverse pattern of toxicities for the different organisms or environmental compartments. The results obtained clearly suggest that the anionic moiety is contributing to the toxicity of ionic liquids, but in comparison with the side chain effect, the influence of the anion is less pronounced. Also, no explicit toxicity patterns were identified. For the majority of the ionic liquids examined, no clear increase in toxicity caused by the anion compared with the reference anion Cl^- were observed with one exception: the $[NTf_2]^-$ anion showed partially drastic toxic effects. Especially for the soil invertebrate *Folsomia candida*, the toxicities were comparable to the effects of an insecticide (carbendazime). Here, the anionic influence was even more distinct than the side chain effect of the cation. For algae and bacteria, the $[NTf_2]^-$ anion generally increased the toxicity in comparison with halides. In general, the results for the other investigated anions bis(trifluoromethyl)imide $\{[(CF_3)_2N]^-\}$, octylsulfate and bis[1,2-benzdiolato(2–)]borate were very heterogeneous. It can be concluded that more complex mechanisms of absorption and distribution – which cannot be directly linked to hydrophobicity and chemical reactivity – might be responsible for these results obtained with organisms of higher biological complexity. The differences in toxicity can be explained by the high structural heterogeneity of the anions investigated resulting in completely different modes of toxic action. The bis(trifluoromethyl)imide anion, being a chemically reactive compound, showed surprisingly low toxicities for all test organisms investigated, whereas for octylsulfate, the recorded effects were drastic for *Vibrio fischeri* and the bis[1,2-benzdiolato(2–)]borate anion showed generally high intrinsic toxicities. The results for both studies can be summed up as follows: chloride, octylsulfate and tetrafluoroborate contribute less to the toxicity of ionic liquids whereas $[NTf_2]^-$ showed partially drastic toxic effects and had a strong influence. This is in contrast to the results of Frade et al. [29], who investigated a large set of ionic liquids on two colon carcinoma cell lines (HT-29 and CaCo-2). In all cases, the toxicity decreased when the $[NTf_2]^-$ anion was used in comparison with $[BF_4]^-$, $[PF_6]^-$ and $[DCA]^-$. This effect occurred independently of the cationic moiety.

Ganske and Bornscheuer [30] investigated the growth of three microorganisms (*E. coli*, *Pichia pastoris* and *Bacillus cereus*) in the presence of imidazolium-based ionic liquids with tetrafluoroborate and hexafluorophosphate anions. Tetrafluoroborate was toxic for all three microorganisms, whereas for hexafluorophosphate differences were observable: *E. coli* reacted very sensitively with growth inhibition (concentration of 0.3–0.7% v/v), whereas for *P. Pastoris* no adverse effects were observed at 10% v/v. The toxicities achieved were comparable to the toxicities caused by the organic solvent dimethyl sulfoxide.

The influence of seven anions on the toxicity of ionic liquids was analyzed for the green alga *Pseudokirchneriella subcapitata* by Cho et al. [22]. This green alga reacted most sensitively to hexafluoroantimonate and less to chloride and bromide. Additionally, the hydrolysis of the fluoride-containing anions was investigated. As expected, an increasing toxicity was observed, but the relatively short test duration of 96 h was not sufficient to enhance the toxicity of the test solutions (owing to the hydrolytic process generating fluoride) but a 6 month old stock solution clearly exhibited increased toxicity. Hydrolysis occurred most for hexafluoroantimonate, less for tetrafluoroborate and not at all for hexafluorophosphate under the applied conditions.

10.2.4
Toxicity of Ionic Liquids as a Function of the Surrounding Medium

Most of the studies on the (eco)toxicity of ionic liquids were performed using standard testing protocols (e.g. OECD Guidelines or ISO Guidelines) or modified standard testing procedures. None of these protocols or procedures incorporates varying environmental conditions (e.g. changing abiotic factors such as pH value, organic matter or clay content for terrestrial systems or dissolved organic matter content for aquatic ecosystems or salinity). However, those factors clearly influence the bioavailability of ionic liquids and therefore their toxicity. Up to now there have been only a few studies that have analyzed environmental key variables with respect to the bioavailability of ionic liquids. The main results are presented in the following.

In the aquatic environment, dissolved organic matter (DOM) is almost ubiquitous. Larson et al. [37] investigated the influence of natural dissolved organic matter and commercially available humic and fulvic acids on the toxicity of different ionic liquids in a test system with the aquatic plant *Lemna minor* (duckweed). The side chain length effect was investigated with imidazolium-based ionic liquids (butyl, hexyl and octyl alkyl side chains) and also the influence of different cationic head groups such as ammonium and pyridinium. The side chain effect was proven even in the DOM-containing test systems for all test compounds. Comparing the influence of DOM on the different head groups, only for [C_4mim]Br did a reduction in toxicity occurred and not for tetrabutylammonium bromide ([$N_{4,4,4,4}$]Br) and *N*-butylpyridinium bromide ([C_4Py]$^+$Br). Hence the first stated hydrophobic interactions being responsible for the reduction in toxicity cannot be the sole mechanism. Additionally, only the natural version of dissolved organic matter reduced the toxicity and not the commercially available mixture of humic and fulvic acids. This can perhaps be explained by the fact that natural organic matter contains a highly diverse mixture of molecules that may alter the permeability and charge of cellular membranes. Commercially available humic matter may not always mimic these effects.

The influence of salinity on the toxicity of ionic liquids was investigated by Latala et al. [38]. They analyzed the influence of imidazolium-based ionic liquids with varying alkyl side chain lengths on the Baltic algae *Oocystis submarina* and *Cyclotella meneghiniana*. In addition to the well-known side chain effect they found additionally that at higher salinities the toxicity of $[C_4mim]^+$ and $[C_6mim]^+$ entities towards *O. submarina* was significantly lower than at low salinities.

The bioavailability of ionic liquids for terrestrial plants was investigated as a function of varying organic matter and clay content and two different clay types (differing in their crystalline structure, which leads to varying sorption capacities) by Matzke et al. [44, 45]. The influence of the structural element's side chain and anion on the toxicity of imidazolium-based ionic liquids was systematically checked in relation to different soil types. In general, the differences in the observed toxic effects can be explained by the different clay minerals and their varying characters [44]. To investigate the influence of the alkyl side chain length on toxicity as a function of the clay type and content, $[C_2mim][BF_4]$, $[C_4mim][BF_4]$ and $[C_8mim][BF_4]$ were tested. The differences in the observed effects between the reference soil and the kaolinite soils were small because the structural formation of the kaolinite mineral allows only for weak adsorption and no absorption of the tested ionic liquids. For some cases, kaolinite even enhanced the toxic effects of $[C_4mim][BF_4]$. Also, there seems to be a critical side chain length regarding the increasing toxicity with increasing alkyl side chain length because for $[C_2mim][BF_4]$ the toxicity was either similar or even slightly increased in comparison with $[C_4mim][BF_4]$. In general, a short side chain increases the mobility of the substances in soils, because here fewer hydrophobic interactions are possible between the soil matrix and the ionic liquid. The influence of smectite on the bioavailability of the substances was strongest for all tested ionic liquids, which was to be expected based on the ability of smectites to (reversibly) incorporate cations, organic molecules or water to a high extent. Here, the ionic interactions dominate. Saturation of the soils with substance occurred much earlier for the short alkyl side chains because for the ionic liquids with longer alkyl side chains more interaction potentials are given between the soil matrix and the substances. When the different soil types were saturated, the influence of the clay minerals on toxicity vanished completely.

In order to check the influence of different soil types (kaolinite- and smectite-containing soils versus a reference soil) on the bioavailability and toxicity of different anions, two biocompatible anions (Cl^- and $[HSO_4]^-$), $[BF_4]^-$ and $[NTf_2]^-$ were investigated [45]. No clear differences in toxicity were observable between the kaolinite-containing soils and the reference soil, but for the smectite-containing soils with $[NTf_2]^-$ the toxic effects doubled in relation to the reference soil. The general assumption that higher clay amounts (kaolinite or smectite) lead to a lower bioavailability was generated in heavy metal soil ecotoxicology and cannot generally be applied to ionic liquids with toxic anionic moieties, as proved with $[NTf_2]^-$.

Hence, especially for the terrestrial environment, the influence of anions on the toxicity of ionic liquids has to be investigated carefully because soils mostly represent negatively charged matrices and therefore smaller interactions between the anionic moiety and the matrices are to be expected, resulting in possibly higher effects for soil organisms.

For [C_4mim]Cl, [C_4mim][HSO_4] and [C_4mim][BF_4], the observed effects are likely to be cationic effects and the differences in toxicity for the different soils can be explained by the influence of the varying clay soils. Additionally, [HSO_4]$^-$ and Cl$^-$ are much more biocompatible: [HSO_4]$^-$ might serve as a fertilizer for plants and can be easily metabolized and chloride is a "natural" anion, which is present anyway in the metabolism of the plants. Therefore, especially in the smectite-containing soils, the interactions probably occurred exclusively between the remaining cationic core and the clay minerals. The cations are totally adsorbed, whereas the [HSO_4]$^-$ anion obviously supported plant development, so that a stimulation of the plant growth occurred in comparison with untreated controls.

In contrast to the partially unexpected results for the clay minerals (increase in toxicity with clay mineral addition), the influence of organic matter content on the toxicity of the selected ionic liquids was consistent and as to be expected. In this study, a reduced toxicity of [C_4mim][BF_4] and [C_8mim][BF_4] was proven for the soils with increased organic matter content. An increase in organic matter content always resulted in a decrease in the detectable toxicities owing to increasing hydrophobic interaction potentials. Comparing the influence of organic matter and smectite on the toxicity of [C_8mim][BF_4], the clay mineral reduced the effects on plant growth to a greater extent. However, it has to be taken into account that organic matter represents a chemically highly diverse material and the influence on toxicity is strongly dependent on its type and quantity.

10.2.5
Combination Effects

Up to now, the issue of mixture toxicity has been neglected within the prospective hazard assessment strategies for ionic liquids. Two types of mixture toxicity have to be taken into account:

1. An ionic liquid itself represents a mixture consisting of a cationic and an anionic moiety.
2. The occurrence of combinations of ionic liquids or ionic liquids and other xenobiotics when accidentally released to the environment is likely.

For both cases, it would be desirable to have strategies for a prediction of the (toxic) effects which are to be expected. Since it is impossible to test all potential mixtures at all possible concentrations, it is necessary to find reliable concepts for predicting the expected mixture toxicities. Two well-established concepts, originating from pharmacological research, were used in the present studies dealing with mixture toxicity: the concept of Concentration Addition (CA) [72, 73], indicating a mixture of similarly acting substances, and the concept of Independent Action (IA) [74], indicating a mixture of dissimilarly acting substances. Stolte *et al.* [66] applied the CA concept to the mixture effects of imidazolium-based ionic liquids (with varying alkyl side chain lengths) in combination with different anions {(Cl$^-$, [HSO$_4$]$^-$, [BF$_4$]$^-$, bis(oxalato) borate, bis[1,2-benzenediolato(2–)]borate, [NTf$_2$]$^-$, [DCA]$^-$, [SbF$_6$]$^-$ and [OTs]$^-$} to compare the predicted values with the experimentally determined values in an assay using the rat leukemia cell line IPC-81. In general, the comparison between the calculated and the measured cytotoxicity values indicated a good correlation. Only one exception occurred, for [NTf$_2$]$^-$ in combination with different imidazolium cations. Here, a significantly higher cytotoxicity was found than calculated from the model of CA. For all imidazolium-based ionic liquids tested, the experimentally derived EC$_{50}$ values were around three times lower than the calculated values. This over-additive effect may be explained with temporary formation of direct ion pairs in aqueous solutions composed of the cationic and anionic moiety. Ion pair formation can result in a significantly higher bioavailability of the particular ionic liquids amplifying membrane interactions entailing stronger cytotoxic effects. These results were underlined by the findings of Matzke *et al.* [12]. They investigated for the marine bacterium *Vibrio fischeri* combination effects of the cationic moiety [C$_4$mim]$^+$ and the anionic moiety [NTf$_2$]$^-$. Again, over-additive effects occurred, because the toxicity of the [C$_4$mim]Cl was moderate (EC$_{50}$ = 2500 µM) and no intrinsic toxicity was measured for the Li[NTf$_2$] (EC$_{50}$ > 20 000 µM), whereas [C$_4$mim][NTf$_2$] showed an EC$_{50}$ value of 300 µM.

Matzke *et al.* [46] investigated the predictability of a mixture of ionic liquids versus a mixture of ionic liquids with the heavy metal cadmium by applying the concepts of CA and IA. Additionally, the performance of an ionic liquid mixture, which is released to the environment and meets a cadmium background polluted aquatic compartment (using aquatic green algae) and a terrestrial environment (using wheat) was analyzed.

In general, the two concepts CA and IA provided no precise prediction of the observed effects. Only for the ionic liquid–cadmium mixtures in the aquatic environment (green algae), was CA approximately able to predict the observed effect. For the pure ionic liquid mixtures both concepts underestimated the observed toxicities, meaning that a higher mixture toxicity was observed than expected from the models. For the mixtures containing ionic liquids and cadmium, toxicities were overestimated, because lower mixture toxicity was measured than predicted. Even if the concepts used strictly exclude interactions of the mixture compounds, the deviations of the experimental data from the predictions lead to the assumption that interactions between the tested substances or interactions with the environmental compartment appeared. This holds true especially for ionic substances such as those investigated in this study. The lack of knowledge about the mode of toxic action of ionic liquids and the variability of the mode of toxic action, which can be caused by cadmium, complicate the interpretation of the results obtained. Also, differences in the strength of the observed mixture effects between the two environmental compartments occurred. The green algae proved to be two orders of magnitude more sensitive than the terrestrial plant, probably based on the higher bioavailability of the pollutants in the aquatic surroundings. The soil matrix represents a good interaction basis for ionic substances to react with and thus reduces the bioavailability of the pollutant or the pollutant mixture, for instance due to sorption processes. As stated above, an ionic liquid itself represents a mixture because it consists of a cationic and an anionic moiety. The above-mentioned studies with the mammalian cell line IPC-81 and the marine bacterium *Vibrio fischeri* disproved the applicability of the CA model for the prediction of the toxicity for a single ionic liquid when the $[NTf_2]^-$ anion was present. This means that in the presence of $[NTf_2]^-$ the observed mixture toxicity was synergistic in comparison with the prediction. Hence a possible explanation for the synergistic effects of the ionic liquid mixture could be the presence of $[NTf_2]^-$. Therefore, the question was raised as to whether concentration addition could precisely predict mixture toxicity for ionic liquid mixtures excluding ionic liquids containing $[NTf_2]^-$, for example a mixture of $[C_4mim][BF_4]$, $[C_6mim][BF_4]$ and $[C_8mim][BF_4]$.

This question was analyzed by Wang *et al.* [70]. They confirmed similar modes of toxic action for bromide salts (choline-, imidazolium- and pyridinium-based ionic liquids with ethyl and butyl, hexyl or octyl side chains) – affirmed by the model of CA. This indicates that the substituent determines the mechanism to a greater extent than the cation class of the compound.

10.2.6
(Quantitative) Structure–Activity Relationships and Modes of Toxic Action

Experimental studies have clearly documented the correlation between increasing hydrophobicity and increasing toxicity and that the hydrophobicity of an ionic liquid is mainly driven by the substituents at the cationic head group (e.g. the alkyl side chain). Even though recent studies indicate that for some cases the anion can play an important role in ionic liquid toxicity (e.g. $[NTf_2]^-$), this is more or less the exception. Especially the structural heterogeneity within the pool of ionic liquids provides an almost unlimited number of compounds. This heterogeneity requires efficient testing strategies to generate data sets leading to a profound insight into modes of toxic action and target sites of chemicals. Since it is impossible to analyze all possible ionic liquid structures according to their hazard potential, the need for toxicity prediction strategies is rising. Several authors have applied QSAR (Quantitative Structure–Activity Relationships) models to data sets of ionic liquids to detect general assumptions on the toxicity and behavior of these compounds. The overarching aim is the development of ionic liquids that are safe for humans and the environment. All authors agree that the cation dominates the toxic characteristics of ionic liquids.

Couling *et al.* [25] applied a QSAR model with four molecular descriptors, log EC_{50} and LC_{50} values for the marine bacterium *Vibrio fischeri* and the freshwater crustacean *Daphnia magna*. They were able to prove with their model the well-established link between side chain length and toxicity for imidazolium, pyridinium and quaternary ammonium compounds; the anion plays a secondary role regarding toxicity. As an outcome, they suggested that choline and quaternary ammonium compounds may be more environmentally friendly alternatives than aromatic compounds and some conventional organic solvents. Putz et al. [54] applied a Spectral Structure–Activity Relationship Model (S-SAR) to the ecotoxicology of ionic liquids with *Daphnia magna* as a test organism. The algorithm employed proved that the cationic effect clearly dominates the anionic effects.

Garcia-Lorenzo et al. [31] developed a QSAR approach with TOPS-MODE on the CaCo-2 carcinoma colon cell line for imidazolium-based ionic liquids to predict side chain effects and anion toxicities. As a result, they recommend methylimidazolium alkylsulfate derivatives with a relatively low toxic potential.

Luis et al. [41] developed a novel group contribution method for estimating EC_{50} values of ionic liquids by means of a QSAR model on the database of *Vibrio fischeri* EC_{50} values. The group contribution method allows for an estimation of different combinations of groups in the toxicity of ionic liquids.

The general correlation between hydrophobicity and cytotoxicity is supported by many publications. The underlying mechanism is based on a general interdependence between hydrophobicity and toxicity of a compound and is well known in ecotoxicology either as "narcosis" (for mammals) [75–77] or as "baseline toxicity" (for aquatic organisms) [78]. With increasing hydrophobicity of a chemical (in this case determined by the length of the alkyl side chain), the uptake into an organism is facilitated, resulting in higher internal concentrations and higher toxic effects. Additionally, the impact on the membrane system is stronger because cations with an increased hydrophobicity adsorb on or intercalate into membranes, resulting in membrane perturbation (expansion or swelling, increase in fluidity, lowering of the phase transition temperature and alteration of the ion permeability of the membrane).

This correlation between hydrophobicity and toxicity was also supported by Ranke and co-workers [13, 58, 65], who determined log k_0 values [via a reversed-phase gradient high-performance liquid chromatographic (HPLC) method], which represented measurements of cationic lipophilicity. Those values were correlated with toxicity data (rat leukemia cells, *Vibrio fischeri* and *Scenedesmus vacuolatus*) and as an outcome they proved that this method can be used to estimate the cytotoxic effects of an ionic liquid. Stepnowski and Skladanowski also applied reversed-phase and immobilized artificial membrane chromatography to determine the lipophilicity of ionic liquids cations [63].

Stolte et al. [13] determined membrane–water coefficients (a commonly used method in toxicology) to investigate interactions between membranes and ionic liquids. They used the Transil technology (porous silica beads covered with a unilamellar lipid bilayer) and found that the coefficients for ionic liquids with longer alkyl side chains were higher, indicating that they interacted much more with the lipid-bilayer. Compared with common organic solvents, the membrane–water coefficients of the ionic liquids tested ranged between 1-hexanol and 2,4,5-trichlorotoluene.

Pretti et al. [53] found for ammonium salts with long cocosyl and stearyl side chains (Amoeng 100 and 130) histological alterations in the tissue of the fish *Danio rerio*.

This is explained by the well-known surfactant action of these ammonium salts on biological membranes, which increases membrane permeability, resulting in an alteration of the physical properties of the lipid -bilayer. Especially the enhanced calcium influx from the external medium into the cytoplasm results in detachment of the membranes.

Ranke et al. [56, 58] proved a correlation between hydrophobicity, cellular sorption and cytotoxicity of imidazolium-based ionic liquids when investigating cellular sorption and distribution among three cellular fractions (cytosol, nuclei and membranes) of rat leukemia cells and rat glioma cells.

Cornmell et al. [24] proved the accumulation of trihexyltetradecylphosphonium and methyltrioctylammonium $[NTf_2]^-$ in the bacterium *Escherichia coli*. This was achieved by Fourier transform infrared spectroscopy (without radio or isotope labeling) and underlines the correlation of toxicity and hydrophobicity because the ionic liquids were found in the cells and lipid membranes of the cells.

For the modes of toxic action for anions, it is suggested from the studies of Stolte et al. [65] that for rat leukemia cells, the mechanisms are based on the hydrophobicity and chemical reactivity. In contrast, for higher testing organisms, this trend was not proven. Here, the authors concluded that more complex mechanisms of absorption and distribution – which cannot be directly linked to hydrophobicity and chemical reactivity – might be responsible for heterogeneous toxicity patterns obtained with organisms of higher biological complexity. The differences in toxicity can be explained by the high structural heterogeneity of the anions investigated, resulting in completely different modes of toxic action.

Regarding mutagenicity, which can result in cancerogenicity in higher organisms, there is only one study available, by Docherty et al. [28]. They investigated the mutagenicity of 10 ionic liquids, including imidazolium, pyridinium and quaternary ammonium compounds, on *Salmonella typhimurium* employing the Amas test. The results indicated that none of the compounds tested met the criteria for mutagenicity (according to the US EPA) except $[C_4mim]Br$ and $[C_8mim]Br$. These showed slight tendencies to mutagenicity but only at the highest doses tested.

Wang et al. [70] observed that HeLa cells undergo apoptosis when they are exposed to $[C_2mim][BF_4]$. This was proven by inverted phase contrast microscopy. The cells had developed vesicles in the cytoplasm and surface blebs but the mechanism is as yet unexplored.

Comparing ionic liquid toxicity with the effects caused by conventional solvents (e.g. methanol, acetone, dimethyl sulfoxide, dimethylfuran, propanol, ethanol, acetonitrile), most studies showed higher toxicities for the ionic liquids (for some cases even by several orders of magnitude).

10.2.7
Conclusion

Here the main results documenting the state of the art of ionic liquid ecotoxicity will be summarized and combined with recommendations for the structural design of inherently safer ionic liquids. Ionic liquids with long alkyl side chains ($>C_4$)

(increasing hydrophobicity) lead to strongly increased toxicities and should therefore be avoided. In contrast, ionic liquids with short or functionalized side chains (reduced hydrophobicity) can result in lower toxicities and should be favored. In general, the hydrophobicity of the cation is mainly dominated by the side chain; the head group itself is of minor importance with one exception: the 1-butyl-4-(dimethylamino)pyridinium head group showed drastic toxicities and should therefore be avoided. In contrast, the morpholinium head group may be very promising in terms of a reduced ecotoxicological hazard potential, especially in combination with functionalized, polar side chains. For many applications, hydrophobic ionic liquids are needed and therefore the anionic moiety is an important element to influence this physicochemical properties. Hence investigations regarding (eco)toxicity and in particular (bio)degradability of the anion species forming hydrophobic ionic liquids (especially if combined with non-toxic polar cations) are of major concern for the structural design and a sound hazard assessment of ionic liquids.

The results obtained for the anionic moieties are heterogeneous and diverse. From an ecotoxicological point of view, Cl^-, Br^-, $[BF_4]^-$ and octylsulfate (with some limitations) can be recommended for application. In contrast, the $[NTf_2]^-$ anion should be avoided owing to the drastic toxicities especially for the terrestrial compartment. The bis(trifluoromethyl)imide anion, $[(CF_3)_2N]^-$, being a chemically reactive compound, showed surprisingly low toxicities for organisms with different biological complexity (algae, bacteria, plants, invertebrates), but it cannot be recommended owing to the hazards emanating for the operator (vulnerability to hydrolytic cleavage and degradation involving HF formation). For all anions containing fluorine (e.g. hexafluorophosphate, hexafluoroantimonate, tetrafluoroborate), hydrolytic effects (and therefore increasing toxicity with increasing time) should be taken into account. In contrast, the bis(1,2-benzenediolato)borate anion generally showed high intrinsic toxicities and therefore cannot be recommended.

In comparison with common organic solvents, ionic liquids do not perform very well. They are in most cases more toxic than or at least equally toxic as conventional solvents. This is probably based on their non-volatility. Nevertheless, compounds with moderate or minor toxic effects have been developed, for example choline derivatives are recommended as promising by several authors. The same holds true for the development of more biocompatible and biodegradable anions. Examples of more biocompatible anions are alkylsulfates (such as methylsulfate, ethylsulfate and octylsulfate), linear alkylsulfonates (methylsulfonate), linear alkylbenzenesulfonates (p-toluenesulfonate) and salts of organic acids (such as acetate or lactate). Additionally, tetrafluoroborate and hexafluorophosphate can be recommended due to their only moderate toxic effects.

Mixture toxicity and combination effects have to be taken into account when evaluating the hazard potential of ionic liquids. Also, the application of traditional prediction concepts for combination effects (for example CA and IA) has to be reconsidered. When applying those concepts to ionic liquid mixtures and mixtures of ionic liquids and cadmium, the prediction quality was only moderate. In consequence, better prediction concepts have to be developed. In general, the development of reliable prediction models (not only for mixture toxicity but also for single ionic liquid toxicities) is favored owing to the high structural diversity of these compounds. Promising tools to fulfill this task are (Q)SARs.

Only a few studies have investigated the influence of environmental key variables (organic matter, clay minerals, pH, salinity, etc.) on the bioavailability and therefore toxicity of ionic liquids. This has to be considered much more in future studies. Especially ionic liquids with short alkyl side chains proved to be less toxic, but highly mobile in soils (low sorption potential) and not biodegradable. Therefore, they could become problematic, because they show a potential for persistence and, due to their high mobility, they could for example reach the ground water and enrich there. In addition to abiotic factors, which influence the bioavailability of ionic liquids, biotic factors (for example community structures, predator–prey relationships and the behavior of organisms) should be taken into account. On the one hand, these parameters can be much more sensitive than, for example, mortality, but on the other hand, the results are much more difficult to interpret.

To sum up the state of the art of ionic liquid ecotoxicity, much work has been carried out and a broad database already exists, which can be used to develop promising prediction tools on expected toxicities. Unfortunately, less is known with regard to factors influencing the bioavailability of ionic liquids. Hence it is necessary to complement prediction tools with information on environmental biotic and abiotic key factors, which do influence the bioavailability of xenobiotics. Altogether, this will help to select (eco)toxicologically favorable structural elements and thus contribute to the design of inherently safer ionic liquids with a reduced hazard for humans and the environment.

10.3
Biodegradability of Ionic Liquids

10.3.1
Introduction

The extensive production and use of synthetic chemicals in almost all fields of human activities results in a continuous release of chemicals into the biosphere. Since the early 1940s, several serious environmental problems have been caused by especially persistent chemicals with high stabilities against biotic and abiotic transformations. For instance, the use of non-biodegradable, synthetic surfactants such as an alkyl arylsulfonate, tetrapropylene benzenesulfonate (TPBS), was responsible for widespread foaming in rivers and coastal waters. Fatal effects for humans and the environment were found for persistent and bioaccumulating substances such as halogenated aromatic hydrocarbons. The most prominent example of such a chemical is the pesticide dichlorodiphenyltrichloroethane (DDT), which was extensively applied (more than 1.5 million tons since 1940) *inter alia* in agriculture. The recalcitrance of DDT, its isomers and its metabolites to biodegradation combined with the high lipophilicity of this compounds led to a strong uptake into organisms (bioaccumulation) and to an enrichment of this chemicals within the food chain (biological magnification) [80]. The American biologist Rachel Carson published in 1962 the book *Silent Spring* [81], which describes the environmental impacts of DDT and particularly the dramatic toxic effects towards birds, which came along with a bird die-off in the USA.

In the 1970s and 1980s, the use of DDT was banned in most countries. However, even many years after the prohibition of DDT, several other species (mammals, reptiles, amphibians, fish) were strongly affected by DDT because of the long half-life of this toxic chemical.

Based on these experiences, the need for biodegradability tests arose and entered into regulatory assessment. Recently, the United Nation enacted the Stockholm Convention on Persistent Organic Pollutants (POPs), adopted by 127 countries in 2001. It was the first global, legally binding instrument of its kind with scientifically based criteria for potential POPs and it is a process that ultimately may lead to global elimination of POP substances [82]. Furthermore, persistent, bioaccumulating and toxic chemicals (PBTs) are covered by different national and international regulations. For instance, by European Union (EU) regulation, the so-called PBT substances are considered to be substances of high concern under REACH (Registration, Evaluation, Authorization and Restriction of Chemicals), demanding particularly rigorous evaluation and an authorization procedure.

To cut a long story short, the environmental legislation demands for non-persistent chemicals. For "sustainable" or "green" chemicals, complete and rapid biotic and/or abiotic degradation is a crucial requirement in accordance with Anastas and Warner's 10th principle of green chemistry: "Design chemicals and products to degrade after use: Design chemical products to break down to innocuous substances after use so that they do not accumulate in the environment" [1].

The following sections will give an overview on existing biodegradation data of ionic liquids and will discuss whether this 10th principle of green chemistry is fulfilled for distinct compounds out of this substance class or not. In addition to biodegradation data, biological degradation products (metabolites) and abiotic degradation processes are also included in the discussion. The collected data should help to improve the prospective design of inherently safer ionic liquids by reducing their risk for humans and the environment.

10.3.2
Testing of Biodegradability

The biodegradability of organic chemicals influences the exposure of substances towards organisms and, hence, it is a key parameter within the risk assessment to estimate environmental concentrations and long-term adverse effects on biota.

In its simplest definition, biodegradation can be described as the breakdown of an organic substance by enzymes produced by living organisms. To simulate biodegradation processes, several test methodologies were developed and standardized by, for instance, the Organization for Economic Cooperation and Development (OECD). General test guidelines describe the biodegradability of chemicals with different concepts and terms such as "ready biodegradability", "inherently biodegradable" and "ultimate biodegradation". For a clearer understanding, some important terms and their relevance within the assessment of biodegradability of chemicals are given in Table 10.3. For every type of biodegradation experiment, a test substance is brought into contact with a mixed population of bacteria – from surface waters, aquatic sediments, soils or from waste water treatment plants – in a solution of inorganic nutrients under standardized conditions (e.g. pH, temperature, substance concentration, test duration).

Non-specific composite parameters such as dissolved organic carbon (DOC), CO_2 production and biochemical oxygen demand (BOD) are used to follow the course of biodegradation. This has the advantage that these integral parameters are applicable to a wide variety of organic substances and they respond to any biodegradation residues or transformation products.

For example: the BOD, the amount of oxygen consumed by microorganisms when metabolizing a test chemical, is measured via manometric respirometry or using an oxygen electrode. Finally, the biodegradation rate is calculated from the

Table 10.3 Aerobic biodegradation testing procedures.

Level 1. Ready biodegradability (e.g. according to OECD test protocols 301 A–F)
Readily biodegradable is an arbitrary classification of chemicals which have passed a screening test for ultimate biodegradation (complete mineralization of the test chemicals to *inter alia* CO_2, H_2O, NO_3^-). A ready biodegradation test is based on stringent conditions[a]. Thus, if a chemical passes the ready test, there is a high probability that it is readily biodegradable in the environment (no further testing is necessary). However, if a chemical fails this test, it does not necessarily follow that it will not degrade in the environment. Further testing is required.

Level 2. Ready biodegradability (e.g. according to OECD test protocols 301 A–F)
Inherently biodegradable is a classification for which there is unequivocal evidence of (partial) biodegradation. An inherent biodegradation test is based on non-stringent conditions. Thus, if a chemical fails the inherent test, there is a high probability that is non-readily biodegradable in the environment. However, if the chemical passes the inherent test, it does not necessarily follow that it will degrade in the environment. Only some inherently biodegradable chemicals will so degrade in the (following) simulation test.

Level 3. Simulation test (e.g. according to OECD test protocol 303)
Simulation tests are designed to determine the elimination and the primary degradation (which means the biological production of organic derivatives which exhibit their own properties and fates) and/or ultimate biodegradation of organic compounds by aerobic microorganisms in a continuously operated test system simulating, for example, the activated sludge process. Some systems closely mimicked full-scale plants, having sludge settlement tanks with settled sludge being pumped. These experiments allow for a more precise and more realistic assessment of the biodegradability of chemicals compared with the above-mentioned screening tests, but they are associated with relatively high cost and effort of applying this simulation test

[a]High test chemical concentration, low cell density of inoculum, inoculum not pre-exposed to test chemical, only inorganics in the medium, 28 day test duration.

ratio between the oxygen consumption that is caused by the degradation of the test item (corrected by the blank values of the inoculum) and the theoretical oxygen demand (ThOD). The ThOD can be calculated from the molecular formula and represents the total amount of oxygen required to oxidize a chemical completely. A degradation extent of at least 60% ThOD within 28 days is defined as the criterion for classifying a particular test substance as "readily biodegradable". Additionally, experiments with sterilized inocula were carried out to check for possible non-biological degradation and for adsorption processes of the test chemical to the organic matter of the inocula (especially when activated sewage sludge is used). Furthermore, to determine if the test chemical inhibits the inoculum, tests with a mixture of a readily biodegradable reference compound and the test chemical can be carried out.

In general, the assessment of biodegradability is a very complex field. The results are strongly dependent on the kind of degradation test performed and on the composite parameters used. Each test procedure has its own intrinsic advantages and drawbacks, which can result in very different degradation rates for one chemical when different methods are applied. Furthermore, the intrinsic variability of bio-degradation test methods can lead to different results for one compound using the same test system. This might lead to the conclusion that results from biodegradability studies have no validity, but these measurements are not comparable to the determinations of simple properties such as melting points or viscosities, for which there is only one correct answer. Especially the nature of the inoculum is a very versatile factor within the assessment of biodegradability. For every experiment a unique microorganism community is used, which comes along with a restricted reproducibility based on the biological variability.

To handle the cascade of complexity – based on different testing methodologies available and on the biological variability – the interpretation of results should address the following question [83]: which test was performed, under what conditions and with which microorganisms?

If biodegradation was observable, two questions arise, namely "at what rate?" and "breakdown to what?" Without careful consideration of these questions, the results of biodegradability studies can lead to false or contradictory conclusions. In particular, the unwary interpretation of empirically preassigned pass levels (>60% biodegradation rate = readily biodegradable) can be misleading in the development of biodegradable chemicals.

The following section is aimed to give the reader the extracted main conclusions of the biodegradation studies conducted with ionic liquids so far.

10.3.3
Results from Biodegradation Experiments

In contrast to the fast growing (eco)toxicological knowledge about ionic liquids, data available in the field of biodegradability are still very restricted. So far, no anaerobic studies and only a few aerobic biodegradation studies have been performed.

Fundamental work in this field was conducted by Scammells and co-workers [32, 84–87]. Guided by the knowledge from the development of biodegradable surfactants, they investigated different imidazolium and pyridinium compounds substituted with side chains containing ester and amide functions combined with several anions in different tests for ready biodegradability. Wells and Coombe [71] investigated the biodegradability of ammonium, imidazolium, phosphonium and pyridinium compounds by measuring the BOD. N-Methylimidazolium and 3-methylpyridinium compounds substituted with butyl, hexyl and octyl side chains and bromide as the anion were examined by Kulpa and co-workers [88] applying DOC die-away tests. Stolte et al. [67] investigated different N-imidazole, imidazolium, pyridinium and 4-(dimethylamino)pyridinium compounds substituted with various alkyl side chains and their analogues containing functional groups in a primary biodegradation test and identified different biodegradation products from [C_8mim]Cl. Recently, Stepnowski and co-workers [62] used the closed bottle test to examine the biodegradability of 1-alkoxymethyl-3-hydroxypyridinium cations combined with acesulfamate, saccharinate and chloride anions.

Based on the various publications referred to above, an overview on the existing aerobic biodegradation data for ionic liquid structures is presented in Tables 10.4–10.6. In the following, the biodegradability of the anion and the cation is discussed separately.

10.3.3.1 Biodegradability of Ionic Liquid Anions

Most of the anions of the ionic liquids investigated so far (Tables 10.5 and 10.6) are inorganic moieties (halides, $[BF_4]^-$, $[PF_6]^-$), which are not relevant for biodegradation tests based on the measurement of oxidizable carbon in a molecule. For inorganic anions, abiotic degradation processes are more important (see Section 10.3.6). However, some ionic liquids anions are accessible to microbial degradation. It is known from research on anionic surfactants [89] that linear alkylsulfates (such as methylsulfate or octylsulfate) exhibit an excellent biodegradability and linear alkylsulfonates (such as methylsulfonate) show good biodegradability in addition to linear alkyl benzenesulfonates (such as p-toluenesulfonate). Good biodegradability was found for the diethylphosphate ionic liquid anion [71]. For different 1-alkoxymethyl-3-hydroxypyridinium acesulfamates and saccharinates, a non-specific influence of the anion compared with the chloride compounds has been suggested [62]. For the food additives potassium acesulfame and saccharin, many toxicological data are available, but no or only a few data regarding biodegradation in the environment could be found. For instance, ready utilization of saccharin by the bacterium Sphingomonas xenophaga, which was enriched and isolated from a communal waste water treatment plant, has been described [90]. Likewise, little information was found on the biodegradation behavior of the typical fluorine-containing ionic liquid anions $[(CF_3SO_2)_2N]^-$, $[(C_2F_5)_2PO_2]^-$, $[(C_2F_5)_3PF_3]^-$, $[(C_3F_7)_3PF_3]^-$ and $[(CF_3SO_2)_3C]^-$. The tendency of these anions to be thermally and chemically very stable is presumably mirrored in their stability towards biological degradation processes [57]. In general, many synthetic halo-organics have been found to be resistant to aerobic biodegra-

Table 10.4 Biodegradation data for imidazolium-based ionic liquids.

No.	Structure			Biodegradation (%)	Classification	Biodegradation test	Inocula	Test duration (days)	Substance concentration	Measured parameter[a]	Ref.
	Head group	Side chain	Anion								
1	N⁺N–R (imidazolium)	C_1-CN	Cl^-	0	Not readily biodegradable	Modified OECD 301D Test	Activated sludge	31	200 µmol l^{-1}	Primary biodegradation	[67]
2		C_2	Cl^-	0	Not readily biodegradable	Modified OECD 301D Test	Activated sludge	31	200 µmol l^{-1}	Primary biodegradation	[67]
3		C_2-OH	I^-	0	Not readily biodegradable	Modified OECD 301D Test	Activated sludge	31	200 µmol l^{-1}	Primary biodegradation	[67]
4		–CH_2–O–C_2H_5	Cl^-	0	Not readily biodegradable	Modified OECD 301D Test	Activated sludge	31	200 µmol l^{-1}	Primary biodegradation	[67]
5		C_2H_5–O–CH_2	Cl^-	0	Not readily biodegradable	Modified OECD 301D Test	Activated sludge	31	200 µmol l^{-1}	Primary biodegradation	[67]
6		C_3-OH	Cl^-	0	Not readily biodegradable	Modified OECD 301D Test	Activated sludge	31	200 µmol l^{-1}	Primary biodegradation	[67]
7		C_4	Br^-	0	Not readily biodegradable	Die-Away Test OECD 301A	Activated sludge	43	40 mg Cl^{-1}	DOC	[88]

10.3 Biodegradability of Ionic Liquids

#	R	Anion	%	Classification	Test	Organisms	Days	Concentration	Measure	Ref.
8	C_4	Br^-	<5	Not readily biodegradable	CO_2 Headspace test (ISO 14593)	Wastewater organisms	28	$40\,mg\,l^{-1}$	TOC	[86]
9	C_4	Br^-	1	Not readily biodegradable	Modified Sturm test (OECD 301B)	Wastewater organisms	28	$2\,mg\,l^{-1}$	BOD	[85]
10	C_4	$[PF_6]^-$	0	Not readily biodegradable	Closed Bottle Test (OECD 301D).	Wastewater organisms	28	$2\,mg\,l^{-1}$	BOD	[85]
11	C_4	$[PF_6]^-$	0	Not readily biodegradable	OECD 301F	Not specified	28	$100\,mg\,l^{-1}$	BOD	[71]
12	C_4	$[BF_4]^-$	0	Not readily biodegradable	Closed Bottle Test (OECD 301D).	Wastewater organisms	28	$2\,mg\,l^{-1}$	COD BOD	[86]
13	C_4	$[BF_4]^-$	<5	Not readily biodegradable	CO_2 Headspace test (ISO 14593)	Wastewater organisms	28	$40\,mg\,l^{-1}$	TOC	[86]
14	C_4	Cl^-	0	Not readily biodegradable	Modified OECD 301D Test	Activated sludge	31	$200\,\mu mol\,l^{-1}$	Primary biodegradation	[67]
15	C_4	Cl^-	<5	Not readily biodegradable	Closed Bottle Test (OECD 301D).	Wastewater organisms	28	$2\,mg\,l^{-1}$	BOD	32
16	C_4	Cl^-	0	Not readily biodegradable	OECD 301F	Not specified	28	$100\,mg\,l^{-1}$	BOD COD	[71]
17	$-C_2H_5-O-C_2H_5$	Cl^-	0	Not readily biodegradable	Modified OECD 301D Test	Activated sludge	31	$200\,\mu mol\,l^{-1}$	Primary biodegradation	[67]

(Continued)

Table 10.4 (Continued)

No.	Structure Head group	Side chain	Anion	Biodegradation (%)	Classification	Biodegradation test	Inocula	Test duration (days)	Substance concentration	Measured parameter[a]	Ref.
18	$-C_3H_7-O-CH_3$		Cl^-	0	Not readily biodegradable	Modified OECD 301D Test	Activated sludge	31	$200\,\mu mol\,l^{-1}$	Primary biodegradation	[67]
19		C_4	$[N(CN)_2]^-$	<5	Not readily biodegradable	Closed Bottle Test (OECD 301D).	Wastewater organisms	28	$2\,mg\,l^{-1}$	BOD	[32]
20		C_4	$[(CF_3SO_2)_2N]^-$	<5	Not readily biodegradable	Closed Bottle Test (OECD 301D).	Wastewater organisms	28	$2\,mg\,l^{-1}$	BOD	[32]
21		C_4	$[C_8H_{17}OSO_3]^-$	25	Not readily biodegradable	Closed Bottle Test (OECD 301D).	Wastewater organisms	28	$2\,mg\,l^{-1}$	BOD	[32]
22		C_6	Cl^-	11	Not readily biodegradable	Modified OECD 301D Test	Activated sludge	31	$200\,\mu mol\,l^{-1}$	Primary biodegradation	[67]
23		C_7-COOH	Br^-	100	Inherently biodegradable	Modified OECD 301D Test	Activated sludge	31	$200\,\mu mol\,l^{-1}$	Primary biodegradation	[67]
24		C_6	Br^-	54	Not readily biodegradable; partially mineralized	Die-Away Test OECD 301A	Activated sludge	37	$40\,mg\,Cl^{-1}$	DOC	[88]

10.3 Biodegradability of Ionic Liquids

#	R	Anion	%	Classification	Test	Medium	Days	Conc.	Measurement	Ref.
25	C_8	Cl^-	100	Inherently biodegradable	Modified OECD 301D Test	Activated sludge	31	$200\,\mu mol\,l^{-1}$	Primary biodegradation	[67]
26	C_8	Br^-	41	Not readily biodegradable; partially mineralized	Die-Away Test	Activated sludge	38	$40\,mg\,Cl^{-1}$	DOC	[88]
27	C_8-OH	Br^-	100	Inherently biodegradable	OECD 301A Modified OECD 301D Test	Activated sludge	31	$200\,\mu mol\,l^{-1}$	Primary biodegradation	[67]
28	C_1	Br^-	~17	Not readily biodegradable	Closed Bottle Test (OECD 301D).	Wastewater organisms	28	$2\,mg\,l^{-1}$	BOD	[85]
29	C_2	Br^-	~22	Not readily biodegradable	Closed Bottle Test (OECD 301D).	Wastewater organisms	28	$2\,mg\,l^{-1}$	BOD	[85]
30	C_3	Br^-	~19	Not readily biodegradable	Closed Bottle Test (OECD 301D).	Wastewater organisms	28	$2\,mg\,l^{-1}$	BOD	[85]
31	C_3	$[BF_4]^-$	<20	Not readily biodegradable	Closed Bottle Test (OECD 301D).	Wastewater organisms	28	$2\,mg\,l^{-1}$	BOD	[32]
32	C_3	$[PF_6]^-$	<20	Not readily biodegradable	Closed Bottle Test (OECD 301D).	Wastewater organisms	28	$2\,mg\,l^{-1}$	BOD	[32]
33	C_3	$[(CF_3SO_2)_2N]^-$	<20	Not readily biodegradable	Closed Bottle Test (OECD 301D).	Wastewater organisms	28	$2\,mg\,l^{-1}$	BOD	[32]
34	C_3	$[N(CN)_2]^-$	<30	Not readily biodegradable	Closed Bottle Test (OECD 301D).	Wastewater organisms	28	$2\,mg\,l^{-1}$	BOD	[32]

(Continued)

Table 10.4 (Continued)

No.	Structure			Biodegradation (%)	Classification	Biodegradation test	Inocula	Test duration (days)	Substance concentration	Measured parameter[a]	Ref.
	Head group	Side chain	Anion								
35		C_3	$[C_8H_{17}OSO_3]^-$	49	not readily biodegradable	Closed Bottle Test (OECD 301D).	Wastewater organisms	28	2 mg l^{-1}	BOD	[32]
36		C_4	Br$^-$	~30	Not readily biodegradable	Closed Bottle Test (OECD 301D).	Wastewater organisms	28	2 mg l^{-1}	BOD	[85]
37		C_5	Br$^-$	~32	Not readily biodegradable	Closed Bottle Test (OECD 301D).	Wastewater organisms	28	2 mg l^{-1}	BOD	[85]
38		C_6	Br$^-$	~26	Not readily biodegradable	Closed Bottle Test (OECD 301D).	Wastewater organisms	28	2 mg l^{-1}	BOD	[85]
39		C_8	Br$^-$	~33	Not readily biodegradable	Closed Bottle Test (OECD 301D).	Wastewater organisms	28	2 mg l^{-1}	BOD	[85]
40		$R_1 + R_2 = C_2$	Br$^-$	<10	Not readily biodegradable	Closed Bottle Test (OECD 301D).	Wastewater organisms	28	2 mg l^{-1}	BOD	[86]
41		$R_1 = H$, $R_2 = C_4$	Br$^-$	<10	Not readily biodegradable	Closed Bottle Test (OECD 301D).	Wastewater organisms	28	2 mg l^{-1}	BOD	[86]

10.3 Biodegradability of Ionic Liquids

Structure: imidazolium cation with R1 on one N, and on the other N a –CH2–C(=O)–O–R2 group.

#	Substituents	Anion	%	Classification	Test	Inoculum	Days	Conc.	Method	Ref.
42	R1 = C1, R2 = C4	Br⁻	<10	Not readily biodegradable	Closed Bottle Test (OECD 301D).	Wastewater organisms	28	2 mg l⁻¹	BOD	[86]
43	R1 = H, R2 = C3	Br⁻	24	Not readily biodegradable	Closed Bottle Test (OECD 301D).	Wastewater organisms	28	2 mg l⁻¹	BOD	[86]
44	R1 = H, R2 = C3	Br⁻	24	Not readily biodegradable	CO2 Headspace test (ISO 14593)	Wastewater organisms	28	40 mg l⁻¹	TOC	[86]
45	R1 = H, R2 = C3	[C8H17OSO3]⁻	49	Not readily biodegradable	Closed Bottle Test (OECD 301D).	Wastewater organisms	28	2 mg l⁻¹	BOD	[86]
46	R1 = H, R2 = C3	[C8H17OSO3]⁻	64	**Readily biodegradable**	CO2 Headspace test (ISO 14593)	Wastewater organisms	28	40 mg l⁻¹	TOC	[86]
47	R1 = C1, R2 = C3	Br⁻	23	Not readily biodegradable	Closed Bottle Test (OECD 301D).	Wastewater organisms	28	2 mg l⁻¹	BOD	[86]
48	R1 = C1, R2 = C3	[C8H17OSO3]⁻	55	Not readily biodegradable	Closed Bottle Test (OECD 301D).	Wastewater organisms	28	2 mg l⁻¹	BOD	[86]

(Continued)

Table 10.4 (Continued)

No.	Structure			Biodegradation (%)	Classification	Biodegradation test	Inocula	Test duration (days)	Substance concentration	Measured parameter[a]	Ref.
	Head group	Side chain	Anion								
49		$R_2 = C_3$ $R_1 = C_1$	$[C_8H_{17}OSO_3]^-$	62	Readily biodegradable	CO_2 Headspace test (ISO 14593)	Wastewater organisms	28	40 mg l^{-1}	TOC	[86]
50		$R_2 = C_3$ $R_1 = H$	Br$^-$	32	Not readily biodegradable	Closed Bottle Test (OECD 301D).	Wastewater organisms	28	2 mg l^{-1}	BOD	[86]
51		$R_2 = C_5$ $R_1 = H$	Br$^-$	41	Not readily biodegradable	CO_2 Headspace test (ISO 14593)	Wastewater organisms	28	40 mg l^{-1}	TOC	[86]
52		$R_2 = C_5$ $R_1 = H$	$[C_8H_{17}OSO_3]^-$	54	Not readily biodegradable	Closed Bottle Test (OECD 301D).	Wastewater organisms	28	2 mg l^{-1}	BOD	[86]
53		$R_2 = C_5$ $R_1 = H$	$[C_8H_{17}OSO_3]^-$	67	Readily biodegradable	CO_2 Headspace test (ISO 14593)	Wastewater organisms	28	40 mg l^{-1}	TOC	[86]

54	$R_2 = C_5$ $R_1 = C_1$	Br^-	33	Not readily biodegradable	Closed Bottle Test (OECD 301D).	Wastewater organisms	28	$2\,mg\,l^{-1}$	BOD	[86]
55	$R_2 = C_5$ $R_1 = C_1$	$[C_8H_{17}OSO_3]^-$	56	Not readily biodegradable	Closed Bottle Test (OECD 301D).	Wastewater organisms	28	$2\,mg\,l^{-1}$	BOD	[86]
56	$R_2 = C_5$ $R_1 = C_1$	$[C_8H_{17}OSO_3]^-$	61	**Readily biodegradable**	CO_2 Headspace test (ISO 14593)	Wastewater organisms	28	$40\,mg\,l^{-1}$	TOC	[86]
	$R_2 = C_5$									

[a]Primary biodegradation: breakdown of test chemical is measured by HPLC.
Biochemical oxygen demand is the amount of oxygen consumed by microorganisms when metabolizng the test chemical. The BOD is expressed in mg oxygen depletion per mg test chemical.
Total organic carbon is the amount of carbon bound in an organic test chemical.
Chemical oxygen demand is the amount of oxygen consumed during the oxidation of a test chemical with hot, acid dichromate. The COD is used to measure indirectly the amount of oxidizable organic matter in solution.

Table 10.5 Biodegradation data for pyridinium-based ionic liquids.

No.	Structure Head group	Side chain	Anion	Biodegradation (%)	Classification	Biodegradation test	Inocula	Test duration (days)	Substance concentration	Measured parameter[a]	Ref.
57	N+–R (pyridinium)	C_2	Cl^-	0	Not readily biodegradable	Modified OECD 301D Test	Activated sludge	31	200 µmol l^{-1}	Primary biodegradation	67
58		C_4	Cl^-	0	Not readily biodegradable	Modified OECD 301D Test	Activated sludge	31	200 µmol l^{-1}	Primary biodegradation	67
59		C_8	Cl^-	100	Inherently biodegradable	Modified OECD 301D Test	Activated sludge	31	200 µmol l^{-1}	Primary biodegradation	67
60		C_4	Br^-	0	Not readily biodegradable	Die-Away Test OECD 301A	Activated sludge	43	40 mg Cl^{-1}	DOC	88
61	N+–R (3-methylpyridinium)	C_4	Cl^-	0	Not readily biodegradable	OECD 301F	Not specified	28	100 mg l^{-1}	BOD COD	71
62		C_6	Br^-	97	Not readily biodegradable; fully mineralized	Die-Away Test OECD 301A	Activated sludge	49	40 mg Cl^{-1}	DOC	88
63		C_8	Br^-	96	Readily biodegradable; fully mineralized	Die-Away Test OECD 301A	Activated sludge	25	40 mg Cl^{-1}	DOC	88

#	Structure	R	Anion	%	Classification	Test	Inoculum	Days	Concentration	Primary biodegradation	Ref
64	(pyridinium ethyl acetate)	C_8	Cl^-	100	Inherently biodegradable	Modified OECD 301D Test	Activated sludge	31	200 μmol l^{-1}	TOC	67
65		C_8	Br^-	87	**Readily biodegradable**	CO_2 Headspace test (ISO 14593)	Wastewater organisms	28	40 mg l^{-1}	TOC	87
66		C_8	$[C_8H_{17}OSO_3]^-$	89	**Readily biodegradable**	CO_2 Headspace test (ISO 14593)	Wastewater organisms	28	40 mg l^{-1}	TOC	87
67	(nicotinate ester N$^+$–R)	C_1	I^-	74	**Readily biodegradable**	CO_2 Headspace test (ISO 14593)	Wastewater organisms	28	40 mg l^{-1}	TOC	87
68		C_1	$[C_8H_{17}OSO_3]^-$	75	**Readily biodegradable**	CO_2 Headspace test (ISO 14593)	Wastewater organisms	28	40 mg l^{-1}	TOC	87
69		C_1	$[(CF_3SO_2)_2N]^-$	~65	**Readily biodegradable**	CO_2 Headspace test (ISO 14593)	Wastewater organisms	28	40 mg l^{-1}	TOC	87
70		C_4	$[C_8H_{17}OSO_3]^-$	82	**Readily biodegradable**	CO_2 Headspace test (ISO 14593)	Wastewater organisms	28	40 mg l^{-1}	TOC	87
71	(nicotinamide N$^+$–R)	C_4	$[C_8H_{17}OSO_3]^-$	30	Not readily biodegradable	CO_2 Headspace test (ISO 14593)	Wastewater organisms	28	40 mg l^{-1}	TOC	87
72	(3-hydroxypyridinium N$^+$–CH$_2$OR)	C_3	Acesulfamates	24	Not readily biodegradable	OECD 301D Test	Wastewater organisms	28	4 mg l^{-1}	BOD	62
73		C_3	Saccharinates	43	Not readily biodegradable	OECD 301D Test	Wastewater organisms	28	4 mg l^{-1}	BOD	62
74		C_3	Cl^-	40	Not readily biodegradable	OECD 301D Test	Wastewater organisms	28	4 mg l^{-1}	BOD	62
75		C_4	Acesulfamates	21	Not readily biodegradable	OECD 301D Test	Wastewater organisms	28	4 mg l^{-1}	BOD	62
76		C_4	Saccharinates	13	Not readily biodegradable	OECD 301D Test	Wastewater organisms	28	4 mg l^{-1}	BOD	62

(Continued)

Table 10.5 (Continued)

No.	Structure			Biodegradation (%)	Classification	Biodegradation test	Inocula	Test duration (days)	Substance concentration	Measured parameter[a]	Ref.
	Head group	Side chain	Anion								
77		C_6	Acesulfamates	39	Not readily biodegradable	OECD 301D Test	Wastewater organisms	28	4 mg l^{-1}	BOD	62
78		C_6	Saccharinates	31	Not readily biodegradable	OECD 301D Test	Wastewater organisms	28	4 mg l^{-1}	BOD	62
79		C_7	Acesulfamates	41	Not readily biodegradable	OECD 301D Test	Wastewater organisms	28	4 mg l^{-1}	BOD	62
80		C_7	Saccharinates	32	Not readily biodegradable	OECD 301D Test	Wastewater organisms	28	4 mg l^{-1}	BOD	62
81		C_7	Cl$^-$	44	Not readily biodegradable	OECD 301D Test	Wastewater organisms	28	4 mg l^{-1}	BOD	62
82		C_{11}	Acesulfamates	49	Not readily biodegradable	OECD 301D Test	Wastewater organisms	28	4 mg l^{-1}	BOD	62
83		C_{11}	Saccharinates	72	**Readily biodegradable**	OECD 301D Test	Wastewater organisms	28	4 mg l^{-1}	BOD	62
84		C_{11}	Cl$^-$	48	Not readily biodegradable	OECD 301D Test	Wastewater organisms	28	4 mg l^{-1}	BOD	62
85		C_{18}	Acesulfamates	32	Not readily biodegradable	OECD 301D Test	Wastewater organisms	28	4 mg l^{-1}	BOD	62
86		C_{18}	Saccharinates	20	Not readily biodegradable	OECD 301D Test	Wastewater organisms	28	4 mg l^{-1}	BOD	62
87		C_{18}	Cl$^-$	25	Not readily biodegradable	OECD 301D Test	Wastewater organisms	28	4 mg l^{-1}	BOD	62

10.3 Biodegradability of Ionic Liquids

#	Structure	R	X⁻	%	Classification	Test	Inoculum	Time (d)	Conc.	Type	Ref
88	pyridinium with N(CH₃)₂	C_2	Br^-	0	Not readily biodegradable	Modified OECD 301D Test	Activated sludge	31	200 μmol l⁻¹	Primary biodegradation	67
89		C_4	Cl^-	0	Not readily biodegradable	Modified OECD 301D Test	Activated sludge	31	200 μmol l⁻¹	Primary biodegradation	67
90		C_6	Cl^-	100	Inherently biodegradable	Modified OECD 301D Test	Activated sludge	31	200 μmol l⁻¹	Primary biodegradation	67

[a]Primary biodegradation: breakdown of test chemical is measured by HPLC.
BOD: biochemical oxygen demand is the amount of oxygen consumed by microorganisms when metabolizing the test chemical. The BOD is expressed in mg oxygen depletion per mg test chemical.
TOC: total organic carbon is the amount of carbon bound in an organic test chemical.
COD: chemical oxygen demand is the amount of oxygen consumed during the oxidation of a test chemical with hot, acid dichromate. The COD is used to measure indirectly the amount of oxidizable organic matter in solution.

Table 10.6 Biodegradation data for ammonium and phosphonium compounds.

No.	Structure			Classification	Biodegradation (%)	Biodegradation test	Inocula	Test duration (days)	Substance concentration	Measured parameter[a]	Ref.
	Head group	Side chain	Anion								
91	$R2{-}\overset{R1}{\underset{R1}{P^+}}{-}R1$	$R_1 = C_4$ $R_2 = C_2$	$[(C_2H_5O)_2PO_2]^-$	Not readily biodegradable	9	OECD 301F	Not specified	28	100 mg l^{-1}	BOD COD	71
92	$R2{-}\overset{R1}{\underset{R1}{N^+}}{-}R1$	$R_1 = C_8$ $R_2 = C_1$	$[(CF_3SO_2)_2N]^-$	Not readily biodegradable	0	OECD 301F	Not specified	28	100 mg l^{-1}	BOD COD	71

[a]BOD: biochemical oxygen demand is the amount of oxygen consumed by microorganisms when metabolizing the test chemical. The BOD is expressed in mg oxygen depletion per mg test chemical.
COD: chemical oxygen demand is the amount of oxygen consumed during the oxidation of a test chemical with hot, acid dichromate. The COD is used to measure indirectly the amount of oxidizable organic matter in solution.

dation processes [91]. These anions possess the potential to persist in the environment and – due to their high lipophilicity – accumulate in tissues of living organisms. The ionic liquid anion trifluoroacetate is known to be widely recalcitrant to biodegradation [92] and a similar behavior can be proposed for the salts of trifluoromethanesulfonic acid. Nevertheless, for both types of compound bioaccumulation processes are thought to be unlikely owing to their low octanol–water coefficients.

In general, when assessing the biodegradability of ionic liquids, it has to be considered that anions (whether they are biodegradable or not) may influence the biodegradability of a cation. For instance, the combination of a ready biodegradable cation (measured as halide) with a toxic anion can lead to decreased biodegradability of the cation. In addition, if the anion reduces the water solubility of the cation, a lower biodegradation rate may occur, because of the decreased availability of the cation to the microorganisms in the surrounding medium.

The biodegradation results of ionic liquid cations presented in Tables 10.4–10.6 are discussed according to the different head groups.

10.3.3.2 Biodegradability of Imidazolium Compounds

Most biodegradation studies have been carried out for compounds with an imidazolium core structure substituted with several different alkyl side chains and side chains with incorporated ester, amide, ether, nitrile, terminal hydroxyl and carboxyl groups (Table 10.4). Especially for compounds substituted with short alkyl side chains ($<C_6$), even if functional groups (such as ether, nitrile and terminal hydroxyl) are introduced, no significant biodegradation was observable in different biodegradation tests. The results from HPLC analysis suggest that the cations investigated remained completely intact (no degradation and no transformation occurred) [67]. The increased level of degradation (25%) of 1-butyl-3-methylimidazolium combined with octylsulfate (Table 10.4, No. 21) is solely due to the degradation of the anion. For imidazolium compounds with elongated alkyl side chains (C_6 and C_8), an increased biodegradability was found (Table 10.4, Nos 22–27). These compounds can be described with terms such as inherently biodegradable and partially mineralizable. However, neither cation cannot be classified as readily biodegradable. The results from primary degradation assays indicated a complete degradation of the $[C_8mim]^+$ (Table 10.4, No. 25), but the HPLC analysis used would not detect to any biodegradation residues or transformation products, which might be resistant to further biodegradation. For long-chain compounds containing hydroxyl (Table 10.4, No. 27) and carboxyl (Table 10.4, No. 23) groups, complete primary biodegradation was also found.

The increased biodegradability of compounds with long side chains can be explained in at least two ways [67, 88]. First, ionic liquids with elongated side chains have been proven to be more toxic (see Section 10.2). Therefore, they produce a selective pressure on the microbial community. Microorganisms capable of degrading ionic liquids with longer alkyl chains are privileged, whereas the others that are not able to metabolize these ionic liquids are eliminated. This assumption has been proven by Docherty et al. [88], who analyzed the structure of the microbial community by DNA polymerase chain reaction (PCR) DGGE (denaturing gradient gel electrophoresis) and found an enrichment of some bacterial species in the samples treated with $[C_8mim]^+$.

A second explanation is based on an increased uptake into the organisms, which is also related to the lipophilicity of the compounds. Owing to this (higher) uptake, the substances can be metabolized by appropriate enzyme systems, for instance by the cytochrome P450 system located in the endoplasmatic reticulum of cells.

However, a successive elongation of the side chain does not represent an opportunity to design biodegradable ionic liquid cations, because increasing the alkyl side chain length is associated with increasing inhibitory effects on the microorganisms induced by the rising cytotoxicity of these cations. This was demonstrated by Wells and Coombe [71] for N-methylimidazolium compounds with C_{12}, C_{16} and C_{18} alkyl side chains.

A long list of imidazolium-based ionic liquids containing amide and ester side chains (Table 10.4, Nos 28–56) were investigated in different biodegradation tests [32, 84–86]. In general, the amide compounds (Table 10.4, Nos 40–42) were found to show poor to negligible biodegradation rates [85], whereas the introduction of ester groups into the side chain resulted in a significant increase in biodegradation, especially for esters with an alkyl side chain length of $\geq C_4$ [86]. These ester groups represent a site of enzymatic hydrolysis (by esterases), which can explain the increased biodegradability of these compounds. The saponification of the ester side chain goes along with the formation of readily biodegradable alcohol moieties (the longer the ester alkyl side chain, the longer is the resulting alcohol). However, none of these cations (investigated as halides) could be classified as readily biodegradable.

Higher levels of biodegradation were found for the ester ionic liquids cations combined with octylsulfate (Table 10.4, Nos 35, 45, 46, 48, 49, 52, 53, 55, 56). In one case, the benchmark of ready biodegradability was passed (Table 10.4, No. 53). Here, the biodegradability is based on a slightly increased biodegradability of the cation, but was mainly driven by the high biodegradability of the anion (for more details, see Section 10.3.4). The data suggest that only the ester side chain and the octylsulfate are degraded in the ready biodegradability test, whereas the imidazolium ring seems to stay intact under these conditions. In biodegradation experiments with partially mineralized $[C_8 mim]^+$, the stability of the core structure was confirmed via NMR measurements [88] (see also Section 10.3.5).

The imidazole ring itself and its ring C-substituted derivatives (methyl, ethyl) are known to be ultimately biodegradable, whereas all N-substituted imidazolium compounds such as methylimidazolium are poorly biodegradable [93]. A possible explanation for the poor biodegradability of N-alkylated imidazoles is that N-substitution apparently blocks the attack by enzymes in the urocanase pathway, which are responsible for the enzymatic degradation of the imidazole-containing amino acid histidine [93, 94]. Imidazolium-based ionic liquids have two N-substituents, which apparently do not enhance the biodegradability.

10.3.3.3 Pyridinium and 4-(Dimethylamino)pyridinium Compounds

Several biodegradation studies have been conducted on ionic liquid cations based on N-alkyl- and N-(ethoxycarbonyl)pyridinium, N-alkylated nicotinic acid ester, N-butylnicotinamide, 1-alkoxymethyl-3-hydroxypyridinium and N-alkyl-4-(dimethylamino)pyridinium.

For *N*-alkyl-3-methylpyridinium compounds (Table 10.5, Nos 57–64), an increased rate of biodegradation with an elongated side chain length was determined – similar to the results found for the imidazolium head group, but the degree of degradation was generally higher for the pyridinium compounds. The *N*-hexyl-3-methylpyridinium ([1-C_6,3-C_1py]$^+$) cation (Table 10.5, No. 62) was fully mineralized (97%) after 35–49 days of incubation. Even if it is not classified as "readily biodegradable", it should not persist in the environment [88]. For the *N*-octyl-3-methylpyridinium compound (Table 10.5, No. 64), nearly complete biodegradation (96%) was found within 25 days, which meets the criterion for being classified as readily biodegradable. The results from ^{1}H and ^{13}C NMR measurements indicate that the pyridinium core structure is completely degraded during the biodegradation experiment with Nos 62 and 64 [88].

In primary biodegradation experiments with different *N*-alkyl-4-(dimethylamino) pyridinium cations (Table 10.5, Nos 88–90), no decomposition for ethyl and butyl-substituted compounds was observable, whereas complete primary degradation was detected for the hexyl derivative [67]. However, the ultimate biodegradability of this structure has not been investigated so far.

For *N*-(ethoxycarbonyl)pyridinium cations (Table 10.5, Nos 65, 66) and for the *N*-alkylated nicotinic acid esters (Table 10.5, Nos 67–70), generally good biodegradability was found (all are classified as ready biodegradable) [87]. Paradoxically, the degradation rates of these cations as halides and as octylsulfate were similar. No explanation for this observation was given. In comparison with the *N*-butylnicotinic acid ester (Table 10.5, No. 70), the structural homologue *N*-butylnicotinamide (Table 10.5, No. 71) exhibited significantly lower biodegradability (81 versus 30%) [87].

The biodegradability of 1-alkoxymethyl-3-hydroxypyridinium salts (Table 10.5, Nos 82–87) is also improved with increasing alkyl chain length, but only from four to 11 carbon atoms. However, for most of these ionic liquids, just a middle-rate biodegradation was found [62]. Only one ionic liquid out of these series (Table 10.5, No. 83) could be classified as readily biodegradable. From a structural point of view, the increased biodegradability for this compound was unexpected, because of the middle-rate biodegradability of the tested cation (as halide No. 84) and the non-specific influence of the saccharinate anion.

Summing up, the ionic liquid cations with a pyridinium core exhibit in general a higher degree of biodegradation than the imidazolium derivatives – especially when an ester function is incorporated into the *N*-side chain or if it is attached to the pyridinium skeleton. Thus, many pyridinium compounds can be classified as "ready biodegradable" and also the core seems to be accessible to aerobic biodegradation processes.

10.3.3.4 Biodegradability of Other Head Groups

Very little biodegradation information for ionic liquids with ammonium and phosphonium head groups has been published (Table 10.6). However, several studies were performed for quaternary ammonium compounds, which are used as cationic surfactants in a number of applications (softeners, emulsifiers, disinfectants, etc.).

The ammonium structures are often substituted with benzyl, hydroxyethyl or unbranched alkyl side chains (often C_8–C_{24}) based on natural fats and oils (coconut, tallow fat or palm oil). Without going into the details of this very wide area, some basic rules regarding biodegradability are presented below according to van Ginkel [95]:

- Alkyltrimethylammonium and benzylalkyldimethylammonium salts are better biodegradable than dialkyldimethylammonium cations.
- An increased alkyl side chain length goes along with a reduced biodegradability (based on inhibitory effects).
- The resistance to biodegradation is largely caused by increasing numbers of long alkyl side chains.
- Ammonium salts with ester- and fatty acid-containing side chains are better biodegradable than those with alkyl side chains.

To the best of our knowledge, no biodegradability data have been published for ionic liquids based on other head groups, such as piperidinium, pyrrolidinium, morpholinium and quinolinium. However, Philipp *et al.* [94] used a large set of aerobic biodegradation data to establish rules for the prediction of *N*-heterocycles. It was concluded that the alkylation of the N-atom of the ring in general reduces the biodegradability of *N*-heterocycles. Even if this does not imply that all *N*-alkylated compounds are not biodegradable, there is a high probability that structural modifications (e.g. the introduction of oxo groups into the ring or side chain modulation) are necessary to achieve biodegradable compounds.

10.3.4
Misleading Interpretation of Biodegradation Data

It was mentioned in Section 10.3.2 that misinterpretation of biodegradation results can lead to false and misleading conclusions – as already occurred for some ionic liquids (Tables 10.4 and 10.5, Nos 46, 49, 53, 56 and 69). These compounds have been classified as readily biodegradable according to a "CO_2 Headspace" ISO test procedure, but this categorization is based on a misleading molecular design and a wrong declaration of chemicals. To give an example (Scheme 10.3), for the 1-(pentoxycarbonyl)-3-methylimidazolium cation (tested as bromide; No. 51) a biodegradation rate of 41% ("CO_2 Headspace" test) was observed. This percentage corresponds fairly well to the oxidizable C content of the ester side chain (Scheme 10.3) or rather to the pentanol formed after the proposed enzymatic hydrolysis. This means that the core structure is not touched and degraded by the microorganisms and is probably resistant to further biodegradation. The authors of this study recommended the combination of this cation with the excellently biodegradable octylsulfate anion (Table 10.4, No. 53). Of course this comes along with an increase in biodegradability (now the C content of the biodegradable parts of the side chain and of the anion exceeds the level of >60%, (Scheme 10.3). However, this denoted "readily biodegradable" compound still contains a non-biodegradable core structure.

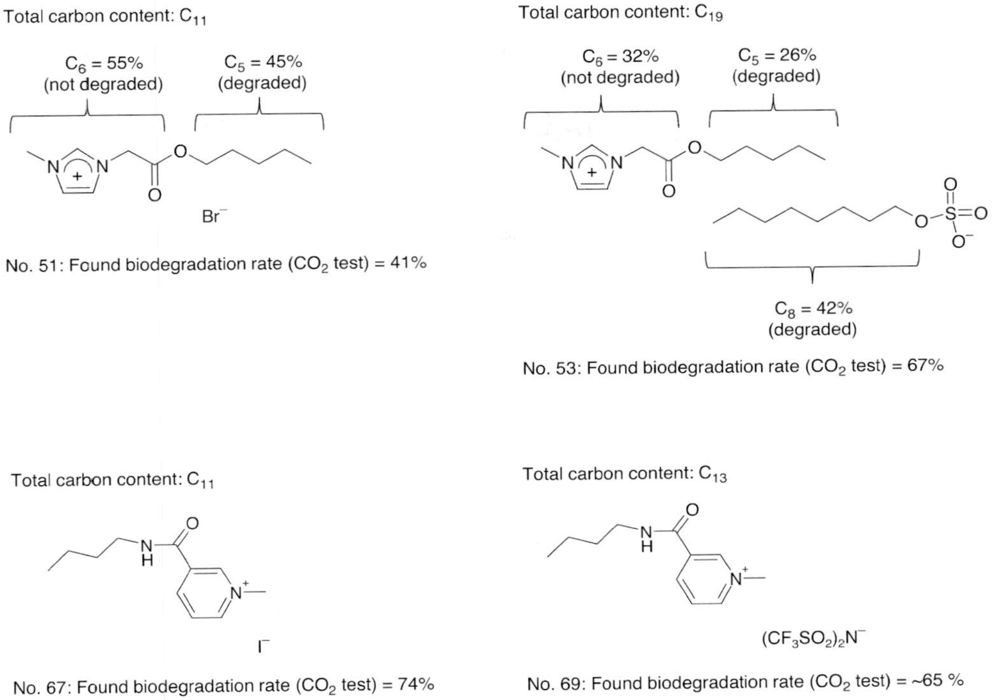

10.3 Biodegradability of Ionic Liquids

Total carbon content: C_{11}

$C_6 = 55\%$ (not degraded) $C_5 = 45\%$ (degraded)

No. 51: Found biodegradation rate (CO_2 test) = 41%

Total carbon content: C_{19}

$C_6 = 32\%$ (not degraded) $C_5 = 26\%$ (degraded)

$C_8 = 42\%$ (degraded)

No. 53: Found biodegradation rate (CO_2 test) = 67%

Total carbon content: C_{11}

No. 67: Found biodegradation rate (CO_2 test) = 74%

Total carbon content: C_{13}

$(CF_3SO_2)_2N^-$

No. 69: Found biodegradation rate (CO_2 test) = ~65 %

Scheme 10.3 Misleading interpretation of biodegradation data.

Even more misguiding is the "ready" classification of substance No. 69 (Table 10.5). This ionic liquid is formed by a readily biodegradable cation and the non-biodegradable $[NTf_2]^-$ anion. In this case, the C content of the cation is overbalanced and the C content of the anion does not contribute significantly to the overall C content of the ionic liquid structural formula (Scheme 10.3). Thus, the measured CO_2 evolution is more or less exclusively related to the cation. The $[NTf_2]^-$-containing ionic liquid No. 69 is thus classified as readily biodegradable, which is a fundamental mistake, because this anion is non-biodegradable and potentially persistent.

In consequence, for the structural design of biodegradable ionic liquids, it has to be considered that metabolizable cations are combined with a biotically or abiotically degradable anion.

The careless interpretation of just biodegradation rates and the misconstruing of pass levels can result in wrong conclusions with respect to the design of non-persistent ionic liquids. Ideally, ionic liquids (and chemicals in general) should not be classified in this arbitrary manner. The aim of biodegradation experiments should not be to pass certain threshold values, but to answer the question of whether a chemical or the combination of chemical entities (such as ionic liquids) and their first transformation products undergo a further and rapid metabolization ending in a complete biodegradation or not.

10.3.5
Metabolic Pathways of Ionic Liquid Cations

There are two reasons to investigate the nature of biological transformation products of ionic liquids substructures: First, transformation pathways have to be known to avoid wrong classifications of ionic liquids, if a pass level is fulfilled, but degradation products are recalcitrant towards biodegradation (see Section 10.3.4). Second, the knowledge of the chemical structure of metabolites formed is crucial with respect to the hazard assessment of these chemicals. Metabolites have their own characteristic ecotoxicological profiles and it may well happen that they are more toxic than their parent compound. Therefore, the classical biodegradation protocols have to be linked to biodegradation studies.

Based on T-SAR, Jastorff, et al. [9, 14] developed a theoretical prediction algorithm to presume metabolites of the $[C_4mim]^+$ and the $[C_8mim]^+$ compounds. Several hydroxylated, carboxylated and dealkylated compounds were proposed and subsequently synthesized to assess their toxicity [15]. In a subsequent study, the same group investigated the metabolic fate of $[C_8mim]^+$ via HPLC–MS analysis in a biodegradation experiment [67]. The degradation products which were identified were found to be in accordance with the presumed structures. The metabolites detected and the proposed breakdown pathway of the $[C_8mim]^+$ compound are shown in Scheme 10.4. In general, the transformation of the alkyl chain seems to start with the oxidation of the terminal methyl group (ω-oxidation) catalyzed probably by monooxygenases (for instance by the cytochrome P450 system). The remaining alcohol is subsequently oxidized by dehydrogenases via aldehydes to carboxylic acids. The resulting carboxylic acids can now undergo β-oxidation and the two carbon fragments released can enter the tricarboxylic acid cycle via acetyl Co-A. The HPLC–MS results also indicate the formation of different non-terminal hydroxyl groups. These secondary alcohol isomers cannot be further degraded via β-oxidation. Their transformation process ended either with the formation of ketones or with additional hydroxylation steps [67].

In general, the transformation products of $[C_8mim]^+$ identified have shorter side chains and/or are functionalized.

In the previous sections it was shown that compounds with short and functionalized side chains exhibit lower toxicities towards, *inter alia*, mammalian cells, marine bacteria, limnic green algae and duckweed. Therefore, a lower hazard potential of the transformation products is proposed as compared with [C₈mim] cation, which has a high aquatic toxicity (especially towards algae). Nevertheless, some restrictions have to be made because the aldehydes, which are intermediates in the oxidation pathway from the –CH$_2$OH group to the –COOH group, have not yet been analyzed regarding their (eco) toxicity and, theoretically, the formation of highly reactive epoxides is also possible.

Docherty et al. [88] used ^{1}H and ^{13}C NMR spectra to analyze initial and final biodegradation samples from different imidazolium ionic liquids. For $[C_6mim]^+$ and $[C_8mim]^+$, a change in the chemical structures was determined indicating a loss of four or five terminal C-atoms from the side chain, while the ring structure remained intact for both compounds and was not used as a carbon source from the microorganisms. In contrast, complete mineralization (including the core) for

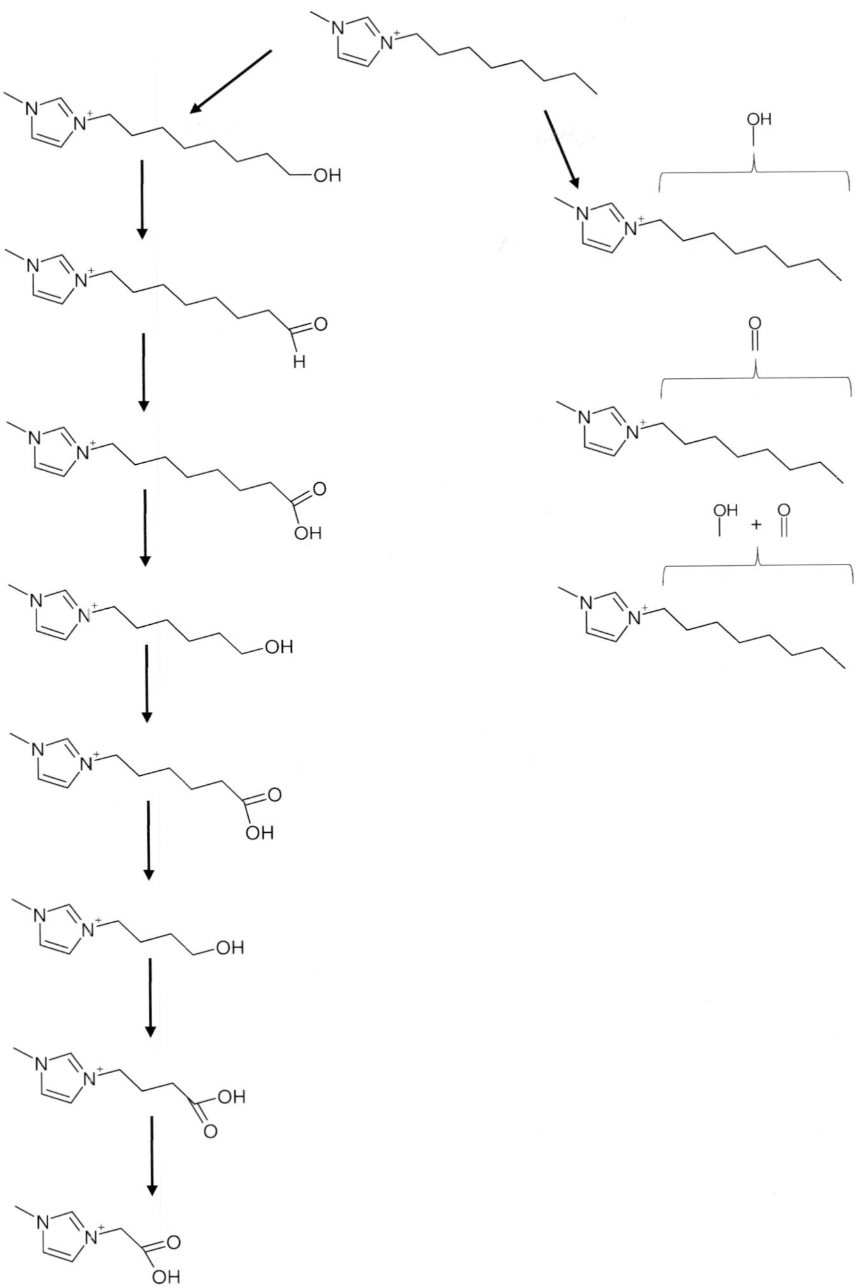

Scheme 10.4 Biodegradation products of [C₈mim]⁺.

the 3-methylpyridinium derivatives with hexyl and octyl side chains was found during the incubation period of 25 and 49 days, respectively [88].

10.3.6
Abiotic Degradation

To describe the environmental fate and the environmental lifetime of ionic liquid structures, abiotic transformation processes also have to be considered. Abiotic mechanisms such as oxidation, reduction, hydrolysis and photolysis represent important pathways for the degradation of synthetic chemicals in the environment. The products formed by such abiotic conversions may be biodegraded further by microorganisms or are found naturally in the environment. For instance, the hydrolysis of $[BF_4]^-$ and $[PF_6]^-$ [96, 97] results in the release of HF. This acutely toxic [98] and corrosive compound is problematic with respect to operational safety and its application in technical processes. Regarding ecotoxicity and bioaccumulation, it is of minor concern because the evolving compounds (fluoride, boric acid and phosphoric acid) are not harmful to the environment in moderate concentrations. The hydrolysis data from the ionic liquid anions trifluorotris(pentafluoroethyl) phosphate and $[NTf_2]^-$ suggest that these anions are resistant to hydrolytic cleavage under environmental conditions. Generally, very limited data sets on the abiotic transformation of cations and anions of ionic liquids are available so far. Photolysis can be the initial transformation step for non-biodegradable cations and anions, which then may lead to biodegradable transformation products.

10.3.7
Outlook

To achieve a valid data pool for the environmental hazard assessment of ionic liquids, the very limited degradation data need to be complemented by data from abiotic degradation studies and by data from both aerobic and anaerobic biodegradation studies. Especially systematic investigations to examine the biodegradability of different head groups are required. Additionally, more detailed knowledge of biodegradation pathways, kinetics and the metabolites of ionic liquids are desirable and would improve the hazard assessment of these chemicals.

To meet the requirements of green chemistry, this should occur before the widespread use and possible release of ionic liquid structures into the environment take place.

10.4
Conclusion

It was the aim of this chapter to review, analyze and discuss the ecotoxicological, toxicological and biodegradation data for ionic liquids available with respect to the following lead questions:

- Is the (eco)toxicological potential of ionic liquids reducible by a directed design of substructural elements?

- What is the impact of each substructural element – head group, side chain and anion – on the (eco)toxicity of the corresponding ionic liquid and, hence, which are the most promising ones to modulate ionic liquid toxicity?
- Which structural factors influence the biodegradability of ionic liquids and how must the substructural elements be designed to be biodegradable?
- Which goal conflicts between technological needs and safety aspects arise in the fields of (eco)toxicity and biodegradability of ionic liquids?
- How can these conflicts be solved or at least be addressed to achieve the goal of inherently safer ionic liquids?

The following subsections provide a concluding summary for the two main issues (eco)toxicity and biodegradation of ionic liquids. The leading questions presented above will be answered and discussed and it will be pointed out where future efforts should be made in research and development for more inherently safe and sustainable ionic liquids.

10.4.1
Toxicity and (Eco)toxicity of Ionic Liquids

Regarding structure–activity aspects in analyzing the (eco)toxicological data from different test systems with varying levels of biological complexity, the following general conclusions can be drawn.

The (eco)toxicitiy of ionic liquids seems to be predominantly determined by the side chains connected to the cationic head group. In several studies it was demonstrated that the (eco)toxicity of an ionic liquid can be modulated over several orders of magnitude by altering the hydrophobicity of the side chain. In accordance with this finding, it was observed that the (eco)toxicity in all test systems investigated so far could be decreased either by reducing the hydrophobicity of the side chain by shortening long alkyl chains ($<C_4$) or by introducing polar functional groups (for example, ether, hydroxyl or nitrile functions) into them.

In contrast to the strong side chain effect, the chemical structure of the cationic head group turned out to be of minor relevance for the (eco)toxicity. Most of the head group structures investigated were found to modulate only slightly the (eco)toxic effects observed. The most prominent exceptions are the N,N-(dimethylamino) pyridinium and quinolinium head groups, for which in the majority of the test systems strong (eco)toxicological effects were observed for the corresponding cations – even when these head groups were linked to short and polar side chains. In contrast, the morpholinium head group was consistently shown to be the least toxicologically active head group in all test systems and hence provides a promising potential for the design of inherently safer ionic liquids – especially when combined with short and polar side chains.

Regarding the anions, it was found that most of the structures tested exhibit no significant (eco)toxicological effects in nearly all test systems investigated so far. Exceptions here are hydrophobic and mostly fluorinated species such as $[(C_2F_5)_3PF_3]^-$ and $[NTf_2]^-$. For these anionic species, strong (eco)toxicological effects were observed in some test systems. These effects were not exclusively linked to the hydrophobicity of the anionic structures, pointing – in addition to unspecific baseline toxicity or narcosis – to some specific modes of toxic action.

Since the anions and in particular the hydrophobic highly fluorinated species represent the most promising candidates for tuning the desired technological properties of an ionic liquid, strong efforts should be made to elucidate their modes of toxic action. Regarding this issue, more biocompatible anions are discussed on the basis of sulfates, sulfonates, phosphates and salts from organic acids (e.g. acetate, lactate and benzoate).

Summing up, one can state that the (eco)toxicity of ionic liquids can be modulated by simple modifications at the side chains connected to the cationic head group and by the choice of a non-toxic anionic moiety. The N,N-(dimethylamino)pyridinium and quinolinium head groups should be avoided.

Regarding the mode of toxic action, the correlation between hydrophobicity and the effects provoked suggests that baseline toxicity or narcosis is most likely the predominant mode of toxic action of ionic liquids. However, the exceptions found for some anionic species and especially the so far unknown long-term effects of ionic liquids in complex biological systems (for example mesocosms) need to be investigated.

Additionally, further studies are needed to analyze the influence of key environmental variables (e.g. organic matter, salinity and pH) on the bioavailability of ionic liquids. In general, the bioavailability of xenobiotics depends strongly on these factors, but regular standard testing protocols for ecotoxicity testing do not consider this. The first studies in this field showed that there is partially a strong influence of organic matter, clay or salinity on the toxicity of selected ionic liquids with sometimes unexpected results (for example ionic liquids with the $[NTf_2]^-$ anion become more bioavailable in soils with increased amounts of smectite). These unexpected results underline the need to consider more the key environmental variables in future testing series aimed at contributing to a prospective hazard assessment.

In addition to the abiotic factors, which influence the bioavailability of ionic liquids, also biotic factors (for example community structures, predator-prey relationships and the behavior of organisms) should be taken into account. On the one hand, these parameters can be much more sensitive than, for example, mortality but on the other hand, the results are much more difficult to interpret.

Ionic liquids with short alkyl side chains proved to be less toxic but much more mobile in soils (low sorption potential) because there is less interaction potential between the matrix and the substance. Therefore, they could become a problem, because they show persistency potential and, due to their high mobility, they could, for example, reach the ground water and enrich there.

Additionally, data on possible carcinogenic, mutagenic or reproduction toxic effects are very limited at the present stage. Only one study has investigated the mutagenic potential of ionic liquids using the Ames test, and another study analyzed the influence of $[C_8mim]Br$ on the development of frog embryos. Therefore, further studies are needed to investigate the structurally highly diverse substance class of ionic liquids for carcinogenic, mutagenic or reproduction toxic effects in more detail.

Also, an analysis for the endocrine disruption potential of ionic liquids seems to be essential, since many industrial chemicals proved to possess this mode of toxic action at a later date (e.g. nonylphenol and bisphenol A).

Furthermore, there is a need to refine the existing models to predict mixture effects of cations and anions of ionic liquids, complex mixtures of two or more ionic liquids and of ionic liquids with environmentally relevant contaminants. Such prediction models represent a crucial tool in the design process and in the ecological hazard assessment of new, inherently safer ionic liquids, since ionic liquids will always be present in the environment in complex composites with other contaminants.

10.4.2
Biodegradability of Ionic Liquids

In general, from screening the literature for biodegradation data, little is known about ionic liquids in this field. However, based on the limited available data, it can be concluded that the design of "readily" or "ultimately" biodegradable ionic liquid cations and anions seems to be possible. With respect to the cation, the pyridinium core exhibits in general a higher degree of biodegradation than the imidazolium derivatives. An increased biodegradability with elongated alkyl side chains ($\geq C_6$) for imidazolium, 4-(dimethylamino)pyridinium and pyridinium cations was found.

10.4.3
The Goal Conflict in Designing Sustainable Ionic Liquids

The results published so far on the effects of an elongating side chain at any type of head group on (eco)toxicity on the one hand and biodegradability on the other create a goal conflict within the design process of inherently safer ionic liquids: The shorter the alkyl side chain, the safer the chemical is with respect to (eco)toxicity issues, but the higher is the risk of persistency and mobility owing to the missing biodegradability and the reduced sorption on organic matter and clay minerals.

Regarding regulatory and legislative issues, especially "persistency" is a key parameter that requires a difficult and complex authorization process under REACH for any substance labeled with it. In consequence, persistent ionic liquids bear the risk of failing during the authorization process or their use will be strongly restricted to only a few applications. Thus, structural design towards biodegradability is an essential prerequisite aiming at the industrial use of ionic liquids.

It was shown that a promising design criterion to achieve biodegradability – and thereby reducing the above-mentioned goal conflict – is the introduction of ester functions into the side chains. Additionally, the use of the pyridinium head group can further improve the biodegradability of the cation. However, especially for ionic liquids which are represented by cationic and anionic structures, it has to be considered that really biodegradable chemical structures have to be designed. This includes not only the biodegradability of the side chains but also of the core structure of the head group. Then, this moiety has to be combined with an (a)biotically degradable anion. Here, different anions such as alkylsulfates (methylsulfate or octylsulfate), linear alkylsulfonates (e.g. methylsulfonate), linear alkyl benzenesulfonates (e.g. *p*-toluenesulfonate) and salts of organic acids (e.g. acetate or lactate) are recommendable with respect to their biodegradability and also from an (eco) toxicological point of view. Typical fluorine-containing ionic liquid anions such as $[(CF_3SO_2)_2N]^-$, $[(C_2F_5)_2PO_2]^-$, $[(C_2F_5)_3PF_3]^-$, $[(C_3F_7)_3PF_3]^-$ and $[(CF_3SO_2)_3C]^-$ should be avoided when designing inherently safer ionic liquids, because these anions come along with a high risk of being persistent in the environment and – based on their high hydrophobicity – of accumulating in tissues of living organisms.

However, for a distinct technological application, it may be necessary to use a non-biodegradable and/or (eco)toxic ionic liquid. The authors would like to point out that also these ionic liquids can be part of sustainable products and processes, but then a comprehensive risk management is needed with respect to operational safety of employees. Furthermore, it has to be ensured that the contamination of the environment is as low as possible. Thus, process strategies are demanded to remove these compounds, for instance from processing effluents via regenerative methods (e.g. membrane filtration) or via advanced destructive oxidation techniques (e.g. electrochemical treatment or UV degradation).

10.4.4
Final Remarks

Not only must sustainable chemistry and its products be non-toxic, readily biodegradable and leave a small ecological footprint, but at the same time sustainable chemicals have to meet their technological features, they must be efficient in a certain process and their production and use must be economically favorable. We have shown that for the case of ionic liquids the major goal conflicts – first between technological needs and (eco)toxicological parameters and second between high biodegradability and low toxicity – on the way towards real sustainable ionic liquids can be addressed and partially overcome by a directed structural design.

Since in most cases the acute toxicity of ionic liquids is low or moderate, with respect to sustainability strong efforts should be made in the future to avoid persistent structural elements in ionic liquids. This means that the head group, side chain and the anion moiety must be intrinsically biodegradable in the environment to avoid the bottleneck of "persistency" in the REACH registration process. Nevertheless, regarding toxicity issues, the wide gap in knowledge in the field of chronic and long-term effects of ionic liquids on humans and the environment needs to be addressed in future research projects.

Finally, it should be stated that even ionic liquids showing a bad performance with respect to safety issues can be sustainable, if their technological, economic and/or socio-economic benefits compensate for the safety drawbacks and if these ionic liquids are used in well-controlled and closed processes. In this context, the chemical industry established the initiative "Responsible Care" in 1984. Among other sustainability-related issues, "Responsible Care" means that for chemicals bearing an intrinsic hazard potential for humans and the environment, the highest technological mode of protection should be applied in industrial processes and products.

However, for processes and applications entailing an uncontrolled or an intended widespread release to humans and the environment, intrinsic safety aspects should be of the highest priority, since "Responsible Care" will not be possible in each case.

In the end, the economic aspects and among them especially the production costs will decide whether ionic liquids will be used for sustainable chemical processes and products in the future – from a structural point of view, they have great potential to contribute to the development of more sustainable products and processes.

References

1 Anastas, P.T. and Warner, J.C. (1998) *Green Chemistry: Theory and Practice*, Oxford University Press, New York.

2 Alaee, M. and Wenning, R.J. (2002) *Chemosphere*, **46**, 579–582.

3 Beach, S.A., Newsted, J.L., Coady, K. and Giesy, J.P. (2006) *Reviews of Environmental Contamination & Toxicology*, **186**, 133–174.

4 Correa-Reyes, G., Viana, M.T., Marquez-Rocha, F.J., Licea, A.F., Ponce, E. and Vazquez-Duhalt, R. (2007) *Chemosphere*, **68**, 662–670.

5 Hekster, F.M., Laane, R.W.P.M. and de Voogt, P. (2003) *Reviews of Environmental Contamination & Toxicology*, **179**, 99–121.

6 Lye, C.M., Bentley, M.G. and Galloway, T. (2008) *Environmental Toxicology*, **23**, 309–318.

7 Preuss, T.G. and Ratte, H.T. (2007) *Umweltschutz Schadst.-Forsch.*, **19**, 227–233.

8 Strack, S., Detzel, T., Wahl, M., Kuch, B. and Krug, H.F. (2007) *Chemosphere*, **67**, 405–411.

9 Jastorff, B., Störmann, R. and Wölke, U. (2004) *Struktur-Wirkungs-Denken in der Chemie*, Universitätsverlag Aschenbeck und Isensee, Bremen, Oldenburg.

10 Jastorff, B., Störmann, R. and Ranke, J., (2007) *CLEAN*, **35**, 399–405.

11 Arning, J., Stolte, S., Böschen, A., Stock, F., Pitner, W.R., Welz-Biermann, U., Jastorff, B. and Ranke, J. (2008) *Green Chemistry*, **10**, 47–58.

12 Matzke, M., Stolte, S., Thiele, K., Juffernholz, T., Ranke, J., Welz-Biermann, U. and Jastorff, B. (2007) *Green Chemistry*, **9**, 1198–1207.

13 Stolte, S., Matzke, M., Arning, J., Böschen, A., Pitner, W.R., Welz-Biermann, U., Jastorff, B. and Ranke, J. (2007) *Green Chemistry*, **9**, 1170–1179.

14 Jastorff, B., Störmann, R., Ranke, J., Molter, K., Stock, F., Oberheitmann, B., Hoffmann, W., Hoffmann, J., Nüchter, M., Ondruschka, B. and Filser, J. (2003) *Green Chemistry*, **5**, 136–142.

15 Jastorff, B., Mölter, K., Behrend, P., Bottin-Weber, U., Filser, J., Heimers, A., Ondruschka, B., Ranke, J., Schaefer, M., Schröder, H., Stark, A., Stepnowski, P., Stock, F., Störmann, R., Stolte, S., Welz-Biermann, U., Ziegert, S. and Thöming, J. (2005) *Green Chemistry*, **7**, 362–372.

16 Balczewski, P., Bachowska, B., Bialas, T., Biczak, R., Wieczorek, W.M. and Balinska, A. (2007) *Journal of Agricultural and Food Chemistry*, **55**, 1881–1892.

17 Bernot, R.J., Brueseke, M.A., Evans-White, M.A. and Lamberti, G.A. (2005) *Environmental Toxicology and Chemistry/SETAC*, **24**, 87–92.

18 Bernot, R.J., Kennedy, E.E. and Lamberti, G.A. (2005) *Environmental Toxicology and Chemistry/SETAC*, **24**, 1759–1765.

19 Cho, C.W., Pham, T.P.T., Jeon, Y.C., Vijayaraghavan, K., Choe, W.S. and Yun, Y.S. (2007) *Chemosphere*, **69**, 1003–1007.

20 Cho, C.W., Pham, T.P.T., Jeon, Y.C., Min, J.H., Jung, H.Y., Lee, D.S. and Yun, Y.S. (2008) *Ecotoxicology*, **17**, 455–463.

21 Cho, C.W., Jeon, Y.C., Pham, T.P.T., Vijayaraghavan, K. and Yun, Y.S. (2008) *Ecotoxicology and Environmental Safety*, **71**, 166–171.

22 Cho, C.W., Pham, T.P.T., Jeon, Y.C. and Yun, Y.S. (2008) *Green Chemistry*, **10**, 67–72.

23 Cieniecka-Roslonkiewicz, A., Pernak, J., Kubis-Feder, J., Ramani, A., Robertson, A.J. and Seddon, K.R. (2005) *Green Chemistry*, **7**, 855–862.

24 Cornmell, R.J., Winder, C.L., Tiddy, G.J.T., Goodacre, R. and Stephens, G. (2008) *Green Chemistry*, **10**, 836–841.

25 Couling, D.J., Bernot, R.J., Docherty, K.M., Dixon, J.K. and Maginn, E.J. (2006) *Green Chemistry*, **8**, 82–90.

26 Demberelnyamba, D., Kim, K.S., Choi, S.J., Park, S.Y., Lee, H., Kim, C.J. and Yoo, I.D. (2004) *Bioorganic &; Medicinal Chemistry Letters*, **12**, 853–857.

27 Docherty, K.M. and Kulpa, C.F. (2005) *Green Chemistry*, **7**, 185–189.

28 Docherty, K.M., Hebbeler, S. and Kulpa, C.F. (2006) *Green Chemistry*, **8**, 560–567.

29 Frade, R.F.M., Matias, A., Branco, L.C., Afonso, C.A.M. and Duarte, C.M.M. (2007) *Green Chemistry*, **9**, 873–877.

30 Ganske, F. and Bornscheuer, U.T. (2006) *Biotechnology Letters*, **28**, 465–469.

31 Garcia-Lorenzo, A., Tojo, E., Tojo, J., Teijeira, M., Rodriguez-Berrocal, F.J., Gonzalez, M.P. and Martinez-Zorzano, V.S. (2008) *Green Chemistry*, **10**, 508–516.

32 Garcia, M.T., Gathergood, N. and Scammells, P.J. (2005) *Green Chemistry*, **7**, 9–14.

33 Grabinska-Sota, E. and Kalka, J. (2003) *Environment International*, **28**, 687–690.

34 Kulacki, K.J. and Lamberti, G.A. (2008) *Green Chemistry*, **10**, 104–110.

35 Lacrama, A.M., Putz, M.V. and Ostafe, V. (2007) *International Journal of Molecular Science*, **8**, 842–863.

36 Landry, T.D., Brooks, K., Poche, D. and Woolhiser, M. (2005) *Bulletin of Environmental Contamination and Toxicology*, **74**, 559–565.

37 Larson, J.H., Frost, P.C. and Lamberti, G.A. (2008) *Environmental Toxicology and Chemistry/SETAC*, **27**, 676–681.

38 Latala, A., Stepnowski, P., Nedzi, M. and Mrozik, W. (2005) *Aquatic Toxicology*, **73**, 91–98.

39 Lee, S.M., Chang, W.J., Choi, A.R. and Koo, Y.M. (2005) *Korean Journal of Chemical Engineering*, **22**, 687–690.

40 Li, X.Y., Zhou, J., Yu, M., Wang, J.J. and Pei, Y.C. (2009) *Ecotoxicology and Environmental Safety*, **72**, 552–556.

41 Luis, P., Ortiz, I., Aldaco, R. and Irabien, A. (2006) *Ecotoxicology and Environmental Safety*, 1–7.

42 Matsumoto, M., Mochiduki, K. and Kondo, K. (2004) *Journal of Bioscience and Bioengineering*, **98**, 344–347.

43 Matsumoto, M., Mochiduki, K., Fukunishi, K. and Kondo, K. (2004) *15–11–Separation and Purification Technology*, **40**, 97–101.

44 Matzke, M., Stolte, S., Arning, J., Uebers, U. and Filser, J. (2008) *Green Chemistry*, **10**, 584–591.

45 Matzke, M., Stolte, S., Arning, J., Uebers, U. and Filser, J., (2008) *Ecotoxicology*, **12**, 199–204.

46 Matzke, M., Stolte, S., Böschen, A. and Filser, J. (2008) *Green Chemistry*, **10**, 784–792.

47 Pernak, J., Rogoza, J. and Mirska, I. (2001) *European Journal of Medicinal Chemistry*, **36**, 313–320.

48 Pernak, J., Kalewska, J., Ksycinska, H. and Cybulski, J. (2001) *European Journal of Medicinal Chemistry*, **36**, 899–907.

49 Pernak, J., Sobaszkiewicz, K. and Mirska, I. (2003) *Green Chemistry*, **5**, 52–56.

50 Pernak, J., Goc, I. and Mirska, I. (2004) *Green Chemistry*, **6**, 323–329.

51 Pernak, J., Goc, I. and Fojutowski, A. (2005) *Holzforschung*, **59**, 473–475.

52 Pham, T.P.T., Cho, C.W., Min, J. and Yun, Y.S. (2008) *Journal of Bioscience and Bioengineering*, **105**, 425–428.

53 Pretti, C., Chiappe, C., Pieraccini, D., Gregori, M., Abramo, F., Monni, G. and Intorre, L. (2006) *Green Chemistry*, **8**, 238–240.

54 Putz, M.V., Lacrama, A.M. and Ostafe, V., (2007) *Res Ecology Letters*, **18**, 197–203.

55 Ranke, J., Molter, K., Stock, F., Bottin-Weber, U., Poczobutt, J., Hoffmann, J., Ondruschka, B., Filser, J. and Jastorff, B. (2004) *Ecotoxicology and Environmental Safety*, **58**, 396–404.

56 Ranke, J., Cox, M., Müller, A., Schmidt, C. and Beyersmann, D. (2006) *Ecotoxicology and Environmental Chemistry*, **88**, 273–285.

57 Ranke, J., Stolte, S., Störmann, R., Arning, J. and Jastorff, B. (2007) *Chemical Reviews*, **107**, 2183–2208.

58 Ranke, J., Muller, A., Bottin-Weber, U., Stock, F., Stolte, S., Arning, J., Störmann, R. and Jastorff, B. (2007) *Ecotoxicology and Environmental Safety*, **67**, 430–438.

59 Romero, A., Santos, A., Tojo, J. and Rodriguez, A. (2008) *Journal of Hazardous Materials*, **151**, 268–273.

60 Samori, C., Pasteris, A., Galletti, P. and Tagliavini, E. (2007) *Environmental Toxicology and Chemistry/SETAC*, **26**, 2379–2382.

61 Skladanowski, A.C., Stepnowski, P., Kleszczynski, K. and Dmochowska, B. (2005) *Environmental Toxicology and Pharmacology*, **19**, 291–296.

62 Stasiewicz, M., Mulkiewicz, E., Tomczak-Wandzel, R., Kumirska, J., Siedlecka, E.M., Golebiowski, M., Gajdus, J., Czerwicka, M. and Stepnowski, P. (2008) *Ecotoxicology and Environmental Safety*, **71**, 157–165.

63 Stepnowski, P., Skladanowski, A.C., Ludwiczak, A. and Laczynska, E. (2004) *Human & Experimental Toxicology*, **23**, 513–517.

64 Stock, F., Hoffmann, J., Ranke, J., Störmann, R., Ondruschka, B. and Jastorff, B. (2004) *Green Chemistry*, **6**, 286–290.

65 Stolte, S., Arning, J., Bottin-Weber, U., Müller, A., Pitner, W.R., Welz-Biermann, U., Jastorff, B. and Ranke, J. (2007) *Green Chemistry*, **9**, 760–767.

66 Stolte, S., Arning, J., Bottin-Weber, U., Matzke, M., Stock, F., Thiele, K., Uerdingen, M., Welz-Biermann, U., Jastorff, B. and Ranke, J. (2006) *Green Chemistry*, **8**, 621–629.

67 Stolte, S., Abdulkarim, S., Arning, J., Blomeyer-Nienstedt, A.K., Bottin-Weber, U., Matzke, M., Ranke, J., Jastorff, B. and Thöming, J. (2008) *Green Chemistry*, **10**, 214–224.

68 Swatlowski, R.P., Holbrey, J.D., Memon, S.B., Caldwell, G.A., Caldwell, K.A. and Rogers, R.D. 2004 *Chemical Communications*, 668–669.

69 Tee, K.L., Roccatano, D., Stolte, S., Arning, J., Jastorff, B. and Schwaneberg, U. (2008) *Green Chemistry*, **10**, 117–123.

70 Wang, X.F., Ohlin, C.A., Lu, Q.H., Fei, Z.F., Hu, J. and Dyson, P.J. (2007) *Green Chemistry*, **9**, 1191–1197.

71 Wells, A.S. and Coombe, V.T. (2006) *Organic Process Research & Development*, **10**, 794–798.

72 Loewe, S. and Muischnek, H. (1926) *Naunyn-Schmiedebergs Arch. Ex.p Pathol. Pharmakol.*, **114**, 313–326.

73 Loewe, S. (1927) *Klinische Wochenschrift*, **6**, 1077–1085.

74 Bliss, C.I. (1939) *The Annals of Applied Biology*, **26**, 585–615.

75 Antkowiak, B. (2001) *Naturwissenschaften*, **88**, 201–213.

76 Meyer, H.H. (1899) *Archives of Experimental Pathology and Pharmacology*, **42**, 109–118.

77 Overton, C.E. (1901) *Studien über die Narkose zugleich ein Beitrag zur allgemeinen Pharmakologie*, Gustav Fischer Verlag, Stuttgart.

78 Escher, B.I. and Schwarzenbach, R.P. (2002) *Aquatic Sciences*, **64**, 20–35.

79 Wells, A.S. and Coombe, V.T. (2006) *Organic Process Research & Development*, **10**, 794–798.

80 GDCh Advisory Committee on Existing Chemicals of Environmental Relevance (BUA), *DDT and Its Derivatives*, BUA Report 216, Wissenschaftliche Verlagsgesellschaft, Stuttgart, 1998.

81 Carson, R. (1962) *Silent Spring*, Houghton Mifflin, Boston.
82 Pavan, M. and Worth, A.P. (2008) *QSAR Combinatorial Science*, **27**, 32–40.
83 Brown, D., (1995) Biodegradability of surfactants, In: *Introduction to Surfactant Biodegradation*, (eds Karsa D.R. and Porter M.R.,) Blackie Academic and Professional, Glasgow, pp. 1–27.
84 Gathergood, N. and Scammells, P.J. (2002) *Australian Journal of Chemistry*, **55**, 557–560.
85 Gathergood, N., Garcia, M.T. and Scammells, P.J. (2004) *Green Chemistry*, **6**, 166–175.
86 Gathergood, N., Scammells, P.J. and Garcia, M.T. (2006) *Green Chemistry*, **8**, 156–160.
87 Harjani, J.R., Singer, R.D., Garcia, M.T. and Scammells, P.J. (2008) *Green Chemistry*, **10**, 436–438.
88 Docherty, K.M., Dixon, J.K. and Kulpa, C.F. (2007) *Biodegradation*, **18**, 481–493.
89 Painter, H.A. (2008) *The Handbook of Environmental Chemistry, Detergents*, Springer, Berlin.
90 Schleheck, D. and Cook, A.M. (2003) *Archives of Microbiology*, **179**, 191–196.
91 Alexander, M. (1994) *Biodegradation and Bioremediation*, Academic Press, New York.
92 US Geological Survey, *Summary of Research Results on Bacterial Degradation of Trifluoroacetate (TFA)*, US Geological Survey, Reston, VA, 1994.
93 Rorije, E., Germa, F., Philipp, B., Schink, B. and Beimborn, D.B. (2002) *SAR QSAR Environmental Research*, **13**, 199–204.
94 Philipp, B., Hoff, M., Germa, F., Schink, B., Beimborn, D. and Mersch-Sundermann, V. (2007) *Environmental Science & Technology*, **41**, 1390–1398.
95 Van Ginkel, C.G., (1995) Biodegradability of surfactants, In: *Biodegradability of Cationic Surfactants*, (eds Karsa D.R. and Porter M.R.,) Blackie Academic and Professional, Glasgow, pp. 183–203.
96 Baker, G.A. and Baker, S.N. (2005) *Australian Journal of Chemistry*, **58**, 174–177.
97 Villagran, C., Deetlefs, M., Pitner, W.R. and Hardacre, C. (2004) *Analytical Chemistry*, **76**, 2118–2123.
98 Smith, F.A. (1993) *Handbook of Hazardous Materials, Fluoride Toxicity*, Academic Press, San Diego.

11
Eco-efficiency Analysis of an Industrially Implemented Ionic Liquid-based Process – the BASF BASIL Process

Peter Saling, Matthias Maase, and Uwe Vagt

11.1
The Eco-efficiency Analysis Tool

11.1.1
General Aspects

Life without chemicals would be inconceivable, but the potential risks and impacts on the environment associated with chemical production and chemical products are viewed critically. Eco-efficiency analysis considers the economic and life-cycle environmental effects of a product or process, giving these equal weighting.

The instrument provides early recognition and systematic detection of economic and environmental opportunities and risks in existing and future business activities. At BASF, this tool has become an integral part of the decision-making process for new investments.

Additionally, innovations can be assessed before realizing them, answering the question of where the new process can be ranked in comparison with the existing process and to process alternatives due to the basis of their sustainability. This information can be used in the decision-making process to reduce the risk of greater investment activities.

Proving early recognition and systematic detection of economic and environmental opportunities for production processes in the chemical industry has been used at BASF since 1996. This powerful eco-efficiency analysis allows for a feasibility evaluation of existing and future business activities. In many cases, decision makers are able to choose among alternative processes for making a product, which means strong support for R&D activities.

11.2
The Methodological Approach

Eco-efficiency has been variously defined and analytically implemented by several workers. In most cases, eco-efficiency is taken to mean the ecological optimization of overall systems while not disregarding economic factors [1]. Eco-efficiency should increase the positive ecological performance of the company in relation to economic value creation – or should reduce negative effects [2].

11.2.1
Introduction

Eco-efficiency analysis compares the economic and environmental pros and cons of each alternative over the whole life-cycle. Thus, eco-efficient solutions are those which provide a specific customer benefit more effectively than others from the financial and environmental point of view. Over 400 eco-efficiency analyses have been conducted at BASF and their results have been used to support strategic decision making, marketing, research and development, and communication with external parties. Eco-efficiency analysis, as one important strategy and success factor in sustainable development, will continue to be a very strong operational tool at BASF.

The major elements of the environmental assessment include energy consumption, resource consumption, emissions to all media, toxicity potential, risk potential and land use. The relevance of each environmental category and also of economic *versus* environmental impacts is evaluated using national emissions and economic data. The eco-efficiency analysis method of BASF is briefly presented and show results for the acid quench of HCl with BASIL in comparison to a chemical process using amines as an auxiliary for acid quench are summarized.

The acronym BASIL™ stands for Biphasic Acid Scavenging utilizing Ionic Liquids. BASF was the first company to transfer ionic liquids from laboratory to commercial dimensions, and the Basil process became the first large-scale industrial process worldwide that uses ionic liquids [3].

The BASIL process is applied for the synthesis of alkoxydiphenylphosphines and dialkoxyphenylphosphines, which are important raw materials in the production of photo initiators to cure coatings and printing inks by exposure to UV light. HCl is formed during the reaction of dichlorophenylphosphine with ethanol (Scheme 11.1).

Scheme 11.1 Synthesis of dialkoxyphenylphosphines.

Acid scavenging with a tertiary amine results in a thick, non-stirrable slurry (Figure 11.1). This problem significantly lowers the yield and capacity of the process. In order to provide a minimum of mixing and heat transfer of the exothermic reaction, a solvent usually has to be added in the conventional process.

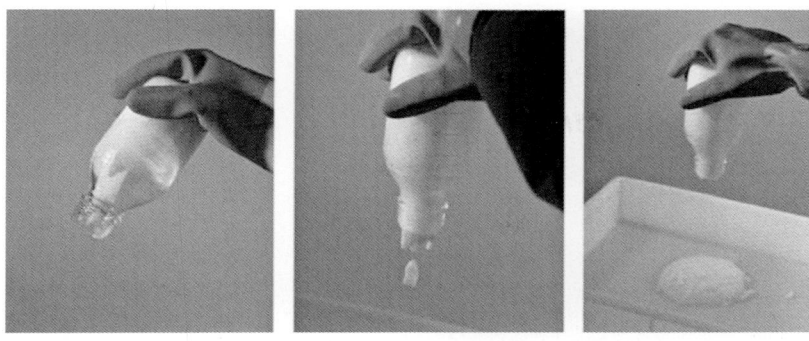

Figure 11.1 Slurry formed when a tertiary amine is used as an acid scavenger (© BASF SE 2003).

In contrast to the conventional technology, the BASF BASIL technology uses an ionic liquid precursor, namely 1-methylimidazole, as an acid scavenger. After the reaction with HCl, an ionic liquid is formed: 1-H–3-methylimidazolium chloride ([H-mim]Cl), which has a melting point of about 75 °C (Scheme 11.2).

Scheme 11.2 Formation of 1-H–3-methylimidazolium chloride ionic liquid in the BASIL process.

During the reaction, two clear liquid phases occur that can easily be separated. The upper phase is the pure product – no solvent is needed any longer – and the lower phase is the pure ionic liquid. In addition to scavenging the acid, 1-methylimidazole also helps in setting the acid free and thereby acts as a nucleophilic catalyst. BASF discovered that the phosphorylation reaction is complete in less than 1 s. Having eliminated the formation of solids and having increased the reaction rate, new reactor concepts were possible. BASF was then able to perform the same reaction in a small jet reactor that has the size of a thumb (Figure 11.2) rather than in a large vessel. In doing so, the productivity of the process was increased by a factor of 8×10^4 to 690 000 kg m^{-3} h^{-1}. At the end of 2004, BASF successfully started a dedicated BASIL plant using this jet stream reactor technology.

BASF has applied its eco-efficiency analysis to the BASIL technology. The eco-efficiency analysis results are presented as aggregated information on costs and environmental impact and show the strengths and weaknesses of a particular product or process. This method uses the main ideas and regulations of the ISO rules for the basic life-cycle analysis (LCA). The ecological calculations of the single results in each category follow the ISO rules 14040 and 14044 in the main points. The quantitative weighting step to obtain the ecological fingerprint and the portfolio are not covered within the ISO rules. Eco-efficiency analysis has more features than are mentioned in the ISO rules.

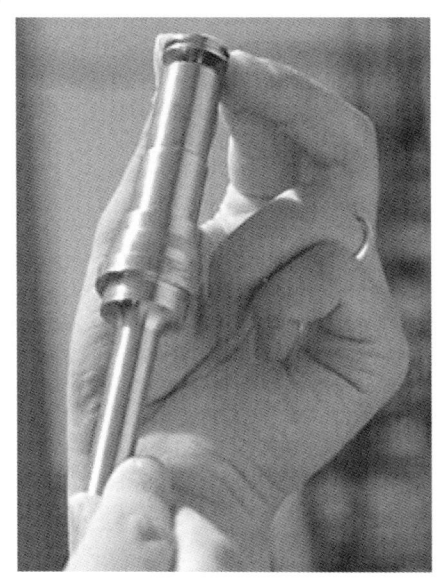

Figure 11.2 Today the BASIL process is run in a small jet reactor, which has a space–time yield of 690 000 kg m^{-3} h^{-1} (© BASF SE 2002).

11.2.2
What is Eco-efficiency Analysis?

The purpose of eco-efficiency analysis is to harmonize economy and ecology. Eco-efficiency analysis is applied in order to use as few materials and as little energy as possible in producing our products and to keep emissions as low as possible. At the same time, sustainable products can help to conserve resources.

The main outline of the BASF eco-efficiency analysis method is provided next; a more detailed discussion is available in literature [4].

Every eco-efficiency analysis passes through several key stages. This ensures consistent quality and the comparability of different studies. Environmental impacts are determined by the LCA method and economic data are calculated using the usual business or, in some instances, national economical models.

The basic preconditions in eco-efficiency analysis are as follows:

- Products or processes studied have to meet the same defined customer benefit.
- The entire life-cycle is considered.
- Both an environmental and an economic assessment are carried out.

The eco-efficiency analysis is worked out by following specific and defined ways of calculations:

- Calculation of total cost from the customer viewpoint.
- Preparation of a specific life-cycle analysis for all investigated products or processes according to the rules of ISO 14040 and 14044.

- Determination of impacts on the health, safety and risks to people, assessing use of area over the whole life-cycle.
- Calculation of relevance and calculation factors for specific weighting.
- Weighting of life-cycle analysis factors with societal factors.
- Determination of relative importance of ecology *versus* economy.
- Creation of an eco-efficiency portfolio.
- Analyses of weaknesses, scenarios, sensitivities and business options.

The specific customer benefit or functional unit always lies at the center of eco-efficiency analysis. In the majority of cases, customers having particular needs and requirements are able to choose between a number of alternative products and processes. In the context of this choice, eco-efficiency analysis compares the economic and environmental data of each solution over the entire life-cycle or within the compartments in which the systems differ in life-cycle.

In this case study for BASIL, the calculation of total cost from the producer's viewpoint was performed. The dimension of the costs, as a part of sustainability, is given equal importance to the environmental dimension in the BASF eco-efficiency method. The relation of both in the calculation is considered only by relevance factors.

11.2.3
Preparation of a Specific Life-cycle Analysis for All Investigated Products and Processes

Once the viable alternatives for a product or process have been identified, data are collected over the life-cycle and impacts in the following environmental categories are determined:

- resource consumption
- energy consumption
- emissions
- risk potential
- toxicity potential
- land use.

11.3
The Design of the Eco-efficiency Study of BASIL

The eco-efficiency study described here for different options for an acid quench in two different chemical reactions compares the use of the ionic liquid precursor 1-methylimidazole (BASIL alternative) with the use of defined amines (AMINE alternative). The target products are ethoxydiphenylphosphine (EDPP) and diethoxyphenylphosphine (DEPP). As described before, the customer benefit is in every case "Providing acid quench chemicals for neutralization of acids in the production of 1000 kg of phenylphosphine products" as described in Figure 11.3. The whole calculation and the impact graphs that follow represent "differences" between the

11 Eco-efficiency Analysis of an Industrially Implemented Ionic Liquid-based Process

```
Customer benefit          Alternative in focus        Alternatives

Providing acid            • BASIL EDPP                • AMINE
quench                                                  EDPP
chemicals for             • BASIL DEPP
neutralization of                                     • AMINE
acids in the                                            DEPP
production of
1000 kg of phenyl-
phosphine
products
```

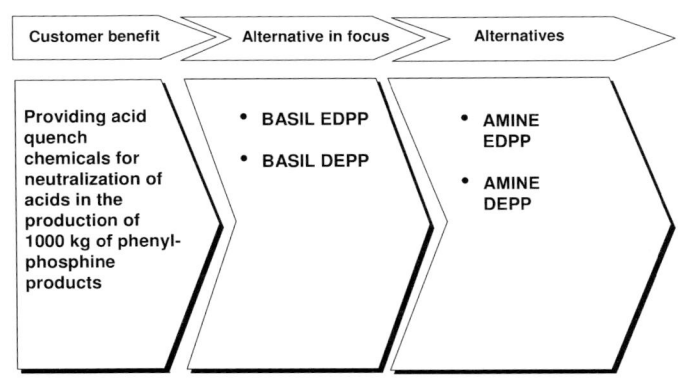

Figure 11.3 Definition of customer or customer benefit and alternatives.

AMINE and BASIL alternatives. The impacts are not "total" impacts. The advantage of assessing only differences between the processes reduces the amount of identical data of the alternatives causing a data ballast that does not help to differentiate between the alternatives. It is a principle of the eco-efficiency analysis to point out significant differences between the chosen alternatives.

In this assessment, the system boundaries show the different steps in the whole life-cycle of the alternative systems compared. As an example, the system boundaries of the EDPP/BASIL process alternative are shown in Figure 11.4. The use of the final product was not considered, so overall it is a cradle-to-gate result for DEPP and EDPP production. The production of 1000 kg of the amines is defined as customer benefit (CB).

11.4
Selected Single Results

11.4.1
Energy Consumption

The energy consumption category of impact includes all energies to fulfill the customer benefit. Fossil energy resources are included before production, as is renewable energy before harvest or use. All impacts leading to a use of energy are considered and are summarized at different important steps of the life-cycle-based view on the processes. Figure 11.5 shows the differences in energy use in the four alternatives assessed. It shows clearly the advantages of the BASIL processes.

The most important impact on the energy consumption are the contributions of the reactants, followed by the acid quench step and the application of additives. A reduction in the use of reactants by increasing the yield would optimize the AMINE processes. Using less additives and 1-methylimidazole will increase the position of the BASIL processes. Due to the calculation of differences in the environment, impacts for steam and electricity are not shown for the BASIL alternatives.

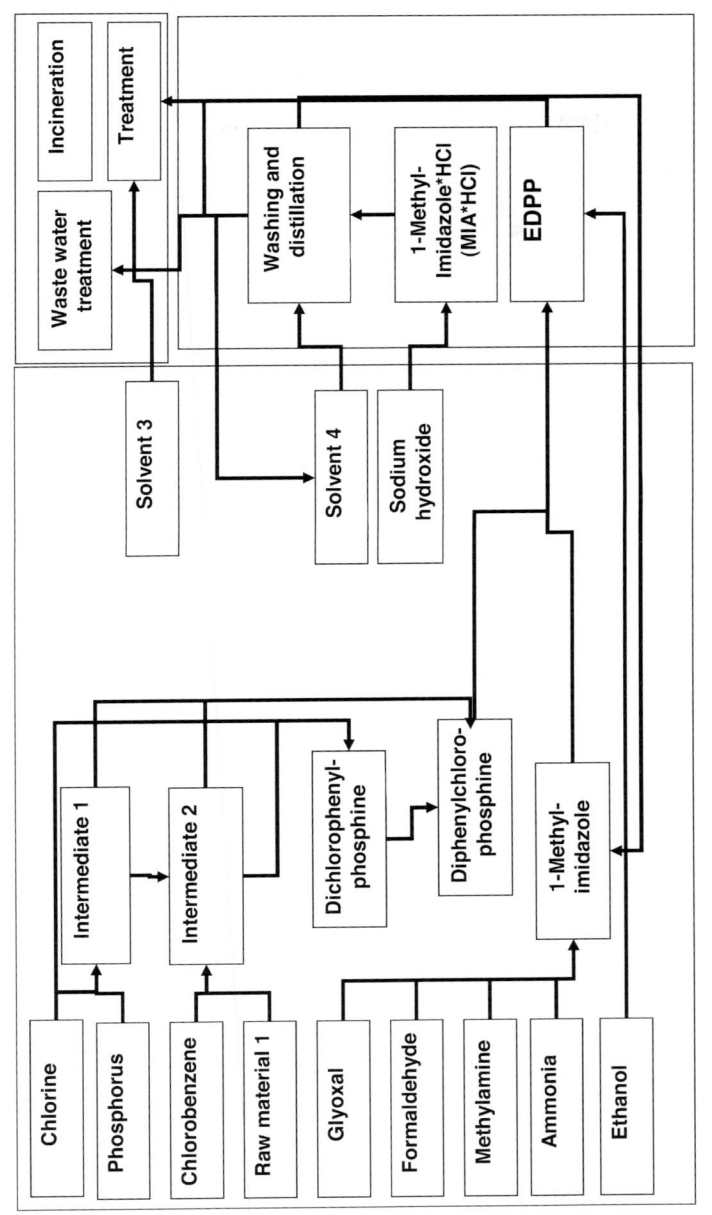

Figure 11.4 System boundaries for the EDPP/BASIL alternative.

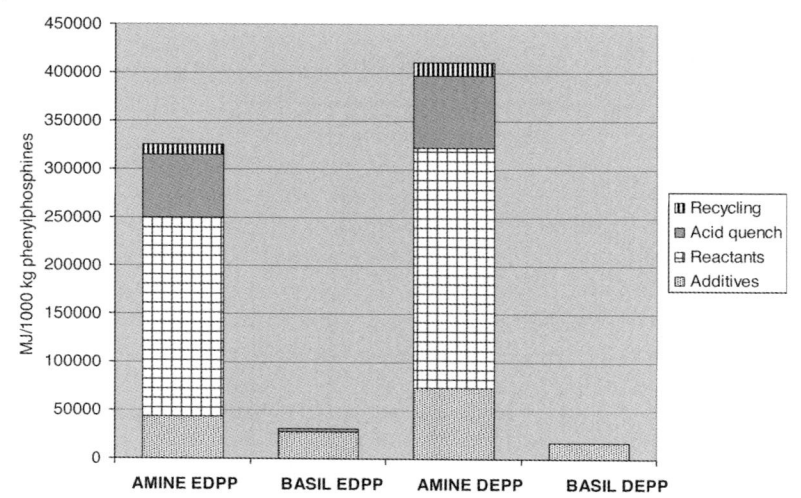

Figure 11.5 Comparison of energy consumption for acid scavenging per customer benefit (CB).

11.4.2
Global Warming Potential (GWP)

Air emissions of different gases are recorded separately and added up to a global warming potential (GWP) over the whole life-cycle for all input materials (Figure 11.6). In most processes, the emission of carbon dioxide is the largest air emission. This emission is typically followed (in terms of quantity) by emissions of N_2O and halogenated hydrocarbons.

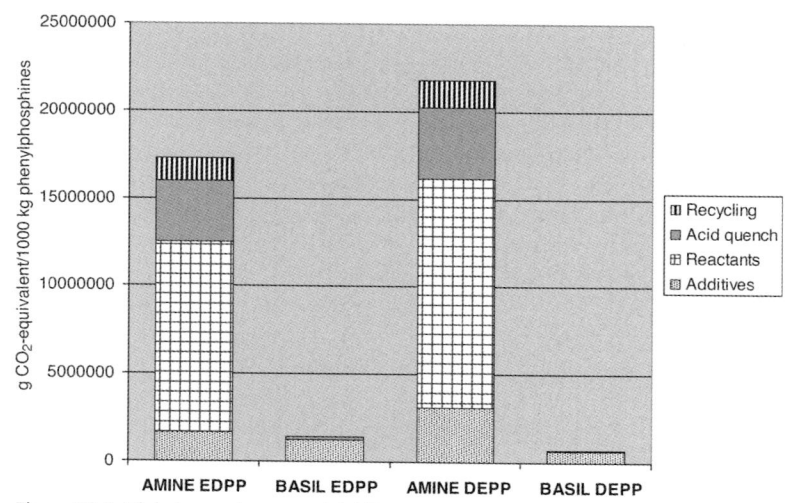

Figure 11.6 Global warming potential (GWP).

The high GWP is linked mainly to the amount of electricity that is used in the pre-chain of the production of the reactants and in the use of primary energy for the production of reactants and the recycling steps.

11.4.3
Water Emissions

The assessment of water pollution is carried out by means of the "critical volume" model. For selected pollutants that enter the water, the theoretical water volume affected by the emission up to the statutory limit value (critical load) is determined. In these factors, eco-toxic effects are also considered to a certain extent. The volumes calculated for each pollutant are added up to yield the "critical volume".

Due to the unavoidable emissions of 1-methylimidazole to the wastewater during the quench process and the worse biodegradation of this compound, the AMINE processes have advantages in this category of environmental assessment (see Figure 11.7). Further reduction of the losses will increase the position of BASIL in the future. This shows that the eco-efficiency analysis helps to find the weaknesses of an alternative and contributes to the further improvement of chemical processes.

11.4.4
The Ecological Fingerprint

Consolidation of the results of the environmental assessment into one overall environmental impact leads to the creation of the environmental fingerprint. This is carried out after normalization of single results via a series of different weighting factors. The fingerprint gives an overview and a summary of the different results of the assessment of the ecological factors (Figures 11.8 and 11.9). In this study, it is shown that only the

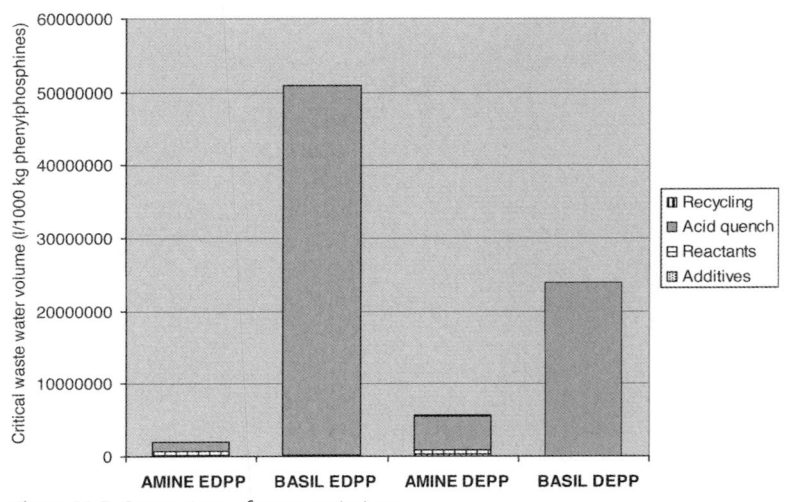

Figure 11.7 Comparison of water emissions.

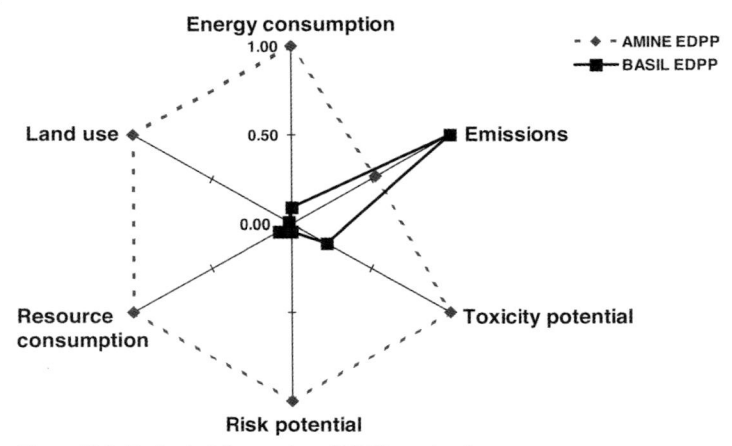

Figure 11.8 Ecological fingerprint of EDPP production.

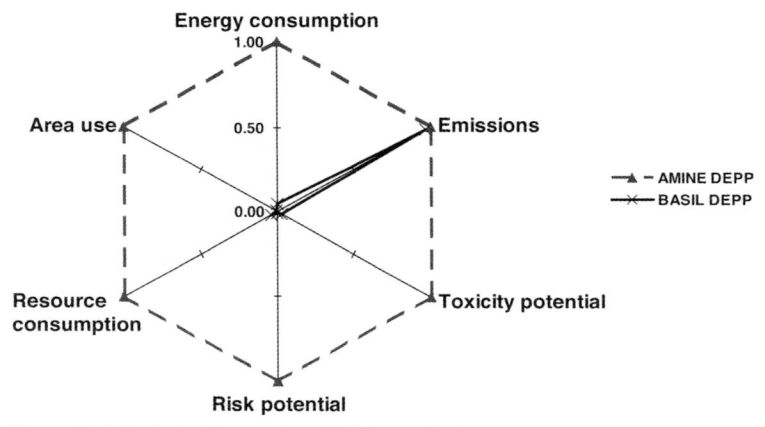

Figure 11.9 Ecological fingerprint of DEPP production.

emissions for the EDPP process are higher for BASIL processes due to the higher water emissions with a relatively high relevance factor. In the DEPP process, the emissions are equal - because of the relatively lower water emissions and the higher GWP. These differences result in a shift of relevance factors compared to the EDPP case. In the category "Emissions" both alternatives are now equal. In all other categories, the BASIL processes are much more favorable. The use of weighting factors is explained in general in the methodology paper [4]. The final calculation factors for this study are shown later in Figures 11.11 and 11.12 together with the eco-efficiency portfolios.

11.4.5
Cost Calculation

Economic viability is crucial for a product to succeed. To address this, an economic analysis is carried out to quantify costs such as investment, maintenance, labor,

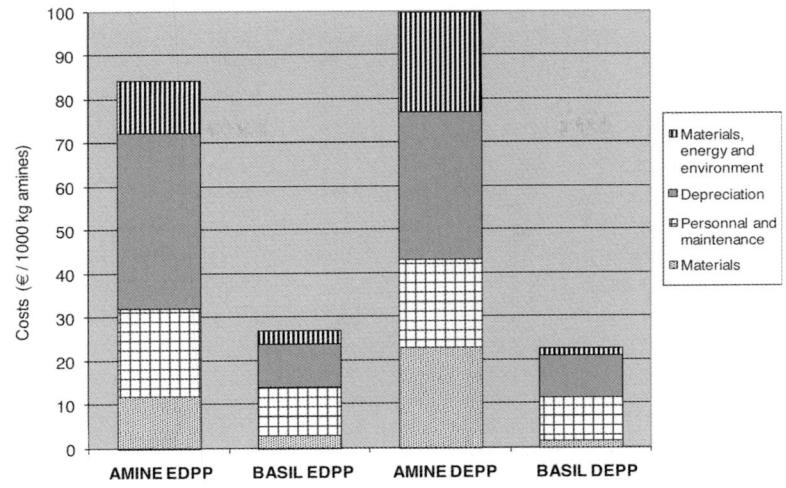

Figure 11.10 Life-cycle costs calculation.

transportation and disposal. Most important are the costs of the reactants, followed by the additives. Process optimization will reduce these costs effectively in the future. The cost evaluation in Figure 11.10 shows clearly that the use of efficient agents such as ionic liquids will increase the capacity of a plant and decrease the capital investment. It also reduces the follow-up costs of chemicals and recycling steps. Higher costs for the quenching agent causes an overall much better cost position.

11.5
The Creation of the Eco-efficiency Portfolio

Finally, the environmental and economic results are portrayed in the eco-efficiency portfolio. In summary, the eco-efficiency portfolio concisely represents the relative overall and economic impact of various alternatives. Thereby eco-efficiency analysis enables the user to understand all effects, both "macroscopic" and "microscopic".

The overall cost calculation and the calculation of the ecological fingerprint constitute independent calculations of the economic and environmental considerations of a complete system with different alternatives. Since ecology and economy are equally important in a sustainability study, a system can compensate for weaknesses in one area by good performance in the other. Alternatives whose sums of ecological and economic performance are equal are considered to be equally eco-efficient.

The values obtained from the ecological fingerprint are multiplied by weighting factors [4] and added up in order to determine the environmental impact of each alternative. The various environmental impact values are normalized by the mean environmental impact and plotted on the eco-efficiency portfolio.

310 | 11 Eco-efficiency Analysis of an Industrially Implemented Ionic Liquid-based Process

Calculation factors for the aggregation of environmental results into the portfolio

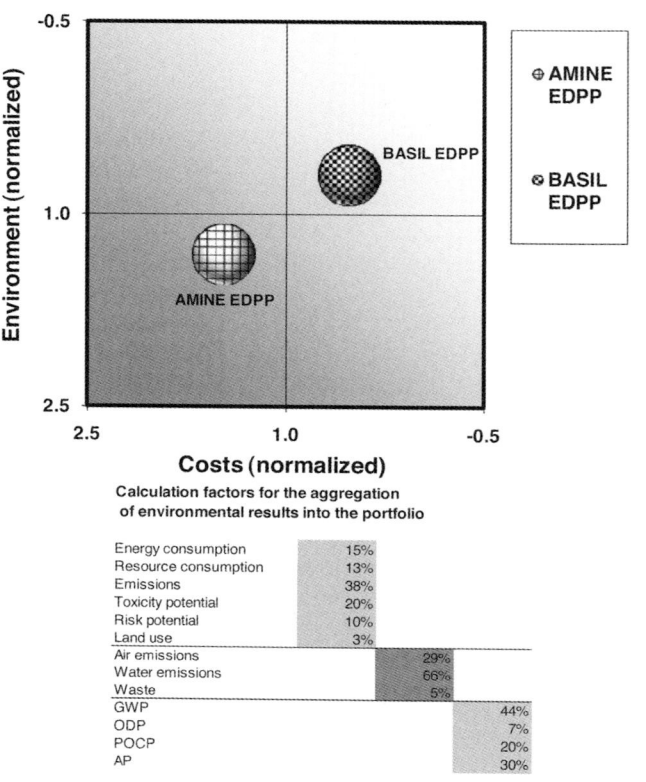

Energy consumption	15%
Resource consumption	13%
Emissions	38%
Toxicity potential	20%
Risk potential	10%
Land use	3%
Air emissions	29%
Water emissions	66%
Waste	5%
GWP	44%
ODP	7%
POCP	20%
AP	30%

GWP: global warming potential; **ODP:** ozone depletion potential; **POCP:** photochemical ozone creation potential, local emissions that lead to an increase in ozone close to the ground; **AP:** abbr. for acidification potential or acid rain. cause forest death.

Figure 11.11 The eco-efficiency portfolio for EDPP.

Beside the fact, that costs and environment are rated with the same societal weighting factor, an additional relevance factor (GDP relevance) assesses the relative importance of costs and environmental burden to each other. This calculation is based on statistical numbers. The relevance factor in this case shows clearly the greater importance of the results evaluating the environmental effects compared with costs. That leads to smaller distances on the costs axis and to larger distances in the environment even if the cost differences of the alternatives are relatively high. For both products, the BASIL processes are the more eco-efficient alternatives, due to the environmental aspects and due to the life-cycle costs (Figures 11.11 and 11.12).

Even if the processes look very similar, they are different processes with different materials, reactions and single impacts. This was shown in the single graphs. The Amine process and the related BASIL processes are different for EDPP and DEPP. That results also in differences of calculated numbers resulting in different relevance

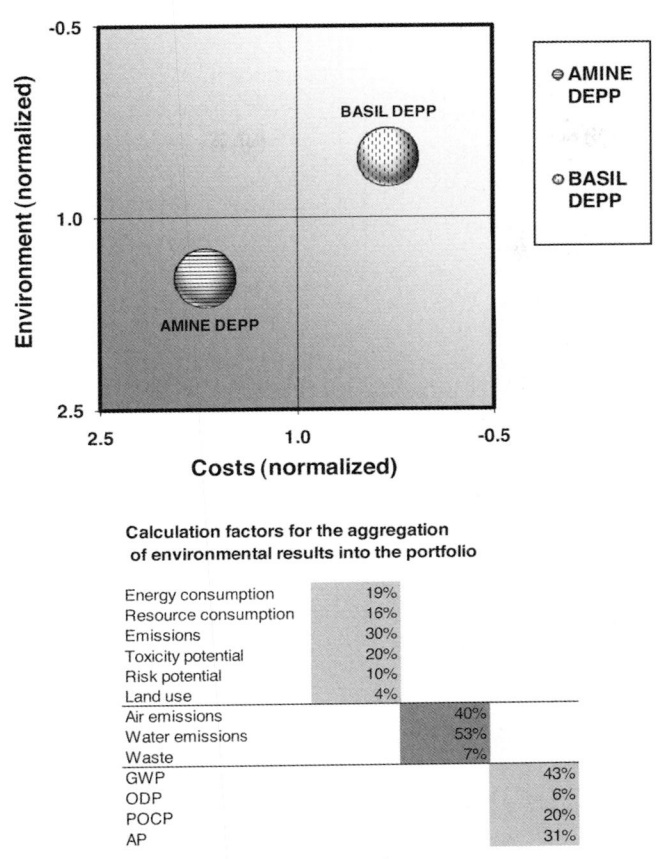

Figure 11.12 The eco-efficiency portfolio for DEPP.

Calculation factors for the aggregation of environmental results into the portfolio:

Energy consumption	19%
Resource consumption	16%
Emissions	30%
Toxicity potential	20%
Risk potential	10%
Land use	4%
Air emissions	40%
Water emissions	53%
Waste	7%
GWP	43%
ODP	6%
POCP	20%
AP	31%

and calculation factors. These differences are expressed in the two portfolios by different positions of the alternatives.

11.6
Scenario Analysis

The value of the eco-efficiency analysis tool, apart from its description of the current state, lies in the recognition of dominant influences and in the illustration of "what if …?" scenarios.

The stability of the results is verified by means of sensitivity analyses in every project. Not only the assumptions made but also the system boundaries and different calculation factors were varied and checked within realistic ranges. It was found in many of the analyses which have been carried out that even substantial changes to the weighting factors have only a very small impact on the eco-efficiency conclusions. True, the relative positions in the portfolio vary (in some cases even their order on the

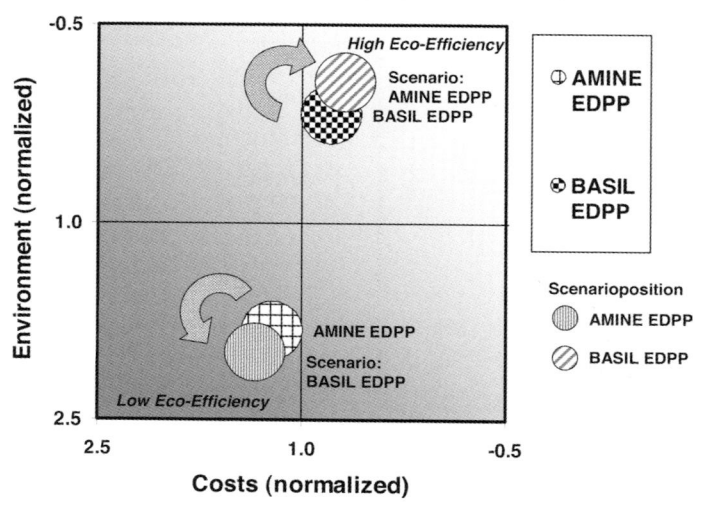

Figure 11.13 The eco-efficiency portfolio scenario for EDPP, showing a reduction in BASIL use of 50% in the process.

ecological axis), but the conclusions with regard to eco-efficiency (environment and costs) change very rarely. In these rare cases, the system is termed unstable and the products are termed similarly eco-efficient. From experience, the largest influence on the result by far is possessed by the input data and the system boundaries.

As an example, it is shown that the impact to the eco-efficiency result is very high if in the BASIL process the amount of ionic liquid used is reduced by 50% (Figure 11.13). Other scenarios show that this impact factor is very important in comparison with other possible optimization options. Eco-efficiency is therefore able to support R&D strategies very efficiently and helps to avoid optimization of low impact factors. The result will be a more sustainable process, developed in shorter times and for lower costs.

In the eco-efficiency analysis, the results are shown relative to each other in the portfolio. They are linked together and if one alternative becomes better in a scenario relative to that also the other alternative moves in the graph even if there is no change in their data in the calculation.

11.7
Conclusion

In summary, the eco-efficiency portfolio concisely represents the relative overall and economic impact of various alternatives. The eco-efficiency enables the user to understand all effects, both "macroscopic" and "microscopic".

The eco-efficiency analysis presented here compares alternative process chains for the production of phenylphosphines, including the use of an ionic liquid in the acid quench step. In the classical approach, an amine forming a solid ammonium salt is

Figure 11.14 The eco-efficiency label.

used. In the new processes BASIL, here 1-methylimidazole is used. It is forming an ionic liquid under the reaction conditions.

The space–time yield of the ionic liquid-based process is higher by a factor of 80 000 compared with the classical process, which leads to a significant improvement in eco-efficiency.

In the base case and in all scenarios, the BASIL alternatives are much more eco-efficient than the classical trialkylamine-based process.

The quench step and the recycling step have important impacts on the overall results.

The results are robust; differences in input data in certain ranges have no influence on the overall result. The scenarios always show advantages of the BASIL processes.

BASF SE has developed a new label for products that have been evaluated by an eco-efficiency analysis (Figure 11.14). The award of this label is dependent on demanding requirements: after realization of the analysis, a third-party evaluation (critical review, CR) is requested. D. R. Shonnard from the Michigan Tech University performed the CR for this study [5]. Furthermore, publication of the results of the analysis will be undertaken via the Internet (www.oeea.de). The label can be carried for 3 years; after that period, a revision of the analysis is required to cover market developments and product diversity.

The acid quench alternatives using BASIL obtained the label and were ranked in the first position of eco-efficiency. The label can be used in marketing for communicating and pushing more sustainable solutions.

11.8
Outlook

Eco-efficiency analysis can be used in a large number of applications and readily yields understandable conclusions in the case of multifactorial problems in relatively short times and at relatively low cost. In the future, eco-efficiency will become more important in the context of sustainability to show which process is more favorable than other alternatives.

The analysis allows for a holistic view of chemical products and processes that combines assessment of life-cycle environmental impacts with economic performance in equal measure to achieve a greater level of sustainability. The relevance of the eco-efficiency analysis on internal strategic decisions is very high within BASF and most analyses are presented to higher management, customers, NGOs and politicians.

References

1 von Weizsäcker, E.U. and Seiler-Hausmann, J.-D. (eds) (1999) *Ökoeffizienz Management der Zukunft*, Birkhäuser Verlag, Switzerland, ISBN 3-7643-6069-0.

2 Schaltegger, S. and Sturm, A. (1998) *Eco-efficiency by Eco-controlling*, vdf, Zürich; Schaltegger, S., Kleiber, O. and Müller, J. (2002) *Nachhaltigkeitsmanagement in Unternehmen. Konzepte und Instrumente zur nachhaltigen Unternehmensentwicklung*, BMU and BDI, Berlin.

3 Swatloski, R.P., Rogers, R.D. and Holbrey, J.D. (2002) US Patent 2003/0157351; Swatloski, R.P., Holbrey, J.D., Chen, J., Daly, D. and Rogers, R.D. (2005) US Patent 2005/0288484;Fort, D.A., Rensing, R.C., Swatloski, R.P., Moyna, P., Moyna, D. and Rogers, R.D. (2007) *Green Chem.* **9**, 63; BASF AG, World Patent WO 03/062171, WO 03/062251, WO 05/061416;Maase, M. (2004) *Chemie in Unserer Zeit* 434; Freemantle, M. (2003) *Chemical & Engineering News*, **81** (13), 9;Rogers, R.D. and Seddon, K.R. (2003) *Inorganic Materials* **2**, 363;Seddon, K.R. (2003) *Science*, **302**, 792–793.

4 Saling, P., Kicherer, A., Dittrich-Krämer, B., Wittlinger, R., Zombik, W., Schmidt, I., Schrott, W. and Schmidt, S. (2002) *International Journal of Life Cycle Assessment* **7** (4), 203–218; Landsiedel, R. and Saling, P. (2002) *International Journal of Life Cycle Assessment* **7** (5), 261–268; Schmidt, I., Meurer, M., Saling, P., Kicherer, A., Reuter, W. and Gensch, C.-O. (2004) *Greener Management International*, (45), 79.

5 Shonnard, D.R., Kicherer, A. and Saling, P. (2003) *Environmental Science & Technology* **37** (23), 5340–5348; Shonnard, D.R., Michigan Technological University, Houghten, MI.

12
Perspectives of Ionic Liquids as Environmentally Benign Substitutes for Molecular Solvents

Denise Ott, Dana Kralisch, and Annegret Stark

12.1
Introduction

Ionic liquids can offer novel, potentially "green" perspectives and considerable advantages. They have been investigated as solvents and also as auxiliaries in a great number of applications, e.g. in organic and catalytic syntheses such as Heck reactions, hydrogenations and Diels–Alder reactions [1–10], and also as solvents for extraction [11–13]. Furthermore, the application potential of ionic liquids in enzymatic reactions [14], electrochemical applications, e.g. the use of ionic liquids as electrolyte material for metal deposition [15, 16] or batteries [17, 18], and also as sensors [19, 20] are some examples of the huge area of potential application, as further highlighted in several chapters of this book.

Most of all, ionic liquids are discussed as substitutes to molecular organic solvents and might have the potential to be a "green" alternative for these, since they offer *inter alia* considerable chemical advantages. Additionally, ionic liquids can feature specific physicochemical properties, in particular non-flammability and high thermal stability. Due to their negligibly low vapor pressure, they cannot emit volatile organic compounds and thus can contribute to safety at the workplace. Against this background, ionic liquids were at first uncritically referred to in the context of green chemistry. Subsequently, first results on their partial toxicity, production effort and environmental impact led to a more differentiated point of view, so that nowadays the assessment of their chemical and biological properties and the resulting environmental impacts have become important research and development topics. In line with the 12 Principles of Green Chemistry, as defined by Anastas and Warner [21], Nelson [22] defined properties and general criteria for a first evaluation of solvents with regard to their potential to act as "green" substitutes in order to indicate environmentally benign products and processes, such as ionic liquids or process technologies using ionic liquids. These criteria are in particular that the solvent/process

- generates less waste (in the solvents' production, use and disposal)
- is innocuous or substantially innocuous
- employs renewable resources in the synthesis of the solvents
- solves other environmental problems
- improves selectivity, reaction efficiency and product separation
- reduces known hazards associated with solvents, such as effects on human and environmental health.

In order to design and apply "green" solvents, information on mechanisms of action, structure–activity relationships and the consideration of the elimination of toxic functional groups, the reduction of bioavailability, the design of innocuous fate and the minimization of the energy consumption are of particular importance [22]. Ionic liquids so far only partly fulfill these requirements. For example, a Strengths, Weaknesses, Opportunities, Threats (SWOT) analysis of ionic liquid preparation is detailed in Chapter 2.

From a more holistic point of view, the environmental friendliness of products and processes generally depends on the environmental impact resulting from all life cycle stages, including upstream processes, the application phase and downstream processes (Figure 12.1).

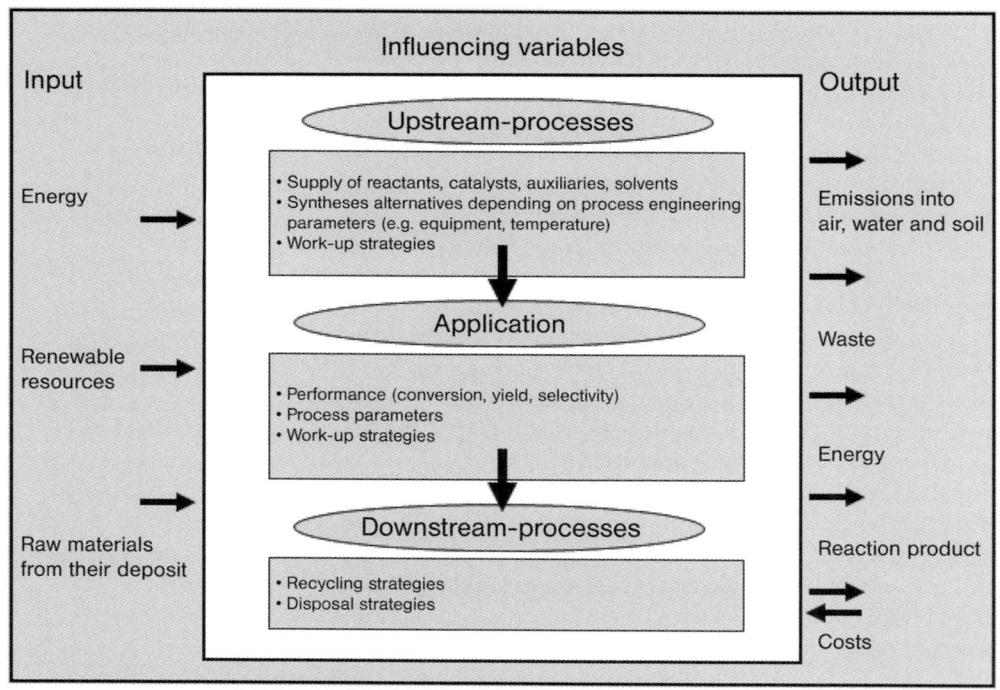

Figure 12.1 Evaluating the environmental/economic impact of a product or process considering all life cycle stages.

In order to integrate these aspects into already ongoing R&D, our group developed the ECO (Environmental and Economic Optimization) method [23, 24]. By means of this screening tool, synthesis pathways for ionic liquids can be optimized and processes using ionic liquids can be evaluated regarding ecological sustainability already at the R&D stage. It can be used to search for optimal configurations in an iterative procedure following a life cycle approach. In terms of a future applicability of these R&D results on an industrial production scale, both economic and ecological aspects are included in the optimization and evaluation strategy. Alternative approaches in the context of ionic liquid evaluation have been developed by Jastorff and co-workers [25, 26] and Zhang et al. [27]. Jastorff's group assessed the sustainability of products or processes by defining risk potentials for humans (*toxicophore*) and the environment (*ecotoxicophore*) regarding technical constraints (*technicophore*). On the example of ionic liquids, they were able to demonstrate that information on resulting risk potentials are already available at the stage of product design and reasoned that it can be used to assist the development of "green" ionic liquids.

Zhang et al. [27] performed a life cycle assessment of the synthesis and application of an ionic liquid compared with selected molecular solvents. They emphasized the challenges and uncertainties of a product or process assessment in early stages of development. However, their studies showed the importance of life cycle analysis for the ecological evaluation of ionic liquids in contrast to other solvent systems.

In this chapter, a brief literature review of different ecological evaluation tools is given, followed by a description of the application of the ECO method to the evaluation and optimization of ionic liquid synthesis. Finally, the solvent performance and ecological (dis)advantages of different solvent systems for a specific R&D application task, i.e. the Diels–Alder reaction of cyclopentadiene and methyl acrylate, is assessed.

12.2
Evaluation and Optimization of R&D Processes: Developing a Methodology

12.2.1
Solvent Selection Tools

A number of groups have already published solvent selection/replacement tools in order to support the decision-making process concerning the selection of "green" and sustainable alternatives [28, 29]. They range from a merely qualitative or semi-quantitative (e.g. ABC/XYZ evaluation [30]) to complex life cycle approaches. Some computer-aided methodologies and software tools, e.g. developed by Gani et al. [31, 32], EPA's SAGE (Solvent Alternatives Guide) for surface cleaning processes [33] and PARIS II (Program for Assisting the Replacement of Industrial Solvents), reflecting solvent properties and environmental issues [34–36], have also been introduced.

In recent activities, solvent selection tools were developed allowing for the concurrent consideration of environmental, health and safety aspects at the R&D

stage, partly under consideration of life cycle aspects/life cycle assessment (LCA) [37–39], and also economic criteria [40, 41]. For instance, the solvent selection guide presented by Capello *et al.* [39] integrates the life cycle assessment method and the EHS (environment, health, safety) method developed by Hungerbühler and co-workers [42, 43]. Further, the ecosolvent tool [38] is used as an LCA tool that facilitates the quantification of the environmental impact of waste solvent treatment. Moreover, it contains life cycle inventories of the petrochemical production of the integrated solvents based on the Ecoinvent database [44].

12.2.2
LCA Methodology

Tools for the assessment of the "greenness" of a chemical substance have not been restricted to solvents but have also been applied to a broad range of chemical synthesis evaluations. The best known, and widely approved as a profound means of ecological evaluation, is the previously mentioned life cycle assessment (LCA) methodology, normalized by the International Organization for Standardization (ISO) [45, 46]. The assessment is based on a compilation of an inventory of relevant inputs and outputs of the system, an evaluation of the potential environmental impacts associated with those inputs and outputs and an interpretation of the results of the inventory analysis and impact assessment phases in relation to the goal and scope of the study. In contrast to a single evaluation of the product/process step under development, it also allows for the consideration of upstream and downstream processes. Thus, the development of products/processes with optimized performance but with increased environmental impact outside the system boundary is avoided. Examples of environmental impact potentials considered within an LCA are resource depletion, global warming, acidification, human and ecotoxicity and land use. Thus, a wide range of actual environmental problems is covered. It can be applied to guide decision making on environmental design and improvement and to rank the environmental burden of alternative products or processes. The level of detail and also the boundaries of an LCA study depend on the subject and intention of the study, but the ISO standards have to be followed. Several LCA studies have been performed in the context of chemical processes [27, 47–50].

The results obtained from the LCA methodology are well founded and comprehensive, but its application is very time consuming and requires an extensive database. However, data and time are often limited during early design stages and a comprehensive LCA study is therefore often not feasible. Nevertheless, the optimization potential is highest and the costs to change the system are lowest at the beginning of the product or process design and development phase [51, 52].

The linkage between the complex LCA methodology on the one hand and simple metrics such as atom efficiency [53, 54], which cover only parts of the system, on the other, can be attained by a simplified or "streamlined" LCA (SLCA). According to the Society of Environmental Chemistry and Toxicology Europe [55], SLCA is an application of the LCA methodology during a comprehensive screening assessment, which covers the whole life cycle, but on a more superficial level. Thus, SLCA involves lower costs, time and effort to run the assessment and allows for the exclusion of certain life cycle stages, system inputs/outputs and impact categories and also the use of qualitative and quantitative generic data modules to fill data gaps. The simplified assessment focuses on the most important environmental aspects with respect to potential environmental impacts, LCA stages and/or phases, thoroughly assessing the reliability of the results with the help of sensitivity and consistency checks [46]. SLCA consists of three steps which are iteratively interlinked: (i) screening, (ii) simplification and (iii) assessment of reliability. SLCA is usually utilized for screening purposes (see, e.g., Fleischer and Schmidt [30] and Gasafi et al. [56]). Typical screening indicators used in an SLCA are the cumulative energy demand (CED) [57], material intensity per service unit (MIPS) [58] and single impact categories such as the global warming potential (GWP) [42]. In conclusion, the simplification of LCA is an approach to deal with data asymmetries, data gaps and inconsistencies in life cycle inventories – essential for applying LCA already at the design stage. Application examples of SLCA have been proposed [59–62]. However, despite the advantages in terms of time, data requirements and costs, a risk remains that not all key issues are identified with SLCA, which may result in incorrect decisions.

LCA or SLCA may be further coupled with (multi-)objective optimization procedures. Kniel et al. [63] were one of the first groups to discuss the potential of combining LCA and multi-objective optimization in a design tool for processes. Further, Azapagic and Clift [64] dealt with the coupling of multi-objective optimization and LCA to facilitate the decision-making process. The system under investigation is simultaneously optimized on a number of environmental objective functions defined and quantified through the LCA approach [65]. The so-called decision-aid tool-optimum LCA performance (OLCAP) methodology based on the results of an LCA study as a starting point of the optimization procedure refers to a predefined optimization problem.

In summary, a wide range of methodologies, which partly overlap and also supplement each other, exist in the context of environmental benign (chemical) product/process design and optimization. However, especially during early stages of development, the influence of single parameter variations on the objectives under consideration is of great importance to guide future activities. At this stage, the analysis of dependencies and proportionalities is much more important for a detailed understanding of the system under investigation than the exact determination of values which may be outperformed during the next step of development.

12.2.3
The ECO Method

The ECO method represents a combination of well-known approaches in the context of ecological evaluation with process optimization tools in order to allow for an ecologically benign design of chemical processes [24]. It was developed to search for optimal configurations of chemical synthesis parameters in an iterative procedure starting during the R&D stage. In order to represent terms of both ecological and economic sustainability, three objective functions which incorporate (i) energy demand (*EF*), (ii) risks concering human health and the environment (*EHF*) and (iii) costs (*CF*) were defined. They have to be minimized during the optimization process to maximize the benefit regarding ecological and economic sustainability. Then, the decision-making process is guided by an outranking of pareto-optimal solution candidates referring to these three key objectives. All of them incorporate a life cycle-based evaluation of the entire process.

As discussed before, the LCA methodology is too complex and is based on data which are partially not available during process development. That is why the determination of the three objective functions is based on the SLCA approach [55] extended by economic issues. In order to limit further the extent of the required database for both the evaluation and the calculation effort, the ECO method utilizes a local optimization of single process steps. A selection of preferred parameter values for each single step results. The additional benefit of this approach is that detailed information about dependencies between the different parameters under investigation is provided. The summation of these results over the entire process chain allows a search for a local optimum regarding ecological and economic sustainability. Finally, a number of preferrable parameter configurations are received for each process step.

12.2.3.1 The Key Objectives

The three life cycle-based key objectives, energy factor (*EF*), environmental and human health factor (*EHF*) and cost factor (*CF*), comprise a wide range of relevant aspects, since their determination is based on data available already during the early stages of development.

The *EF* is based on the calculation of the cumulative energy demand (CED) tailored to the evaluation of chemical synthesis strategies. With the help of this metric, the variation of the energy demand of different alternatives concerning all life cycle stages can be assessed (Equation 12.1). This also allows for an implicit evaluation of the variations of other impact categories of LCA such as abiotic resource depletion, global warming, stratospheric ozone depletion, tropospheric photooxidant formation, acidification and eutrophication, as demonstrated elsewhere (e.g. [23, 66]) and in the following.

$$EF = \frac{\sum_{i=1}^{x_S} E_i^S + \sum_{i=1}^{x_R} E_i^R + \sum_{i=1}^{x_W} E_i^W + \sum_{i=1}^{x_A} E_i^A + \sum_{i=1}^{x_D} E_i^D}{n_{\text{product}}} \quad (12.1)$$

The *EHF* allows for a comparison of different chemical substances used, e.g. reactants, solvents or auxiliaries, regarding the resulting risks for humans and the environment. It is based on the remaining potential of danger ($RPoD_{ij}$), calculated according to the EHS method suggested by Koller *et al.* [42]. The *EHF* is divided into three sub-objectives: *EHF*(AcT), *EHF*(ChrT) and *EHF*(WmE), referring to the categories acute toxicity, chronic toxicity and water-mediated effects, respectively. Their calculation is demonstrated using the example of *EHF*(AcT) in Equation 12.2.

$$EHF(AcT) = \frac{\sum_{i=1}^{x_S} RPoD(AcT)_i^S + \sum_{i=1}^{x_R} RPoD(AcT)_i^R}{n_{product}}$$
$$+ \frac{\sum_{i=1}^{x_W} RPoD(AcT)_i^W + \sum_{i=1}^{x_A} RPoD(AcT)_i^A + \sum_{i=1}^{x_D} RPoD(AcT)_i^D}{n_{product}}$$
(12.2)

Since ecologically advantageous alternatives will only be adopted into industrial processes if they are economically competitive, the cost factor *CF* has been chosen as a the third criterion of the ECO method (Equation 12.3). The calculation is similar to current approaches of life cycle cost (LCC) analyses, again tailored to the evaluation of chemical synthesis strategies.

$$CF = \frac{\sum_{i=1}^{x_S} C_i^S + \sum_{i=1}^{x_R} C_i^R + \sum_{i=1}^{x_W} C_i^W + \sum_{i=1}^{x_A} C_i^A + \sum_{i=1}^{x_D} C_i^D}{n_{product}}$$
(12.3)

In all cases, the effort is summed up for the

1. the supply of the reactants, solvents and auxiliaries (*S*)
2. the performance of the reaction (*R*)
3. the energy demand necessary for the work-up (*W*)
4. the application of the products (*A*)
5. the disposal of waste (*D*)

and is related to a product-based benefit, e.g. the product molarity or the product mass. The units of the key objectives described in Equations (12.1–12.3) are *EF* in MJ mol^{-1}, *EHF(AcT,WmE,ChrT)* in kg mol^{-1} and *CF* in € mol^{-1}. Alternatively, they can be related to the product mass.

12.2.3.2 The Evaluation and Optimization Procedure

To start the evaluation acompaning the ongoing R&D, a preselection of optimization parameters and a range of parameter variations is made and all experiments required to characterize the system under investigation are performed.

Parallel to this, freely accessible or commercial databases are explored in inventory databases (e.g. the Ecoinvent database [44], SimaPro [67] or PROBAS [68]) for energy issues, from safety data sheets [69] or from software tools such as EPIWIN [70] for toxicological criteria and from market prices for cost issues. Having the experimental data resulting from a parameter variation at hand, the calculation of the sustainability factors can be executed. Finally, the values of all factors are collected in a performance matrix. Its compilation is an iterative process and it may change significantly during the optimization process. The ranking of alternatives is executed using a partial ranking procedure.

Partial ranking is realized by using, for example, the software Decision Lab 2000 [71] [the outranking results presented in this chapter were determined using the software Decision Lab 2000 and are valid under the following regulations: minimize EF, EHF(AcT), EHF(ChrT), EHF(WmE), CF, weight: 33:11:11:11:33, preference function: linear, threshold unit: percent]. This decision support tool allows for the partial ranking of the parameter alternatives under investigation according to the three key objectives *EF*, *EHF* and *CF* by means of the PROMETHEE method [72]. Thus, unfavorable (dominated) alternatives can be excluded in ongoing research work to restrict the experimental effort. The emphasis is to range the choices from a set of pareto-optimal solutions, rather than to define the preferences explicitly before analyzing the alternative trade-offs.

The optimization step discussed above results in a range of parameter values referring to a chemical compound, a synthesis pathway, a work-up procedure or another process which is optimized regarding the objective functions, leading to new insights, which can be used to optimize further and/or identify new areas of process/product development. If this procedure is executed along the process chain, a local optimization of the entire process regarding ecological and economic sustainability can be approximated.

In the case of data uncertainties or existing gaps, an assessment by means of similar parameter characteristics or other generic data can be executed. Should this be impossible, a worst-case scenario is assumed. This happens against the background that incalculable environmental impacts, toxicological risks or costs should be pointed out.

Whereas the approach is not suited to provide specific values describing, for example, quantitative environmental impacts or costs, it is responsible for relative evaluations of a broad range of parameter variations based on small-scale preparation with the help of simple determination methods. It allows for the detection of sensitive parameters and of hot-spots along the entire life cycle chain. In order to quantify the progress in ecological and economic sustainability achievable by the newly designed product or process compared with its alternatives, the reliability of the results needs to be proven by a detailed LCA/LCC analysis following the optimization process.

12.3
Assessment of Ionic Liquid Synthesis – Case Studies

As discussed before, ionic liquids have some outstanding advantages over conventional organic solvents. However, their declaration as a "green" solvent alternative requires more than the investigation of their performance in application [73, 74]. Additionally, important aspects are environmental impacts resulting from upstream and downstream processes (such as starting materials, solvents and auxiliaries), energy supply, work-up, solvent recycling and waste treatment, process engineering aspects, the influence of different reaction conditions and comparison with alternative solvents.

Figure 12.2 represents the generalized system boundary of the investigation and assessment of a chemical reaction with respect to the parameter variations

12.3 Assessment of Ionic Liquid Synthesis – Case Studies

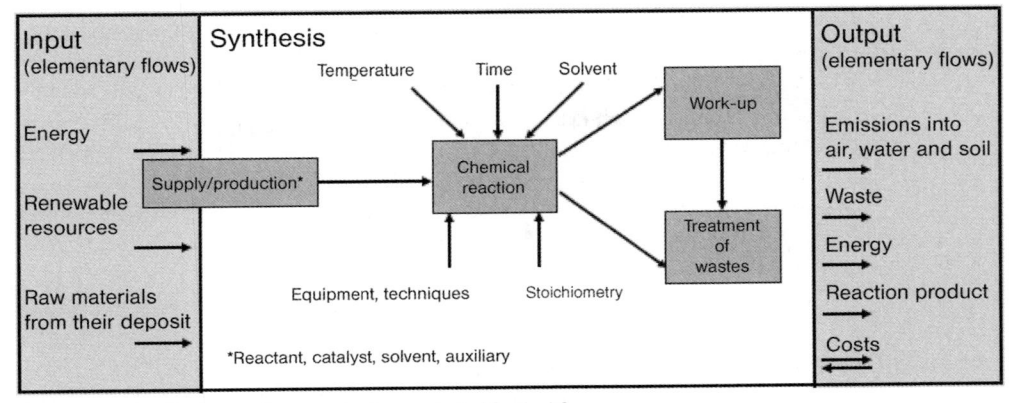

Figure 12.2 "Cradle-to-grave" approach, demonstrated for the life cycle stage of the synthesis of the product under investigation.

undertaken. As pointed out, a plethora of parameters can influence the ecological and economic impact of a chemical reaction during reaction engineering, summarized as inputs and outputs.

In the following, some selected results for the evaluation of the ionic liquid supply via a commonly applied two-step pathway will be discussed.

The first step of ionic liquid synthesis is commonly the alkylation of an N-, P- or S-containing organic compound, for instance N-methylimidazole, pyridine, trialkylphosphine or dialkyl sulfide, followed by an anion exchange as the second step. Both steps, particularly the alkylation step, involve energy- and time-consuming synthesis and work-up procedures. Especially the work-up (extraction) often requires a high input of organic solvents, resulting in energy-consuming distillation steps.

The development of environmentally benign synthesis routes for ionic liquids will be demonstrated on the example of the alkylation step (Menschutkin reaction) in more detail [75]. The preparation of the ionic liquid 1-hexyl-3-methylimidazolium chloride ([C_6mim]Cl) is taken as a representative experiment (Scheme 12.1).

The following process parameters were investigated: temperature, solvent, concentration, molar ratio and reaction time. In addition, the N-base was altered in order to prove the transferability of the reaction parameters.

Scheme 12.1 Preparation of 1-alkyl-3-methylimidazolium-based ionic liquids via alkylation of N-methylimidazole.

Table 12.1 summarizes the experimental results. The experiments represented in bold type will be discussed in more detail below, dealing with the influence of different solvents and of the reaction temperature.

Table 12.1 Alkylation of N-methylimidazole, variation of stoichiometry $n_{\text{N-base}}/n_{\text{C}_6\text{H}_{13}\text{Cl}}$, reaction temperature T, reaction time t, solvent and concentration of N-methylimidazole $c_{\text{N-base}}$; $n_{\text{N-base}} = 0.21$ mol.

Experiment No.	$n_{\text{N-base}}/n_{\text{C}_6\text{H}_{13}\text{Cl}}$	T (°C)	t (h)	Solvent	$c_{\text{N-base}}$ (mol l^{-1})	Yield (%)
1	1:1.6	80	30	n-Heptane	1.6	27
2	1:1.6	80	30	o-Xylene	1.6	16
3	1:1.6	80	30	Cyclohexane	1.6	20
4	1:1.6	80	30	Ethanol	1.6	25
5	1:1.6	80	30	Ethanol	3.0	61
6	1:1.6	80	30	Ethanol	1.0	14
7	1:1.2	80	30	n-Heptane	2.5	46
8	1:1.6	80	30	–	3.3	73
9	1:1.0	80	30	–	4.6	77
10	1:0.8	80	30	–	5.3	74
11	1:0.5	80	30	–	6.7	50
12	1:3.0	80	30	–	2.0	52
13	1:4.0	80	30	–	1.6	40
14	1:1.0	70	30	–	4.6	49
15	1:1.2	70	30	–	4.1	45
16	1:2.0	70	30	–	2.8	37
17	1:1.0	90	30	–	4.6	94
18	1:2.0	90	30	–	2.8	92
19	1:1.0	100	30	–	4.6	98
20	1:1.0	70	10	–	4.6	18
21	1:1.0	70	19	–	4.6	35
22	1:1.0	70	72	–	4.6	78
23	1:1.0	70	144	–	4.6	87

12.3.1
Synthesis of Ionic Liquids: Extract from the Optimization Procedure

The assessment and optimization procedure will be discussed by means of experiments 1–8 in Table 12.1 (conversion of N-methylimidazole in different solvents and solvent free). In a baseline experiment, n-heptane was applied as a solvent for the alkylation of N-methylimidazole and 1-chlorohexane (experiment 1). Some molecular solvents also tested, namely o-xylene, cyclohexane and ethanol, resulted in lower yields. However, by increasing the molarity of N-methylimidazole in these solvents, an increased yield was obtained. Consequently, the reaction was carried out under solvent-free conditions. In these experiments, the solvent-free alkylation resulted in the highest yields.

The variance from the baseline experiment 1 is shown exemplarily for the *EHF (AcT)* and *EF* in Figure 12.3. Due to the lower conversions in o-xylene, cyclohexane and ethanol under the same reaction conditions, the reactant and solvent requirements increase, resulting in a higher toxicity potential [*EHF(ChrT,WmE)* indicate the

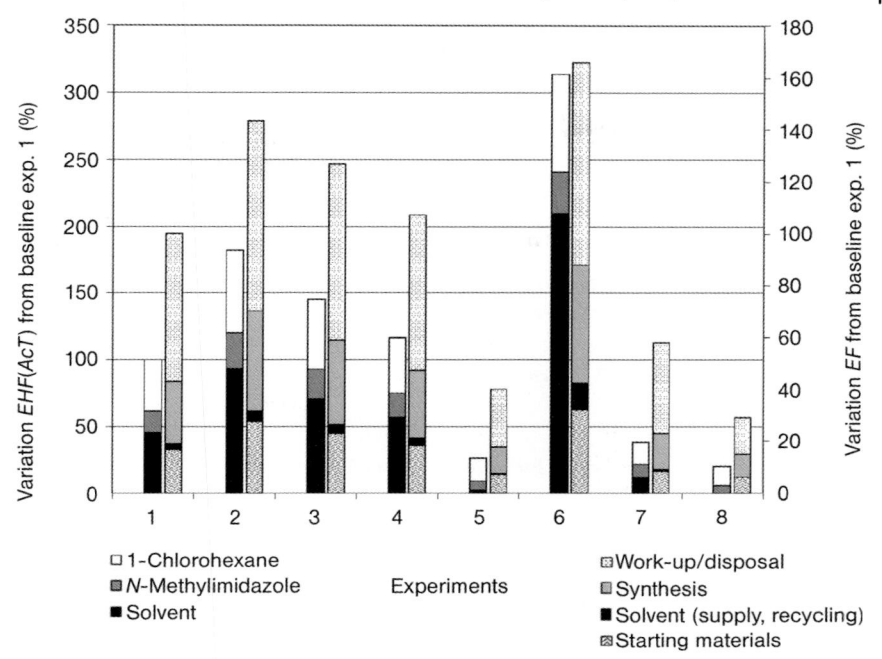

Figure 12.3 Variation of EHF(AcT) (left bars) and EF (right bars) from baseline experiment 1; for reaction conditions, see Table 12.1.

same trend] and a lower energy efficiency. In highly concentrated solutions (see experiments 5 and 7), the conversion increases after a specific reaction time, resulting in a lower reactant requirement. Therefore, the resulting $EHF(AcT)$ is comparable to that calculated for the solvent-free experiment. Thus, solvents which possess a comparably high toxicity potential should not be excluded *a priori*, since their performance significantly influences the efficiency of the synthesis. Working in closed systems has to be guaranteed in this context. In the case of the example discussed earlier, the synthesis under solvent-free conditions is more energy effective and less problematic in terms of the environment and human health. This can be traced back to the avoidance of an additional work-up step and higher conversions (Figure 12.3).

Further experiments indicate an equimolar or approximate amount of starting materials to be the best reactant ratio. Adopting this information, the reaction was carried out under equimolar conditions at temperatures of 80–100 °C, resulting in increasing yields and energy efficiencies with increase in temperature.

Finally, the evaluation of all experimental results and the outranking of the different solution candidates via an outranking procedure resulted in the following optimal synthesis prescription: $T = 100\,°C$, $t = 30\,h$, $n_{MIM}/n_{C_6H_{13}Cl} = 1:1$, solvent free. This parameter configuration has been found to be the best trade-off regarding EF, EHF and CF, determined using the software Decision Lab 2000 [71].

Since all results of the assessment concerning "synthesis" and "work-up" are based on the energy demand determined on a laboratory scale, relative statements between synthesis alternatives are easy to apply and approvable. Since the work-up procedure is taken into account in addition to the synthesis of 1-hexyl-3-methylimidazolium chloride itself, *EF* is mainly influenced by the supply of the solvent for extraction and its recycling. In this case, the efficiency of the work-up procedure seems to be an essential performance criterion for an environmentally benign supply of pure $[C_6mim]Cl$.

To simplify the optimization procedure and to gain more information on the effects of single- and multi-parameter variations and also prognoses on the target values such as yield, energy efficiency and toxicity, statistical methods such as Design of Experiments (DoE) can be used.

12.3.2
Validation of *EF* as an Indicator for Several Impact Categories of the LCA Methodology

As pointed out, *EF* is an appropriate indicator that reflects most other LCA impact categories. To visualize this coherence, a comparative LCA for alternative synthesis pathways of $[C_6mim]Cl$ was performed. To start with, the baseline experiment of the optimization procedure, (experiment 1, Table 12.1) was chosen and compared with the best trade-off solution candidate (experiment 19). Figure 12.4 represents the reduction of the cumulative energy demand (CED) in contrast to the reduction of other life cycle impact categories [abiotic resource depletion potential (ADP), global warming potential (GWP), ozone depletion potential (ODP), acidification potential (AP), eutrophication potential (EP), photochemical ozone creation potential (POCP),

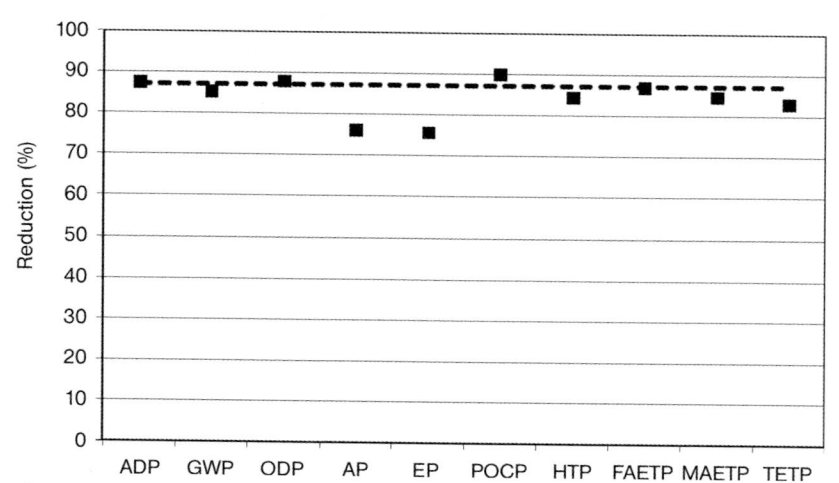

Figure 12.4 Validation of *EF* as an indicator for variations within other LCA impact categories (reduction of *EF*: dashed line).

human toxicity potential (HTP), freshwater aquatic ecotoxicity potential (FAETP), marine aquatic ecotoxicity potential (MAETP) and terrestrial ecotoxicity potential (TETP)] when favoring the conditions of experiment 19.

A reduction of 87% was achieved concering the *EF* (Figure 12.4). The reductions within the other impact categories diverge from the *EF* reduction with a median deviation of 4%. The standard deviation relating to the *EF* is 6%, thus fulfilling the requirements for a screening method during R&D. Therefore, the energy factor *EF* can be regarded as a sufficient screening factor reflecting a couple of LCA impact categories which are dominated by emissions from energy generation processes.

Nevertheless, *EF* is not applicable in all cases to reflect the impacts concerning human and ecotoxicity. In Figure 12.4, these factors seem to be suitable for this task, but, at this stage of development, the release of the ionic liquid into the environment is not integrated and quantified, because information on industrial disposal routes and the resulting emission pathway into the environment are not available. Thus, the criterion "human health and environment", based on the risk estimation procedure of the EHS tool [42, 43], constitutes an important supplement to *EF*. In this way, risks of any release of chemical substances at the workplace or into the environment can be pointed out by means of data which are at hand at the R&D stage or can be covered by a worst-case scenario, respectively.

12.3.3
Comparison of the Life Cycle Environmental Impacts of the Manufacture of Ionic Liquids with Molecular Solvents

In the following, the optimization considerations of the manufacture of the precursor [C_6mim]Cl were included in the evaluation of the effort to supply the ionic liquid [C_6mim][BF_4]. [C_6mim][BF_4] was compared with the effort of supplying an assortment of conventional solvents, namely methanol, acetone, toluene, benzene and water, via life cycle assessment (Figure 12.5). The resulting environmental impact potentials were normalized to those of [C_6mim][BF_4] in order to obtain a clear comparison.

One of the main requirements for a well-founded comparison is to ensure comparability. In this case, a direct comparison of the alternative solvents was difficult, since the same scale of production data for all solvents considered is necessary. The supply of methanol, acetone, toluene, benzene and deionized water can be evaluated on an industrial production scale using the Ecoinvent database [44], but such data are not available for the supply of [C_6mim][BF_4]. The energy demand for a laboratory-scale synthesis (batchwise) of the ionic liquid, known from the above-described experiments, allows no energetic comparison with the industrial supply of the other solvent systems. To guarantee comparability, the energy requirements for heating, stirring and distillation processes within upstream chains were calculated theoretically by means of thermochemical data. The amount of energy which is theoretically required (calculated by means of heat capacity, enthalpy of vaporization, standard enthalpy change of formation, standard enthalpy change of reaction; assumed efficiency for heating = 80%) amounts to less than 10% of the overall

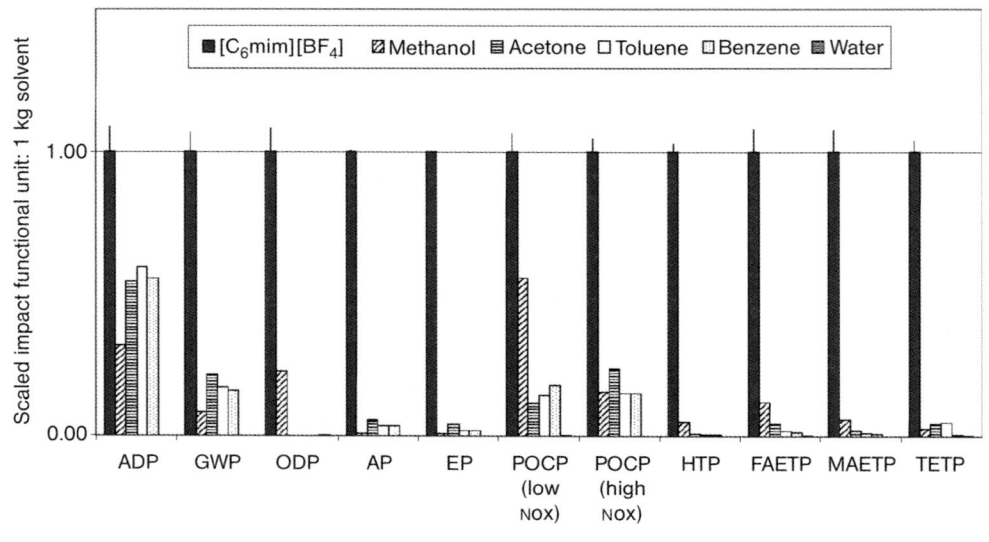

Figure 12.5 Comparison of LCA impact categories [abiotic resource depletion potential (ADP), global warming potential (GWP), ozone depletion potential (ODP), acidification potential (AP), eutrophication potential (EP), photochemical ozone creation potential (POCP), human toxicity potential (HTP), freshwater aquatic ecotoxicity potential (FAETP), marine aquatic ecotoxicity potential (MAETP), terrestrial ecotoxicity potential (TETP)] for the supply of [C$_6$mim][BF$_4$], methanol, acetone, toluene, benzene and water (scaled to the results of [C$_6$mim][BF$_4$]).

result (Figure 12.5, error bar). Against this background, the approach is therefore sufficient regarding the goal of these investigations.

The life cycle impact of the manufacture of the ionic liquid is higher than for all molecular solvents investigated, primarily due to the material intensity, namely organic agents and solvents, along the extensive pathway of synthesis and work-up. Zhang et al. [27] performed a similar LCA comparison considering the supply of [C$_4$mim][BF$_4$]. They attributed the higher impact of the ionic liquid in most LCA categories also to the length of the pathway and large amounts of organic solvents and materials required during synthesis.

In conclusion, the ionic liquid investigated features no relevant vapor pressure and hence does not emit volatile organic compounds (VOCs) – nevertheless, claiming it *a priori* as a "green" substitute for conventional solvents is too narrowly considered. Potentially, the environmental impact resulting from the manufacturing phase of [C$_6$mim][BF$_4$] can be compensated through an advantageous application featuring higher performances in combination with efficient recycling and reuse.

These insights gained became available only under consideration of upstream- and downstream processes. Thus, they are an essential part of a holistic evaluation and classification of a solvent concerning its environmental impact and indicate the need for the assessment of ionic liquids via a life cycle approach.

12.4
Assessment of the Application of Ionic Liquids in Contrast to Molecular Solvents

As dicussed before, choosing a suitable solvent for a process or searching for alternative technologies, environmental, health and safety criteria should be considered in addition to physical and chemical properties of a solvent. With regard to this, the efforts to eliminate, replace, recycle or minimize the use of solvents should start in the earliest stage of product/process development.

Again, the ECO method was applied in order to compare the solvent performance of the ionic liquid [C$_6$mim][BF$_4$] with that of molecular solvents in a holistic approach. For this purpose, the Diels–Alder reaction was chosen as a potential application at the chemical R&D stage.

12.4.1
Case Study: Diels–Alder Reaction

Ionic liquids as reaction media and (acidic) catalysts for Diels–Alder reactions have been the topic of numerous studies (e.g. [5–10]). Several groups [6, 10, 76] have investigated the effect of hydrogen bonds between starting material and solvent molecules (ionic liquid) in terms of selectivity and reaction rate. The C-2 imidazolium proton shows a significant Lewis acid character and is able to coordinate to the carbonyl oxygen of the methyl acrylate molecule during the reaction, stabilizing the transition state of the cycloadduct and hence leading to the preferential formation of the *endo* product. To compare the performance of alternative solvents, the reaction of cyclopentadiene and methyl acrylate (Scheme 12.2) was chosen as an example, leading to a mixture of *endo*- and *exo*-bicyclo[2.2.1]hept-5-ene-2-carboxylic acid methyl esters [74]. [C$_6$mim][BF$_4$] was compared with an assortment of conventional organic solvents and also a solvent-free version (without additional solvent applied in the Diels–Alder reaction).

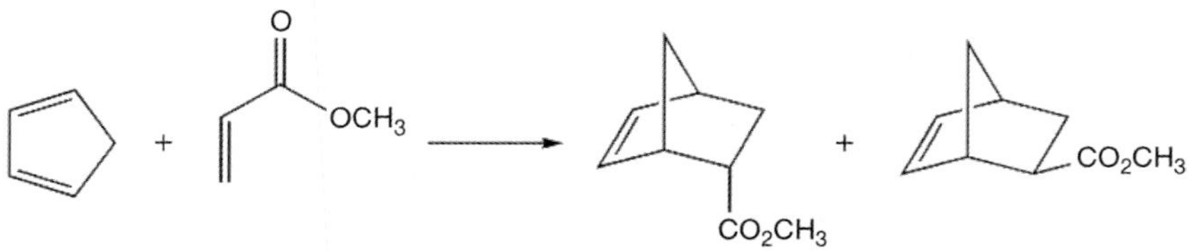

Scheme 12.2 Diels–Alder reaction of cyclopentadiene and methyl acrylate.

Some significant results are discussed in detail below. The process parameter, described in Table 12.2, represented the basis for the evaluation, since in all cases similar and nearly quantitative conversions were reached. Furthermore, the conversions and yields reached a plateau during this time [74].

Table 12.2 Diels–Alder reaction of cyclopentadiene and methyl acrylate. Solvents used and their performances (59 mmol of methyl acrylate per 15 ml of solvent, $n_{\text{methyl acrylate}}$: $n_{\text{cyclopentadiene}} = 1:1.2$, reaction temperature $T = 25\,°C$, reaction time $t = 48\,h$).

Experiment No.	Solvent	Methyl acrylate conversion (%)[a]	Endo:exo ratio[a]
1	Methanol	95	4.9
2	Methanol–water (1 : 1 v/v)	98	5.5
3	Acetone	83	3.3
4	Cyclohexane	90	2.6
5	[C$_6$mim][BF$_4$]	92	3.8
6	Solvent free	98	2.9

[a]Determined by gas chromatography.

12.4.1.1 Evaluation of the Solvent Performance

In general, the ionic liquid [C$_6$mim][BF$_4$] showed a similar chemical performance in the Diels–Alder reaction to conventional solvent systems at room temperature. The solvents methanol and methanol–water give the highest *endo:exo* ratios at room temperature. In the case of the solvent-free route, good conversions are also obtained, although the *endo:exo* ratio is lower than for the reactions with added solvents. From a purely chemical point of view, the methanol-based systems perform better in terms of conversion and selectivity. The question now arises of how the different solvent systems compare with regard to ecological performance.

12.4.1.2 Evaluation of the Energy Factor *EF*

Figure 12.6 demonstrates the *EF* for the synthesis of 1 kg of methyl *endo*-bicyclo[2.2.1]-hept-5-ene-2-carboxylate. Energy requirements for work-up and stirring were measured on a laboratory scale, whereas the energy demand for the supply of solvents for synthesis and extraction and also starting materials were determined using the Ecoinvent database [44]. The best results with regard to the energy consumption were obtained for methanol, methanol–water and the solvent-free system. This can be explained on the one hand by the comparatively low energy demand for the supply of the solvents and on the other by the solvent performance during the Diels–Alder reaction. Due to the lower selectivity, the *EF* for the supply of the reactants for the solvent-free route is higher than for methanol and methanol–water. In spite of this, the *EF* is comparably low, since no additional energy demand for the solvent supply arises. Furthermore, the work-up procedure involves no additional solvent distillation steps increasing the *EF*. The *EF* reflects two weak points of the application of [C$_6$mim][BF$_4$] for this type of Diels–Alder reaction: first, the supply of [C$_6$mim][BF$_4$] is energetically disadvantageous in comparison with the conventional solvent systems, primarily due to the material intensity along the extensive pathway of the ionic liquid manufacture, as discussed above, and second, the work-up procedure is more complex and results in a higher *EF*, since additional organic solvents become necessary to extract the product from the ionic liquid phase, whereas in the case

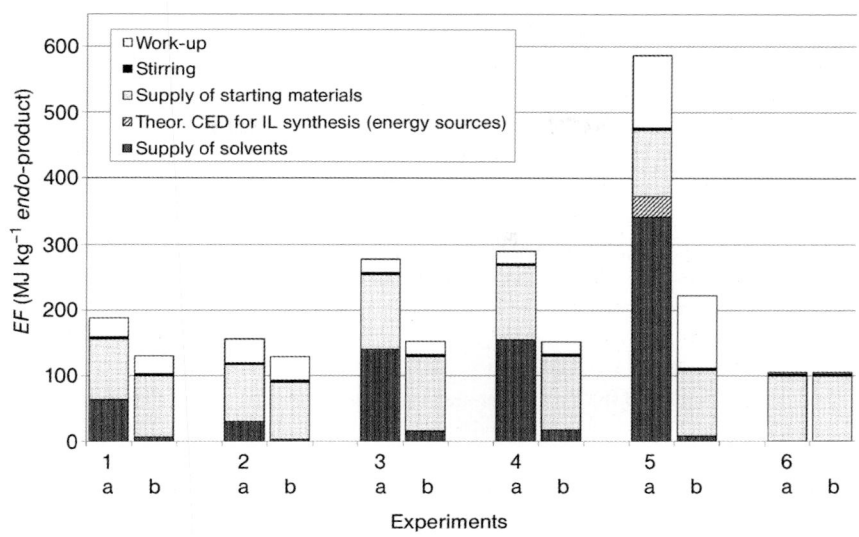

Figure 12.6 EF for the Diels–Alder reaction of methyl acrylate and cyclopentadiene. Solvents: methanol, methanol–water, acetone, cyclohexane, [C$_6$mim][BF$_4$] and solvent-free. For reaction conditions, see Table 12.2. (a) No recycling considered; (b) 100-fold use considered.

of organic solvents, solvent recovery and therefore product isolation are achieved by distillation.

One possibility for increasing the energy efficiency of the ionic liquid in the Diels–Alder synthesis could be an appropriate solvent recycling strategy. Following this approach, [C$_6$mim][BF$_4$] was recycled and reused three times. After three recycling steps, a mass loss of 5% was determined, while the solvent performance remained unchanged (Table 12.3).

If ionic liquids are to represent an ecological and further economic alternative to conventional solvent systems, their process advantages such as solvent recovery and

Table 12.3 Diels–Alder reaction of cyclopentadiene and methyl acrylate. Reaction conversions and selectivities with recycled [C$_6$mim][BF$_4$] (59 mmol of methyl acrylate per 15 ml of ionic liquid, $n_{\text{methyl acrylate}}:n_{\text{cyclopentadiene}} = 1:1.2$, reaction temperature $T = 25\,°C$, reaction time $t = 48$ h).

Cycle	Total mass loss (%)	Endo:exo ratio[a]	Methyl acrylate conversion (%)[a]
1		3.8	92
2	3.1	3.7	96
3	4.3	3.7	97
4	5.0	3.7	98

[a]Determined by gas chromatography.

solvent/catalyst recycling are of particular importance and processes using ionic liquids should be characterized by long-term use. Therefore, a 100-fold recycling of the solvents was assumed in order to demonstrate the impact on ecological and economic criteria (the authors assumed that the solvent performance does not diminish within 100 runs and the mass loss of 5% per four runs does not increase. In the case of the molecular solvents, due to their vapor pressure and type of use on laboratory scale, a total mass loss of 10% was assumed). Figure 12.6 shows a significant improvement in energy efficiency concerning the supply of [C_6mim][BF_4], resulting in a lower EF (reduction of ~98%, black bars) associated with a reduced life cycle impact. However, the work-up procedure seems to become a new bottleneck (white bars).

12.4.1.3 Evaluation of the Environmental and Human Health Factor *EHF* - Examples

The investigations initially addressed toxicological aspects of the solvent choice by means of criteria concerning aspects of mobility, acute toxicity and chronic toxicity for humans, acute toxicity for aquatic organisms, persistence in the environment and bioaccumulation, using the data given in safety data sheets [69]. Quantitative data regarding toxicity for humans, ecological effects and information about the accumulation and biodegradability of [C_6mim][BF_4] are hardly or not available. Due to insufficient data concerning environmental and human health concerns, a worst-case scenario was assumed and the risks were classified as high.

Figure 12.7 demonstrates exemplarily the criterion *EHF(AcT)* regarding the alternative solvents used. Methanol is classified as a toxic substance and therefore has a significant impact on human health. Although the risk is assumed to be high in the case of [C_6mim][BF_4], the resulting acute toxicity for humans is lower than for methanol-based systems, since [C_6mim][BF_4] has no relevant vapor pressure and thus represents a low imminent hazard. However, due to the additional solvent (diethyl ether) for extraction in this case, the performance of the ionic liquid is worse than with the use of solvents with low boiling points, which can be removed from the product phase simply by distillation.

12.4.1.4 Evaluation of the Cost Factor *CF*

Figure 12.8 displays the cost factor *CF* for the synthesis of 1 kg of methyl *endo*-bicyclo [2.2.1]hept-5-en–2-carboxylate under consideration of a 100-fold solvent reuse strategy [77]. Here, the costs of starting-materials, solvents for reaction and work-up and energy costs were considered. Personnel costs were not included, since their calculation on a laboratory scale would be not representative.

The conventional solvents are readily available and relatively cheap and hence the cost factor is mainly defined by the supply of the starting materials. Therefore, in the case of 100-fold use, the *CF* depends mainly on the performance of the different reaction media. In the case of the ionic liquid, the additional solvent demand for the extractant and its provision lead to a comparably higher *CF* than for the synthesis in conventional solvents and the solvent-free synthesis route.

For the present assessment of the *CF*, an ionic liquid price of 667 € kg^{-1} was taken into account [77]. Interestingly, this cost assessment shows that a reduction of the ionic

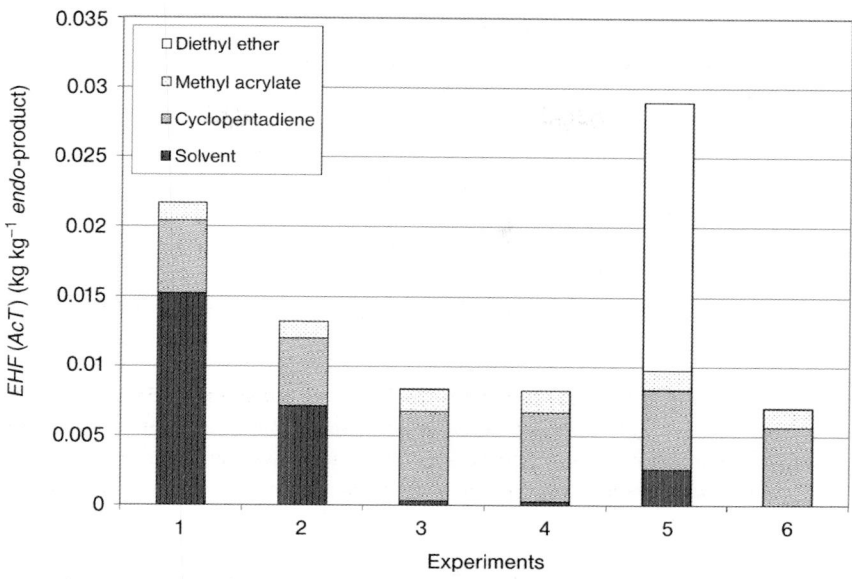

12.4 Assessment of the Application of Ionic Liquids in Contrast to Molecular Solvents | 333

Figure 12.7 Dependence of $EHF(AcT)$ on the choice of the solvent for the Diels–Alder reaction of methyl acrylate and cyclopentadiene. Solvents: methanol, methanol–water, acetone, cyclohexane, [C$_6$mim][BF$_4$] and solvent free. For reaction conditions, see Table 12.2.

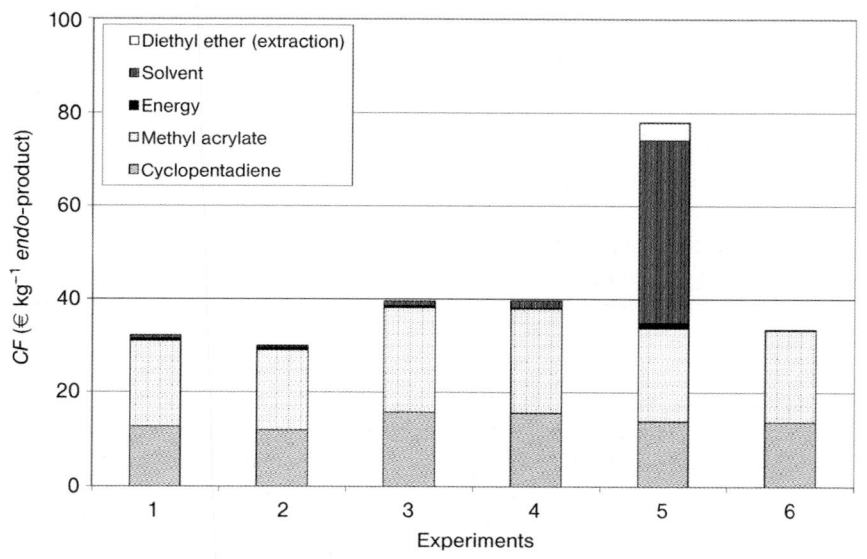

Figure 12.8 Dependence of CF on the choice of the solvent system for the Diels–Alder reaction at 100-fold use. Solvents: methanol, methanol–water, acetone, cyclohexane, [C$_6$mim][BF$_4$] and solvent free. For reaction conditions, see Table 12.2.

liquid price to 22 € kg^{-1} would allow for the production of 1 kg of the *endo* product (assumption: production price of the *endo* product = 40 € kg^{-1}) at a similar price to that with the organic solvents, with 100-fold use of the solvent. This is in accordance with predictions of ionic liquid manufacturers, who expect that a range of ionic liquids will become commercially available at €25–50 per liter on a ton scale [78].

12.4.1.5 Alternative Ionic Liquid Choices

As pointed out above, several authors have found that the outcome of the Diels–Alder reaction is highly dependent on the choice of the cation and the anion of the ionic liquid. Summarizing the evidence presented in the literature, it can be said that the activating interaction of the cation is via hydrogen bonding with the carbonyl moiety of the dienophile. Such an interaction can be achieved using either 1,3-dialkylimidazolium- [10, 76] or 1*H*-3-alkylimidazolium-based ("protonated") ionic liquids [6, 7], given that the hydrogen atoms on the cation are available for interaction with the substrate (i.e. they are not involved in strong interactions with the ionic liquid anion). Hence high *endo:exo* ratios (up to 6.0 for cyclopentadiene and methyl acrylate) have been reported when using anions with little coordinating ability, such as [Tf$_2$N]$^-$ [6].

In a best-case scenario for the ionic liquid [C$_6$mim][BF$_4$], the selectivity and conversion resulting from the solvent system methanol–water were used to determine the parameter sensitivity of *EF*, *CF* and *EHF*. In accordance with a yield improvement of the desired *endo* product and by considering 100-fold use, the parameter values decrease, but only by 14%, indicating the great ecological and economic impact of the supply and use of the ionic liquid and also the ecological impact of additional work-up steps. Hence, further improvement of the performance of ionic liquids with regard to both chemical and ecological aspects point to the use of protonated ionic liquids. Protonated ionic liquids are produced by simply adding an amine, such as 1-methylimidazole, to equimolar amounts of acid, which reduces the *EHF*, *EF* and *CF* of the ionic liquid production process considerably. Although the use of these ionic liquids containing non-fluorinated anions leads to lower *endo:exo* ratios compared with fluorinated anions with lower anion–cation interaction, an overall lower *EHF*, *EF* and *CF* of the process is expected. Furthermore, the selectivity may be improved by using catalysts [5, 8]. An additional optimization potential lies in the optimization of the temperature and reactant concentrations, which are factors well known to affect both yield and selectivity [7], and also the optimization of the product isolation and solvent recycling.

12.4.1.6 Decision Support

In the case of the Diels–Alder reaction investigated, the solvent system methanol–water and the solvent-free synthesis were indicated by the ECO method to be the most ecologically sustainable alternatives. The outranking of the different solution candidates with consideration of 100-fold use was determined using the software Decision Lab 2000 [71] and resulted in the following order of preference: solvent-free ≥ methanol–water > methanol > acetone > cyclohexane > [C$_6$mim][BF$_4$].

Nevertheless, the use of ionic liquids in the Diels–Alder reaction instead of water-containing systems may be preferable if moisture-sensitive reactants are used. In addition, the use of catalysts in ionic liquids might be advantageous compared with conventional solvents if the catalyst remains in the reaction medium after work-up. Furthermore, the ecological and economic impact might be reduced by applying protonated ionic liquids. However, reactions in media with no relevant vapor pressure often require additional solvents during work-up, which might affect environmental aspects adversely. Therefore, multiphasic systems, which often can be established in ionic liquid processes, may in certain instances provide more efficient pathways for practical applications. As a consequence, processes should be designed in which the ionic liquid (and a potential catalyst) forms a stationary phase remaining in the reactor, so that the energetic, toxicological and cost aspects discussed above will become less relevant.

Finally, it should be pointed out that the ionic liquids considered here represent only a few selected examples of the huge class of ionic liquids and there is great potential for further optimization by changing the ionic liquid, for instance by considering the environmental and human health factor or by considering structure–activity relationships during product/process development. An evaluation broader in scope has to be the focus of future research to close this gap of knowledge.

In fact, a synthetic application that works very well in cheap and ecologically benign solvents such as water and alcohols may not be the ideal field of green ionic liquid application (see Chapter 3). This is simply due to the fact that ionic liquids are complex and sophisticated solvents and their use has a much higher potential to be green if a performance can be realized that is not possible with water or alcohols (see Chapter 3 for many examples). This should be kept in mind in all cases where ionic liquids are investigated as solvent substitutes. In this respect, the example of the Diels–Alder reaction turns out to be particularly unfavorable for the use of ionic liquids.

12.5
Conclusions

To evaluate ionic liquid life cycle stages in the context of application-oriented chemical R&D, methods such as the ECO method can serve as an instrument for decision support and optimization, in order to develop environmentally benign and cost-optimized products and processes. To meet the limitations of data and time during the early stages of product and process development, the application of such a simplified life cycle approach has been proven to be suitable for screening and hot-spot detection along the entire process chain. The results obtained from this should be scrutinized regarding validity and correctness by iteratively increasing the detail while the knowledge increases with ongoing development progress so that a comprehensive LCA is finally performed, quantifying the environmental impact potentials related to this process in a holistic way.

The implementation of this method in ionic liquid R&D demonstrates the high optimization potential for common synthetic strategies for these compounds and emphasizes the need for a critical evaluation already at early process development stages. Furthermore, the life cycle impact of the ionic liquid investigated is higher than for the selected molecular solvents, primarily due to the extensive pathway of its manufacture. Potentially, the ecological and economic impacts resulting from the manufacture can be compensated within the application phase.

The application of ionic liquids in the Diels–Alder reaction was chosen exemplarily from the scientific literature investigating the reaction performances of ionic liquids. In spite of the actually still low commercial relevance, it reflects the advantages, disadvantages and challenges of ionic liquid technology.

Ionic liquids can become attractive alternatives to conventional solvents if their use results in an essential improvement in a specific field of application (e.g. acid scavengers, electrolytes for dye-sensitized solar cells, absorption chillers, lubricants and solvent applications, as demonstrated in other chapters of this book). When used as solvents, the property of low volatility can be used to enhance the separation efficiency and recyclability, if the product phase forms a second layer and a potentially used catalyst remains selectively in the ionic liquid phase. By this means, the higher ecological and economic impact of ionic liquid manufacture in comparison with conventional solvents can be reduced or even over-compensated in selected examples.

References

1 Dyson, P.J. and Zhao, D., (2005) Hydrogenation, in *Multiphase Homogeneous Catalysis*, vol. 2, (eds B. Cornils, W. A. Herrmann, I.T. Horvath, W. Leitner, S. Mecking, H. Olivier – Bourbigou and D. Vogt), Wiley-VCH Verlag GmbH, Weinheim, pp. 494–511.

2 Seddon, K.R. and Stark, A. (2002) Selective catalytic oxidation of benzyl alcohol and alkylbenzenes in ionic liquids. *Green Chemistry*, **4** (2), 119–123.

3 Sheldon, R. (2001) Catalytic reactions in ionic liquids. *Chemical Communications*, 2399–2407.

4 Welton, T. (1999) Room-temperature ionic liquids. Solvents for synthesis and catalysis. *Chemical Reviews*, **99** (8), 2071–2083.

5 Earle, M.J., McCormac, P.B. and Seddon, K.R. (1999) Diels–Alder reactions in ionic liquids. *Green Chemistry*, **1** (1), 23–25.

6 Janus, E., Goc-Maciejewska, I., Lozynski, M. and Pernak, J. (2006) Diels–Alder reaction in protic ionic liquids. *Tetrahedron Letters*, **47** (24), 4079–4083.

7 Fischer, T., Sethi, A., Welton, T. and Woolf, J. (1999) Diels–Alder reactions in room-temperature ionic liquids. *Tetrahedron Letters*, **40** (4), 793–796.

8 Silvero, G., Arevalo, M.J., Bravo, J.L., Avalos, M., Jimenez, J.L. and Lopez, I. (2005) An in-depth look at the effect of Lewis acid catalysts on Diels–Alder cycloadditions in ionic liquids. *Tetrahedron*, **61** (39), 7105–7111.

9 Bartsch, R.A. and Dzyuba, S.V. (2003) Polarity variation of room temperature ionic liquids and its influence on a Diels–Alder reaction, in *Ionic Liquids as Green Solvents: Progress and Prospects*, vol. 856 (eds R.D. Rogers and K.R. Seddon), ACS Symposium Series, American Chemical Society, Washington, DC, pp. 289–299.

10 Aggarwal, A., Lancaster, N.L., Sethi, A.R. and Welton, T. (2002) The role of hydrogen bonding in controlling the selectivity of Diels–Alder reactions in room-temperature ionic liquids. *Green Chemistry*, **4** (5), 517–520.

11 Visser, A.E., Swatloski, R.P. and Rogers, R.D. (2000) pH-Dependent partitioning in room temperature ionic liquids. *Green Chemistry*, **2** (1), 1–4.

12 Dietz, M.L. (2006) Ionic liquids as extraction solvents: where do we stand? *Separation Science Technology*, **41** (10), 2047–2063.

13 Alonso, L., Arce, A., Francisco, M. and Soto, A. (2008) Solvent extraction of thiophene from n-alkanes (C_7, C_{12} and C_{16}) using the ionic liquid [C_8mim][BF_4]. *Journal of Chemical Thermodynamics*, **40** (6), 966–972.

14 Sheldon, R.A., Lau, R.M., Sorgedrager, M.J., van Rantwijk, F. and Seddon, K.R. (2002) Biocatalysis in ionic liquids. *Green Chemistry*, **4** (2), 147–151.

15 Abbott, A.P. and McKenzie, K.J. (2006) Application of ionic liquids to the electrodeposition of metals. *Physical Chemistry Chemical Physics*, **8** (37), 4265–4279.

16 Endres, F. (2007) Ionic liquids for metal deposition. *Nachrichten Chemie*, **55** (5), 507–511.

17 Sakaebe, H. and Matsumoto, H. (2003) N-Methyl-N-propylpiperidinium bis (trifluoromethanesulfonyl)imide (PP13-TFSI) – novel electrolyte base for Li battery. *Electrochemistry Communications*, **5** (7), 594–598.

18 Howlett, P.C., MacFarlane, D.R. and Hollenkamp, A.F. (2004) High lithium metal cycling efficiency in a room-temperature ionic liquid. *Electrochemistry Solid-State Letters*, **7** (5), A97–A101.

19 Silvester, D.S. and Compton, R.G. (2006) Electrochemistry in room temperature ionic liquids: a review and some possible applications. *Zeitschrift Fur Physikalische Chemie*, **220** (10–11), 1247–1274.

20 Wei, D. and Ivaska, A. (2008) Applications of ionic liquids in electrochemical sensors. *Analytica Chimica Acta*, **607** (2), 126–135.

21 Anastas, P.T. and Warner, J. (1998) *Green Chemistry: Theory and Practice*, Oxford University Press, New York.

22 Nelson, W.M. (2002) Are ionic liquids green solvents? in *Ionic Liquids; Industrial Applications to Green Chemistry*, vol. 818 (eds R.D. Rogers and K.R. Seddon), ACS Symposium Series, American Chemical Society, Washington, DC, pp. 30–41.

23 Kralisch, D. (2006) Ökologische Nachhaltigkeit im Fokus der chemischen Forschung und Entwicklung, PhD thesis, Friedrich-Schiller University, Jena.

24 Kralisch, D., Reinhardt, D. and Kreisel, G. (2007) Implementing objectives of sustainability into ionic liquids research and development. *Green Chemistry*, **9** (12), 1308–1318.

25 Jastorff, B., Stoermann, R., Ranke, J., Moelter, K., Stock, F., Oberheitmann, B., Hoffmann, W., Hoffmann, J., Nuechter, M., Ondruschka, B. and Filser, J. (2003) How hazardous are ionic liquids? Structure–activity relationships and biological testing as important elements for sustainability evaluation. *Green Chemistry*, **5** (2), 136–142.

26 Jastorff, B., Moelter, K., Behrend, P., Bottin-Weber, U., Filser, J., Heimers, A., Ondruschka, B., Ranke, J., Schaefer, M., Schroeder, H., Stark, A., Stepnowski, P., Stock, F., Stoermann, R., Stolte, S., Welz-Biermann, U., Ziegert, S. and Thoeming, J. (2005) Progress in evaluation of risk potential of ionic liquids-basis for an eco-design of sustainable products. *Green Chemistry*, **7** (5), 362–372.

27 Zhang, Y., Bakshi, B.R. and Demessie, E.S. (2008) Life cycle assessment of an ionic liquid versus molecular solvents and their applications. *Environmental Science & Technology*, **42** (5), 1724–1730.

28 Curzons, A.D., Constable, D.J.C. and Cunningham, V.L. (1999) Solvent selection guide: a guide to the integration of environmental, health and safety criteria into the selection of solvents. *Clean Products Processes*, **1**, 82–90.

29 Alfonsi, K., Colberg, J., Dunn, P.J., Fevig, T., Jennings, S., Johnson, T.A., Kleine, H.P., Knight, C., Nagy, M.A., Perry,

D.A. and Stefaniak, M. (2008) Green chemistry tools to influence a medicinal chemistry and research chemistry based organisation. *Green Chemistry*, **10** (1), 31–36.

30 Fleischer, G. and Schmidt, W.P. (1997) Iterative screening LCA in an eco-design tool. *International Journal of Life Cycle Assessment*, **2** (1), 20–24.

31 Gani, R. (2004) Chemical product design: challenges and opportunities. *ComputerS & Chemical Engineering*, **28** (12), 2441–2457.

32 Gani, R., Jimenez-Gonzalez, C. and Constable, D.J.C. (2005) Method for selection of solvents for promotion of organic reactions. *ComputerS & Chemical Engineering*, **29** (7), 1661–1676.

33 Darvin, C.H. and Monroe, K. (1997) SAGE Solvent Alternatives GuidE: system improvements for selecting industrial surface cleaning alternatives. *Metal Finishing*, **95** (3), 24–25.

34 Cabezas, H., Zhao, R., Bare, J.C. and Nishtala, S.R. (1999) Designing environmentally benign solvent substitutes. *NATO Science Ser. 2: Environment Security*, **62**, 317–331.

35 Cabezas, H., Harten, P.F. and Green, M.R. (2000) Designing greener solvents. *Chemical & Engineering*, **107** (3), 107–109.

36 Li, M., Harten, P.F. and Cabezas, H. (2002) Experiences in designing solvents for the environment. *Industrial & Engineering Chemistry Research*, **41** (23), 5867–5877.

37 Jimenez-Gonzales, C., Curzons, A.D., Constable, D.J.C. and Cunningham, V.L. (2005) Expanding GSK's Solvent Selection Guide – application of life cycle assessment to enhance solvent selections. *Clean Technology Environment Policy*, **7** (2), 42–50.

38 Capello, C., Hellweg, S. and Hungerbühler, K. (2006) The Ecosolvent Tool, ETH Zurich, Safety and Environmental Technology Group, Zurich, http://www.sust-chem.ethz.ch/tools/ecosolvent.

39 Capello, C., Fischer, U. and Hungerbühler, K. (2007) What is a green solvent? A comprehensive framework for the environmental assessment of solvents. *Green Chemistry*, **9** (9), 927–934.

40 Elgue, S., Prat, L., Cognet, P., Cabassud, M., Le Lann, J.M. and Cezerac, J. (2004) Influence of solvent choice on the optimization of a reaction-separation operation: application to a Beckmann rearrangement reaction. *Separation and Purification Technology*, **34** (1–3), 273–281.

41 Elgue, S., Prat, L., Cabassud, M. and Cezerac, J. (2006) Optimisation of solvent replacement procedures according to economic and environmental criteria. *Biochemical Engineering Journal*, **117** (2), 169–177.

42 Koller, G., Fischer, U. and Hungerbühler, K. (2000) Assessing Safety, health and environmental impact early during process development. *Industrial & Engineering Chemistry Research*, **39** (4), 960–972.

43 Sugiyama, H., Fischer, U. and Hungerbühler, K. (2006) The EHS Tool, ETH Zurich, Safety and Environmental Technology Group, Zurich, http://sust-chem.ethz.ch/tools/EHS.

44 Ecoinvent database by Frischknecht, R., Jungbluth, N., Althaus, H.-J., Bauer, C., Doka, G., Dones, R., Hischier, R., Nemecek, T., Primas, A. and Wernet, G. Swiss Centre for Life Cycle Inventories, Dübendorf. Versions used here: v.1.3 (2006), v.2.0 (2007).

45 ISO 14040:2006 (2006) *Environmental Management – Life Cycle Assessment – Principles and Framework*. European Committee for Standardization, Brussels.

46 ISO 14044:2006 (2006) *Environmental Management – Life Cycle Assessment – Requirements and Guidelines*. European Committee for Standardization, Brussels.

47 Burgess, A.A. and Brennan, D.J. (2001) Application of life cycle assessment to chemical processes. *Chemical Engineering Science*, **56** (8), 2589–2604.

48 Jödicke, G., Zenklusen, O., Weidenhaupt, A. and Hungerbühler, K. (1999) Developing environmentally-sound

processes in the chemical industry: a case study on pharmaceutical intermediates. *Journal of Cleaner Production*, **7**, 159–166.

49 Hellweg, S., Fischer, U., Scheringer, M. and Hungerbühler, K. (2004) Environmental assessment of chemicals: methods and application to a case study of organic solvents. *Green Chemistry*, **6** (8), 418–427.

50 Kralisch, D. and Kreisel, G. (2007) Assessment of the ecological potential of microreaction technology. *Chemical Engineering Science*, **62** (4), 1094–1100.

51 Biwer, A. and Heinzle, E. (2001) Prozesssimulation zur frühen ökologischen Bewertung biotechnologischer Prozesse: Beispiel Zitronensäure. *Chemie Ingenieur Technik*, **73** (11), 1467–1472.

52 Hoffmann, V.H., Hungerbühler, K. and McRae, G.J. (2001) Multiobjective screening and evaluation of chemical process technologies. *Industrial & Engineering Chemistry Research*, **40** (21), 4513–4524.

53 Trost, B.M. (1991) The atom economy: a search for synthetic efficiency. *Science*, **254** (5037), 1471–1477.

54 Trost, B.M. (1995) Atom economy – a challenge for organic synthesis: homogeneous catalysis leads the way. *Angewandte Chemie (International Edition in English)*, **34** (3), 259–281.

55 SETAC, (1997) *Simplifying LCA: Just a Cut? – Final Report of the SETAC–Europe Screening and Streamlining Working Group*, Society of Environmental Chemistry and Toxicology (SETAC), Brussels.

56 Gasafi, E., Meyer, L. and Schebek, L. (2003) Using life-cycle assessment in process design. *Journal of Industrial Economics*, **7** (3–4), 75–91.

57 VDI, (1997) *VDI-Richtlinie 4600: Cumulative Energy Demand – Terms, Definitions, Methods of Calculation*. Verein Deutscher Ingenieure, Düsseldorf.

58 Schmidt-Bleek, F. (1993) MIPS – a universal ecological measure? *Fresenius' Environment Bulletin*, **2**, 306–311.

59 Fleischer, G., Gerner, K., Kunst, H., Lichtenvort, K. and Rebitzer, G. (2001) A semi-quantitative method for the impact assessment of emissions within a simplified life cycle assessment. *International Journal of Life Cycle Assessment*, **6** (3), 149–156.

60 Weidenhaupt, A. and Hungerbühler, K. (1997) Integrated product design in chemical industry. A plea for adequate life-cycle screening indicators. *Chimia*, **51** (5), 217–221.

61 Hochschorner, E. and Finnveden, G. (2003) Evaluation of two simplified life cycle assessment methods. *International Journal of Life Cycle Assessment*, **8** (3), 119–128.

62 Hur, T., Lee, J., Ryu, H. and Kwon, E. (2005) Simplified LCA and matrix methods in identifying the environmental aspects of a product system. *Journal of Environmental Management*, **75** (3), 229–237.

63 Kniel, G.E., Delmarco, K. and Petrie, J.G. (1996) Life cycle assessment applied to process design: Environmental and economic analysis and optimisation of a nitric acid plant. *Environmental Progress*, **15** (4), 221–228.

64 Azapagic, A. and Clift, R. (1999) The application of life cycle assessment to process optimisation. *Computers & Chemical Engineering*, **23** (10), 1509–1526.

65 Azapagic, A. (1999) Life cycle assessment and its application to process selection, design and optimization. *Biochemical Engineering Journal*, **73** (1), 1–21.

66 Huijbregts, M.A.J., Rombouts, L.J.A., Hellweg, S., Frischknecht, R., Hendriks, A.J., Van de Meent, D., Ragas, A.M.J., Reijnders, L. and Struijs, J. (2006) Is cumulative fossil energy demand a useful indicator for the environmental performance of products? *Environmental Science & Technology*, **40** (3), 641–648.

67 PRé Consultants (2006) SimaPro v.7, PRé Consultants, Amersfoort, The Netherlands.

68 PROBAS (Prozessorientierte Basisdaten für Umweltmanagement-Instrumente),

http://www.probas.umweltbundesamt.de/php/index.php, (2008).

69 Merck KGaA (2008) Safety Data Sheets, Merck KGaA, Darmstadt.

70 EPA . (2000) EPIWIN, v.3.11, Office of Pollution Prevention and Toxics, US Environmental Protection Agency, Washington, DC.

71 Visual Decision Decision Lab 2000, v. 1.01.0386, Visual Decision, Montreal.

72 Brans, J.P., Vincke, P. and Mareschal, B. (1986) How to select and how to rank projects: the PROMETHEE method. *European Journal of Operational Research*, 24 228–238.

73 Kralisch, D., Stark, A., Koersten, S., Kreisel, G. and Ondruschka, B. (2005) Energetic, environmental and economic balances: spice up your ionic liquid research efficiency. *Green Chemistry*, **7** (5), 301–309.

74 Reinhardt, D., Ilgen, F., Kralisch, D., Koenig, B. and Kreisel, G. (2008) Evaluating the greenness of alternative reaction media. *Green Chemistry*, **10** (11), 1170–1181.

75 Menschutkin, N. (1890) *Zeitschrift fur Physiologische Chemie*, **6**, 41–57.

76 Nobuoka, K., Kitaoka, S., Kunimitsu, K., Iio, M., Harran, T., Wakisaka, A. and Ishikawa, Y. (2005) Camphor ionic liquid: correlation between stereoselectivity and cation–anion interaction. *The Journal of Organic Chemistry*, **70** (24), 10106–10108.

77 Chemical prices: online quote request, www.merck.de, February 2008; in the case of [C_6mim][BF_4] online quote request, www.solvent-innovation.com, April 2008.

78 Hilgers, C. and Wasserscheid, P. (2003) Quality aspects and other questions related to commercial ionic liquid production, in *Ionic Liquids in Synthesis* (eds P. Wasserscheid and T. Welton), Wiley-VCH Verlag GmbH, Weinheim, p. 21.

Index

a

abiotic degradation 290
– anions 287
absorption chillers 221, 222
– single-effect 222
abstraction
– thermal 67
acidic impurities 117
acids
– acidic oxides 55
– biphasic acid scavenging utilizing ionic liquids 300
– Brønsted 53, 101
– catalysts 55, 329
– Lewis 67, 101
– methanesulfonic 110
– polylactic 93
– protonated acidic ILs 128
– protonic 55
– supported 55
acrylate
– methyl 329, 330
actinides 142
activation
– transition metal complexes 68
active enzymes 153
activity
– structure–activity relationships 236, 261
aerobic biodegradation
– testing procedures 267
air gap spinning 132
(S)-(+)-alapyridain 96
aliphatic hydrocarbons 145
alkoxides
– metal 55
1-alkoxymethyl-3-hydroxypyridinium salts 285

1-alkyl-3-methylimidazolium-based ionic liquids 323
1-alkyl-3-methylimidazolium halide salts 8, 11, 12
1-alkyl-3-methylimidazolium ionic liquids 194
alkylation
– Friedel–Crafts 54
– N-methylimidazole 323, 324
alkylimidazolium cations 227
alkyls
– metal 55
alkyltrimethylammonium salts 286
allyl chloride 74
alumina membranes
– porous 167
amines
– tertiary 300
Ammoeng 110, 153
ammonia
– solubility 227
ammonium compounds 282
ammonium salts 254, 286
anion hydrolysis 81
anionic moieties 264
anions
– biotically degradable 287
– dimethylphosphate 227
– ionic liquids 241
– non-halide 28
antimicrobial properties
– ionic liquids 254
applications
– COSMO-RS 173
– HMF 98
– IL lubricants 217
– ILs in electrolyte systems 191
aquatic test systems 243

aqueous nitrate media 142
aqueous two-phase system (ATPS) 151
aromatic–aliphatic separation 145, 156
aromatic hydrocarbons
– extraction/separation 145, 156
atom economy/efficiency 7, 318
ATPS, see aqueous two-phase system
auxiliary substances
– 12 principles of green chemistry 10

b

bacteria 243
ball-on-disc tribotester 204
baseline toxicity 292
BASF BASIL process 49, 299
batch synthesis 17
batches
– technical-grade 111
bath
– electroplating 195
benzene
– monoalkyl- 57
– 1,2,4-trichloro- 78
benzylalkyldimethylammonium salts 286
bilayer
– lipid- 262
bioavailability 258
– pollutants 261
– xenobiotics 292
biochemical oxygen demand (BOD) 267
biodegradability
– aerobic testing procedures 267
– ammonium compounds 282
– IL anions 269
– imidazolium-based ionic liquids 270
– imidazolium compounds 283
– ionic liquids 34, 265, 293
– phosphonium compounds 282
– pyridinium-based ionic liquids 278
– tested ionic liquids 244
– testing 266
biodegradation 235
– primary 277, 281
bioethanol
– cellulosic 135
biological test systems 240
biomass fractionation 134
biotically degradable anion 287
biphasic acid scavenging utilizing ionic liquids (BASIL) 300
biphasic catalysis 72
– liquid–liquid 56
biphasic reaction systems 46
– IL–scCO$_2$ 75

biphasic separation
– hydrophobic compounds 141
bis(trifluoromethanesulfonyl)amide ionic liquids 197
BOD, see biochemical oxygen demand
borate
– trimethyl 157
boundary friction 209
bromide
– lithium 222
Brønsted acids
– homogeneous 101
– ionic liquids 53
bubbler
– oil 9
1-butene hydroformylation 84
1-butyl-4-(dimethylamino)pyridinium head group 264
1-butyl-3-methylimidazolium halide salts 9

c

cancerogenicity 263
capital-dependent costs 116
carbene complexes 68
carbon dioxide
– CO$_2$ production (biodegradability parameter) 267
– Henry constant 174
– separation 165
– supercritical
  – scCO$_2$ 75
carbon disulfide 125
catalysis
– biphasic 72
– heterogeneous 77
– hydroformylation 79
– liquid–liquid biphasic 56
– multiphase 75
– nanoparticle and nanocluster 75
– SILP 46
– SILP transition metal 71
– transition metal 65
catalysts
– acidic 55, 329
– immobilization 72
– interaction with ionic liquids 67
– ionic liquids 51
– ionic solutions 70
– palladium 76
– platinum 73
– recycling 72
– SCILL 78
– solubility and immobilization 80
cationic head groups 240, 257

cations
- alkylimidazolium 227
- 1,3-dialkylimidazolium 69
- ionic liquid 288
- metabolizable 287
- organic 195
CED, *see* cumulative energy demand
cells
- solar cells 198
cellulose
- derivatives 134
- dissolution and processing 123
- global production 99
- pulp 127
- regeneration 131
cellulose solutions
- rheological behavior 129
cellulosic bioethanol 135
cellulosic fibers 131, 133
centrifugal extractor 168
CF, *see* cost factor
chelate extraction 144
chemical feedstock
- cellulose 125
chemical reactions
- reactions 18
chemical stability 44
chemical structures
- T-SAR analysis 237
chemicals
- natural organic 123
- non-persistent 266
- platform 93
- REACH legislation 15, 235, 293
chillers
- absorption 221, 222
chlorides
- allyl 74
- hemihydrate 12
- 1-H–3-methylimidazolium 301
chloroaluminate 51, 68
clay minerals 258
"CO_2 Headspace" test 286
coefficient of performance (COP) 223, 226
colorless ionic liquids 31
commercial lubricants 210
complex viscosity 129
complexes
- Cu^{2+} 144
- metal–carbene 68
- transition metal 65, 68
compressibility
- ionic liquids 210

Concentration Addition (CA) concept 259
concentration effect 107
condensation
- fructose 100
conductive heating preparation 8, 9
conductivity
- electric 215
- specific 194, 196, 197
conductor-like screening model for real solvent (COSMO-RS) method 47, 173
contact measurement
- steel–steel 204
contactor
- rotating disc 169
continuous flow synthesis 19
continuous processing 114
continuous single-phase process 101
conventional heating 11
conversion efficiency 199
conversion of renewable resources 93
cooling efficiency 223, 226
copper complexes 144
corrosion 215, 225
cost assessment
- BASF BASIL process 308
- continuous processing of HMF 115
- extraction of aromatic hydrocarbons 146
cost factor (CF) 320, 332
cotton 133
coupling reactions
- Heck 76
- Suzuki 76
"cradle-to-grave" life-cycle assessment (LCA) 317, 323
cross-solubility 162
crown ether 143
crystal structures
- dimethylimidazolium chloride hemihydrate 13
crystallization behavior 224
cumulative energy demand (CED) 319
customer benefit 303
cyclopentadiene 329, 330
cytotoxicity 263
- and hydrophobicity 262

d

DDT (dichlorodiphenyltrichloroethane) 265
decolorization 5
- ionic liquids 31
- SWOT Analysis 31
decomposition temperature 225

degradation
– abiotic 290
– anions 287
– biodegradation, biodegradability 277
– UV 294
dehydration 105
denitrogenation 158
density 196
DEPP, see diethoxyphenylphosphine
derivatization 6
design
– inherently safe ionic liquids 235
– separation processes 172
– sustainable ionic liquids 238
desulfurization
– fuels 158
– oxidative 162
device stability 192
dialkoxyphenylphosphines 300
1,3-dialkylimidazolium cation 69
dialkylimidazolium salt 111
diazabicycloundecane 128
dichlorodiphenyltrichloroethane (DDT) 265
dichloromethane 31
die-away tests 269
Diels–Alder reaction 329, 330
diesel
– extraction 159
diethoxyphenylphosphine (DEPP) 303
diethylphosphate 269
dimer selectivity 73
N,N-dimethylacetamide 104
4-(dimethylamino)pyridinium compounds 284
dimethylcyclohexylammonium hydrogensulfate 78
dimethylimidazolium chloride hemihydrate 13
dimethylphosphate anion 227
dissolution
– cellulose 127, 137
dissolved organic carbon (DOC) 267
dissolved organic matter (DOM) 257
distillation 44
– extractive 153
– product isolation 70
distribution coefficient
– aromatic–aliphatic separation 147
disulfide
– carbon 125
DOM, see dissolved organic matter
drop size 169
drugs 152

dry/wet spinning 132
DSSC, see dye-sensitized solar cells
dye-sensitized solar cells (DSSC) 198
dynamic viscosity 194, 196

e
E-factor 7, 18
eco-efficiency
– analysis 299
– label 313
– portfolio 309
ECO method 317, 319, 334
ecological fingerprint 307
economic evaluation
– cost assessment 146
ecotoxicity
– ionic liquids 238, 239, 291
– surrounding medium 257
EDPP, see ethoxydiphenylphosphine
EF, see energy factor
efficiency
– atom economy 7
– atom 318
– cooling 223
– eco- 299, 309
– energy 18
– extraction 18
– power conversion 199
EHF, see environmental and human health factor
elastic liquids 20
elastohydrodynamic friction 209
electric conductivity 215
electrochemical stability 45, 196
electrochemical treatment 294
electrolyte properties
– ionic liquids 193
electrolyte systems 191
electron-rich heteroatoms 51
electrophilic substitution 54
electroplating bath 195
endo:exo ratio 330
energy consumption 304
energy demand
– cumulative 319
energy efficiency 18
energy factor (EF) 320, 330
environment 233
environmental and human health factor (EHF) 320, 332
environmental separation 158
enzymes 242, 255
– active 153
equilibrium

– vapor–liquid   227
ethoxydiphenylphosphine (EDPP)   303
1-ethyl-3-methylimidazolium ionic liquids   196
evaluation   321
– chemicals
  – REACH legislation   15
– economic
  – cost assessment   146
exchange resins   55
extraction   141
– aromatic hydrocarbons   145
– chelate   144
– efficiency   142
– extractive distillation   153
– four-stage   159
– liquid–liquid   141
– metals   141
– proteins   151
– radioactive metals   142
– reactive   116
extractor
– centrifugal   168

f

feedstock
– chemical   124
– petrochemical   93
– solubility   79
fibers
– cellulosic   131, 133
filtration
– membrane   294
– nano-   164
fingerprint
– ecological   307
fluidity   197
fluoride-free ionic liquids   208, 209
four-stage extraction   159
fractionation
– biomass   134
friction   206, 209
friction coefficient   205, 206
Friedel–Crafts alkylation   54
Friedel–Crafts reactions   47, 55
– SILP catalysis   57
fructose   104
– concentration effect on HMF manufacture   107
– condensation   100
– global production   99
fuels
– denitrogenation/desulfurization   158
functionality

– ionic liquids   143
functionalized ionic liquids   165

g

gases
– separation   164
gasoline
– extraction   159
global warming potential (GWP)   306, 319
glucose
– global production   99
– reactivity   112
glycol
– oligoethylene   167
greases
– ionic   216
green chemistry
– 12 principles of   3, 235
green engineering
– ionic liquids   137
green separation processes   139
green synthesis   1
– organic   41
groups
– head   197
GWP (global warming potential)   306, 319

h

halide salts
– 1-alkyl-3-methylimidazolium   8, 12
– 1-butyl-3-methylimidazolium   10
halides   74. *see also* chlorides
– metal   55
halogen-containing ionic liquids   81
HDS, *see* hydrodesulfurization
head groups
– biodegradability   269, 283, 285
– 1-butyl-4-(dimethylamino)pyridinium   255
– cationic   197, 240
heat exchanger   223
heat transfer   224
heating
– conductive   8
– conventional   11
– microwave   15
Heck coupling reactions   94
hemihydrates
– dimethylimidazolium chloride   13
Henry constant   174
Hertz contract stress   207
heteroatoms
– electron-rich   51
heterogeneous catalysis   77

hexafluorophosphate ionic liquids   208, 215
hexoses   105
HMF, see 5-hydroxymethylfurfural
homo-substituted dialkylimidazolium salt   111
homogeneous Brønsted acids   101
homogeneous Lewis acids   101
human health
– environmental and human health factor   320
humines   99
hydrocarbons
– aliphatic   145, 156
– aromatic   145, 156
hydrodesulfurization (HDS)   158
hydrodynamic lubrication   209
hydroformylation   81
– 1-butene   84
– catalysis   79
– 1-octene   82
– SILP catalysis   83
hydrogensulfate
– dimethylcyclohexylammonium   78
hydrolysis
– anion   81
hydrophilic ILs   5
– purification   28
hydrophobic compounds
– biphasic separation   141
hydrophobic ILs   5, 28
hydrophobicity and cytotoxicity   262
hydrosilylation   73
5-hydroxymethylfurfural (HMF)   93, 94
– applications   98
– continuous processing   114
– derivatives   96
– fructose concentration effect   107
– manufacture   99, 105
– oxidation products   97
– reduction products   98
– temperature effect on HMF manufacture   106
– water effect on HMF manufacture   108
hygroscopic ionic liquids   223

i

IL, see ionic liquids
imidazole ring   284
imidazolium
– biodegradability   283
imidazolium-based ionic liquids   256
– biodegradability   270
– fluoride-free   208
immobilization
– catalysts   72, 80

– transition metal complexes   65
impurities
– acidic   117
in situ formation
– carbene complexes   68
Independent Action (IA) concept   259
industrially implemented IL processes   299
inherently safe ionic liquids   235
intrinsic safety aspects   295
inulin   99, 112
ionic catalyst solutions   70
ionic greases   216
ionic ligands   66
– 1-alkyl-3-methylimidazolium   194, 323
– anions   241, 265
– antimicrobial properties   254
– bioavailability   258
– biodegradability   34, 265, 293
– biphasic acid scavenging   300
– biphasic reaction systems   46
– bis(trifluoromethanesulfonyl)amide   197
– Brønsted acids   52
– catalytic   51
– cellulose dissolution   123, 127
– cellulose processing   123, 129
– chemical and thermal stability   44
– colorless   31
– compressibility   210
– conversion of renewable resources   93
– corrosion behavior   225
– crystallization behavior   224
– desulfurization   158
– ecotoxicity   239, 291
– electrochemical stability   45
– electrolyte properties   191, 193
– environmental aspects   233
– 1-ethyl-3-methylimidazolium   25, 195, 208
– extractive distillation   155
– fluoride-free   208, 209
– fractionation of biomass   134
– Friedel–Crafts reactions   55
– functionalized   45, 165
– green engineering   137
– green organic synthesis   41
– green separation processes   139
– 1-H–3-methylimidazolium chloride   301
– halogen-containing   81
– heterogeneous catalysis   77
– hexafluorophosphate   208, 215
– HMF manufacture   99, 105
– hydroformylation catalysis   79
– hydrophilic   5
– hydrophobic   5
– hygroscopic   223

– IL-based spinning process   132
– IL–scCO$_2$ biphasic systems   75, 82
– IL-supported membranes   166
– imidazolium-based   257, 270
– inherently safe   235
– interaction with catalysts   53
– LCA   327
– lubricants   203
– metabolic pathways   288
– 1-methylimidazole   303
– microwave irradiation-promoted synthesis   15
– nanoparticle catalysis   75
– non-toxic   34
– pH values   109
– phase behavior of IL–water systems   226
– phosphite ligands   81
– polarity   65
– precursors   303
– product isolation   49
– protonated   128, 334
– pure   214, 225
– purification   28, 109
– pyridinium-based   160, 278
– reactive   51
– recycling   131
– SCILL   78
– SILP   46
– solubility   77
– structural moieties   117
– substitutes for molecular solvents   315
– sustainable   238
– synthesis   5, 322
– task-specific   143
– tested for biodegradablity   244
– tetrafluoroborate   208
– thermomorphic   75
– transition metal catalysis   65
– ultrasonic irradiation-promoted synthesis   20
– "unusual"   111
– vapor pressure   43
– water-insoluble   111
ionic liquids   3
irradiation
– microwave   15
– ultrasonic   20
isobutyl ketone
– methyl   101
isopropylation   54, 58
isothermal temperature stability test   212

j

jet reactor   302

k

kaolinite-containing soils   258
ketones
– methyl isobutyl   101

l

label
– eco-efficiency   313
lactose   113
lanthanides   142
large-scale processing   123
LCA, *see* life-cycle assessment
lead structures   238
legislation
– REACH   15
Lewis acids   67
– homogeneous   101
life-cycle assessment (LCA)   300, 312, 316, 318
– "cradle-to-grave"   317, 323
– methodology   318
ligands
– ionic   66
– neutral   143
– phosphite   81
lipid bilayer   262
liquids
– elastic   20
– ionic liquids   20
– liquid–liquid biphasic catalysis   56
– liquid–liquid biphasic organic reactions   46
– liquid–liquid extraction   141
– non-Newtonian   129
– separation   141, 144
– vapor–liquid equilibrium   227
lithium bromide   222
long side chains   283
loop reactor   74
loss modulus   130
lubricants   203
– commercial   211
– hydrodynamic   209
– thin films   210, 217
Lyocell   133
– process   126

m

magnetic follower   9
maltose   113
mannose   113
manufacture
– HMF   99, 105
market prices
– HMF starting materials   99

mass transfer 171, 224
melting point 214, 224
membranes
– combination with separation 163
– filtration 294
– IL-supported 166
– membrane–water coefficients 262
– porous alumina 167
MEMS (microelectrochemical–mechanical systems) 217
metabolizable cations 287
metals
– alkoxides 55
– alkyls 55
– extraction 141
– halides 55
– metal–carbene complexes 68
– radioactive 142
– transition metal 65
metathesis 5, 70
methanesulfonic acid (MSA) 110
methanol
– separation from TMB 156
methyl acrylate 329, 330
methyl isobutyl ketone 101
1-methylimidazole 303
N-methylimidazole
– alkylation 323, 324
1-H–3-methylimidazolium chloride ionic liquid 301
N-methylmorpholine N-oxide (NMMO) 126
– spinning process 127
microelectrochemical–mechanical systems (MEMS) 217
microwave heating 15
microwave irradiation
– IL synthesis 15
– simultaneous use with ultrasonic irradiation 23
microwave-promoted synthesis
– SWOT analysis 22
microwave reactors 17
mixed film friction 209
mixture toxicity 264
modes of toxic action 261
moieties
– anionic 264
molecular interaction 65
molecular solvents
– LCA 327
– substitutes 315
molten organic salts 126
monoalkylbenzene 57
monosaccharides 99

Monsanto/Kellog technology 54
MSA, see methanesulfonic acid
multiphase catalysis 75
multiphasic systems 82
mutagenicity 263
MX waste 18, 28

n

nanofiltration 164
nanoparticle and nanocluster catalysis 75
naphtha cracker 146
narcosis 292
natural organic chemicals 123
neutral ligand 143
nitrate
– aqueous media 142
"non-green" attributes 3
non-halide anions 28
non-Newtonian liquids 129
non-persistent chemicals 266
non-toxic ionic liquids 34
nonane 46
nucleophilic attack 226

o

1-octene
– hydroformylation 82
ODP (ozone depletion potential) 310
OH shift 128
oil bubbler 9
olefin–paraffin separation 155, 168
oligoethylene glycol 167
oligomerization 73
"one-pot" synthetic route 5
opportunities
– SWOT analysis 8
optimization 321
organic carbon
– dissolved 267
organic cations 195
organic chemicals
– natural 123
organic compound–water separation 156
organic matter
– dissolved 257
organic reactions
– liquid–liquid biphasic 46
– product isolation 49
organic salts
– molten 126
oxidation
– degradation techniques 292
– products of HMF 97
oxidative desulfurization 162

oxygen demand
– biochemical 267
– theoretical 268
ozone depletion potential (ODP) 310

p

paraffin
– olefin–paraffin separation 155, 168
PCR (polymerase chain reaction) 283
Pd-catalyzed Suzuki coupling 76
penicillin 152
performance
– coefficient of 223, 226
persistency 293
personnel-dependent costs 116
petrochemical feedstock 93
pH value 109
phase behavior
– IL–water systems 226
phase change 222
phosphines
– dialkoxyphenyl- 300
phosphite ligands 81
phosphonium compounds 282
platform chemicals 93
polarity
– ionic liquids 65
pollutants
– bioavailability 261
polylactic acid 93
polymerase chain reaction (PCR) 283
porous alumina membranes 167
portfolio
– eco-efficiency 309
pour point 213
power conversion efficiency 199
pressure
– vapor pressure 43
pretreatment
– supports 58
primary biodegradation 277
12 principles of green chemistry 3, 235
processes
– BASF BASIL 49, 299
– cellulose dissolution and processing 123
– continuous 114
– continuous single-phase 101
– green separation 139
– industrially implemented 299
– large-scale 123
– Lyocell 126
– NMMO 126
– separation (desgin) 172
– spinning process 132

– viscose 125
product isolation 49, 70
"PRODUCTIVELY" 6
propionoxymethylfurfural 96
proteins
– extraction 151
protonated acidic ionic liquids 128
protonated diazabicycloundecane 128
"protonated" ionic liquids 334
protonic acids 55
Pt-catalyzed hydrosilylation 73
pulp
– cellulose 127
pure ionic liquids 214
– viscosity 225
purification
– 1-alkyl-3-methylimidazolium halide salts 12
– hydrophilic/hydrophobic ILs 28
– ionic liquids 28
– routes 5
– SWOT analysis 29
purity
– ionic liquids 109
pyridinium 284
pyridinium-based ionic liquids 278
pyridinium-containing ILs 160

q

quantitative structure–activity relationships (QSAR) 261
quasi-Fermi level 200
quaternization 5, 18

r

radioactive metals 142
rayon 133
RDC, see rotating disc contactor
REACH (Registration, Evaluation, Authorization and restriction of CHemicals) legislation 15, 235
reactions
– alkylation 323, 324
– anion hydrolysis 81
– biphasic 46
– condensation 100
– coupling reactions 76
– decolorization 5, 31
– dehydration 105
– denitrogenation 158
– desulfurization 158, 162
– Diels–Alder 329, 330
– electrophilic substitution 54
– extraction 116

- Friedel–Crafts 55
- Friedel–Crafts alkylation 54
- hydroformylation 79
- hydrosilylation 73
- isopropylation 54, 58
- metathesis 70
- oligomerization 73
- organic reactions 49
- oxidation 97
- oxidative desulfurization 162
- purification 31
- quaternization 5, 18
- reduction 97
- ring-closing metathesis 70
- separation 139
- synthesis 31
- thermal abstraction 67

reactive extraction 116
reactive ionic liquids 51
reactivity
- glucose 112

reactors
- jet 302
- loop 74
- microwave 17

recycling
- catalysts 72
- ionic liquids 131
- solvent 116

reduction products of HMF 97
regeneration
- cellulose 131

renewable resources
- conversion 93

resins
- exchange 55

restriction
- chemicals
   - REACH legislation 15

rheological behavior
- cellulose solutions 129

ring-closing metathesis 70
rotating disc contactor (RDC) 169

S

saccharides 93, 112, 113, *see also* sugars
- fructose 105
- glucose 112
- hexoses 105
- inulin 112
- mono- 99

safety aspects
- intrinsic 295

salinity 258

salts
- 1-alkoxymethyl-3-hydroxypyridinium 285
- 1-alkyl-3-methylimidazolium halide 9, 12
- alkyltrimethylammonium 286
- ammonium 254, 286
- benzylalkyldimethylammonium 286
- 1-butyl-3-methylimidazolium halide 10
- dialkylimidazolium 111
- molten organic 126

scar diameter 206
scavenging
- biphasic acid 300

$scCO_2$ 75, 82
scenario analysis 311
SCILL (solid catalyst with ionic liquid layer) 78
selectivity
- aromatic–aliphatic separation 147
- dimer 73

separation
- aromatic–aliphatic 145, 156
- biphasic 141
- CO_2 165
- combination with membranes 163
- environmental 158
- functionalized ILs 165
- gases 164
- liquids 141
- olefin–paraffin 155, 168
- organic compound–water 156
- process design 172
- processes 139
- TMB–methanol 156

shear rate 129
shear viscosity 129
SHX, *see* solution heat exchanger
side chain effect 243
side chains
- long 283

SILP catalysis 46, 70
- Friedel–Crafts 57
- hydroformylation 83
- transition metal 71

single-effect absorption chiller 222
single-phase process
- continuous 101

smectite-containing soils 258
soil types 258
solar cells 198
- dye-sensitized 198

solid catalyst with ionic liquid layer (SCILL) 78
solid-phase batch synthesis 17
solubility
- ammonia 227

– catalysts 80
– cross- 162
– feedstock 79
– ionic liquids 77
– transition metal complexes 65
– tunable 47
solution heat exchanger (SHX) 223
solution-phase batch synthesis 17
solutions
– cellulose 129
– ionic catalyst 70
solvent-free synthesis 334
solvent recycling 116
solvent-to-feed ratio (S/F) 154
solvents
– COSMO-RS method 47
– molecular 315, 327
– performance 330
– selection tools 317
specific conductivity 194, 197
spinning process 123
– IL-based 132
– NMMO 127
stability
– chemical/thermal 44, 210, 225
– device 192
– electrochemical 45, 196
starch 99
starting materials
– HMF manufacture 99
steel–steel contact measurement 205
storage modulus 130
strengths, weaknesses, opportunities, threats (SWOT) analysis, see SWOT analysis
stress
– Hertz contract 207
Stribeck curve 209
structural moieties
– ionic liquids 117
structure–activity relationships 236
– quantitative 261
substitution
– electrophilic 54
substructural elements 291
sugars 104, see also saccharides, fructose
– global production 99
sulfate
– hydrogen- 78
sulfolane 146, 147, 170
supercritical CO_2, see $scCO_2$
supported acids 55
supported ionic liquid phase, see SILP
supports
– pretreatment 58

sustainable ionic liquids 238
SWOT analysis 8, 14
– IL decolorization 31
– microwave-promoted synthesis 18
– purification of hydrophilic/hydrophobic ILs 29
– simultaneous microwave and ultrasonic irradiation 23, 24
– ultrasonic-promoted synthesis 22
synthesis
– batch 17
– continuous flow 19
– dialkoxyphenylphosphines 300
– green 3
– green organic 41
– ionic liquids 322
– microwave irradiation-promoted 15
– microwave-promoted 18
– solvent-free 334
– ultrasonic irradiation-promoted 20, 22
synthetic routes
– ionic liquids 5
– "one-pot" 5

t
T-SAR concept 236
task-specific ionic liquids (TSILs) 143
technical-grade batches 111
temperature effect on HMF manufacture 106
temperature of thermal decomposition 225
temperature stability test
– isothermal 212
tertiary amine 300
"Test Kit" concept 236
test organisms 242
test systems
– aquatic 243
– biological 240
tests
– biodegradability 266, 267
– "CO_2 Headspace" 286
– die-away 269
tetrafluoroborate ionic liquids 208
theoretical oxygen demand (ThOD) 268
thermal abstraction 67
thermal decomposition temperature 225
thermal stability 210, 225
– ionic liquids 44
thermomorphic ionic liquids 75
thin films
– lubricating 210, 217

"Thinking in terms of Structure–Activity Relationships" concept   236
threats
– SWOT analysis   8
TMB, see trimethyl borate
toluene
– isopropylation   54, 58
toxicity
– baseline   292
– cyto-   256
– dichloromethane   31
– ecotoxicity   257
– ionic liquids   291
– mixture   264
– modes of toxic action   261
toxicology   235
transition metal catalysis   65
– SILP   71
transition metal complexes   65, 68
tribotester
– ball-on-disc   204
1,2,4-trichlorobenzene   78
trimethyl borate   156
triphasic systems   46
TSILs, see task-specific ionic liquids
tunable solubility properties   47
two-phase system
– aqueous   151

u
ultrasonic irradiation   20
– simultaneous use with microwave irradiation   23
ultrasonic-promoted synthesis
– SWOT analysis   22
UNIFAC   173
"unusual" ionic liquids   111
UV degradation   294

v
vapor–liquid equilibrium (VLE)   227
vapor pressure
– ionic liquids   43
viscose process   125
viscosity
– complex   129
– dynamic   194
– pure ionic liquids   214, 225
– zero shear   129
viscosity index   213, 214
VLE, see vapor–liquid equilibrium

w
waste
– MX   18, 28
water
– effect on HMF manufacture   108
– membrane–water coefficients   262
– pollution   307
– separation from organic compounds   156
– water-insoluble ionic liquids   111
weaknesses
– SWOT analysis   8
wear   204
wear scar diameter   206
wear volume   205
WEEE logo   193
wet/wet spinning   132
working pairs   221

x
xenobiotics   259
– bioavailability   292

z
zeolites   55
zero shear viscosity   129